Geospatial Analysis Applied to Mineral Exploration

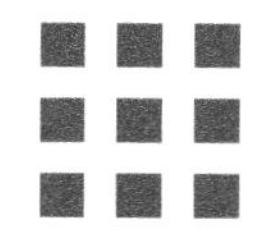

Geospatial Analysis Applied to Mineral Exploration

Remote Sensing, GIS, Geochemical, and Geophysical Applications to Mineral Resources

Edited by

Amin Beiranvand Pour
Institute of Oceanography and Environment (INOS), University Malaysia Terengganu (UMT), Kuala Nerus, Terengganu, Malaysia

Mohammad Parsa
Mineral Exploration Research Centre, Harquail School of Earth Sciences, Laurentian University, Sudbury, ON, Canada; Department of Earth Science, University of New Brunswick, Fredericton, NB, Canada; Metal Earth Research Center, Harquail School of Mines, Laurentian University, Sudbury, ON, Canada

Ahmed M. Eldosouky
Geology Department, Faculty of Science, Suez University, Suez, Egypt

ELSEVIER

Elsevier
Radarweg 29, PO Box 211, 1000 AE Amsterdam, Netherlands
The Boulevard, Langford Lane, Kidlington, Oxford OX5 1GB, United Kingdom
50 Hampshire Street, 5th Floor, Cambridge, MA 02139, United States

ISBN: 978-0-323-95608-6

For Information on all Elsevier publications
visit our website at https://www.elsevier.com/books-and-journals

Publisher: Candice Janco
Acquisitions Editor: Jennette McClain
Editorial Project Manager: Aleksandra Packowska
Production Project Manager: Sruthi Satheesh
Cover Designer: Christian Bilbow

Typeset by MPS Limited, Chennai, India

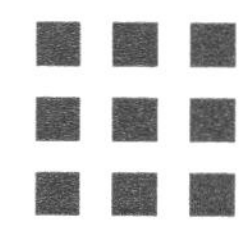

Dedication

To my beloved father (Mashallah Beiranvand) and mother (Delbar Aali Nezhad)

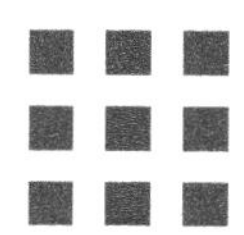

Contents

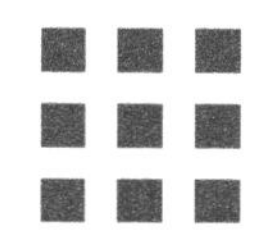

List of contributors

Mohamed Abd El-Wahed Geology Department, Faculty of Science, Tanta University, Tanta, Egypt

Ahmed Moustafa Abdel-Rahman Geology Department, Faculty of Science, Al-Azhar University, Cairo, Egypt

Hamada El-Awny Geology Department, Faculty of Science, Al-Azhar University, Cairo, Egypt

Hatem Mohamed El-Desoky Geology Department, Faculty of Science, Al-Azhar University, Cairo, Egypt

Ahmed M. Eldosouky Geology Department, Faculty of Science, Suez University, Suez, Egypt

Reda A.Y. El-Qassas Exploration Division, Nuclear Materials Authority, Maadi, Cairo, Egypt

Wael Fahmy Geology Department, Faculty of Science, Al-Azhar University, Cairo, Egypt

Jeff Harris Mineral Exploration Research Centre, Harquail School of Earth Sciences, Laurentian University, Sudbury, ON, Canada; Metal Earth Research Center, Harquail School of Mines, Laurentian University, Sudbury, ON, Canada

Mazlan Hashim Geoscience and Digital Earth Centre (INSTeG), Research Institute for Sustainable Environment, Universiti Teknologi Malaysia, Johor Bahru, Malaysia

Ahmed Henaish Department of Geology, Faculty of Science, Zagazig University, Zagazig, Egypt

Mohammad Shawkat Hossain Institute of Oceanography and Environment (INOS), University Malaysia Terengganu (UMT), Kuala Nerus, Terengganu, Malaysia

Thong Duy Kieu Hanoi University of Mining and Geology, Hanoi, Vietnam

Hassan Mohamed Exploration Division, Nuclear Materials Authority, Maadi, Cairo, Egypt

Aidy M. Muslim Institute of Oceanography and Environment (INOS), University Malaysia Terengganu (UMT), Kuala Nerus, Terengganu, Malaysia

Mohammad Parsa Mineral Exploration Research Centre, Harquail School of Earth Sciences, Laurentian University, Sudbury, ON, Canada

Luan Thanh Pham Faculty of Physics, University of Science, Vietnam National University, Hanoi, Vietnam

Amin Beiranvand Pour Institute of Oceanography and Environment (INOS), University Malaysia Terengganu (UMT), Kuala Nerus, Terengganu, Malaysia

Hojjatollah Ranjbar Mining Engineering Department, Shahid Bahonar University of Kerman, Kerman, Iran

Milad Sekandari Mining Engineering Department, Shahid Bahonar University of Kerman, Kerman, Iran

Adel Shirazi Faculty of Mining Engineering, Amirkabir University of Technology, Tehran, Iran

Aref Shirazi Faculty of Mining Engineering, Amirkabir University of Technology, Tehran, Iran

Cuong Van Anh Le University of Science, Vietnam National University Ho Chi Minh City, Ho Chi Minh, Vietnam

Jonas Didero Takodjou Wambo Division of Geological and Planetary Sciences, California Institute of Technology, Pasadena, CA, United States

Mastoureh Yousefi Department of Mining Engineering, Isfahan University of Technology, Isfahan, Iran

Basem Zoheir Geology Department, Faculty of Science, Benha University, Benha, Egypt; Institute of Geosciences, University of Kiel, Kiel, Germany

Renguang Zuo State Key Laboratory of Geological Processes and Mineral Resources, China University of Geosciences, Wuhan, P.R. China

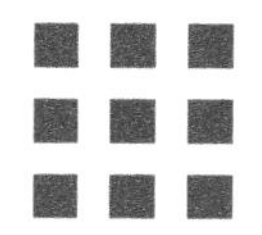

Preface

Nowadays, many surficial mineral deposits are being mined out, leaving only deep-seated mineral deposits for feeding raw materials into the industry. Therefore techniques applied to mineral exploration need to be revisited for discovering new mineral resources, which may be located in harsh and remote regions. Over the past decades, remote sensing technology and geographic information system (GIS) techniques have been incorporated into several mineral exploration projects worldwide. This aim is to bridge the knowledge gap for the geospatial-based discovery of buried, covered, and blind mineral deposits. This book details the main aspects of the state-of-the-art remote sensing imagery, geochemical data, geophysical data, geological data, and geospatial toolbox required to explore ore deposits. It covers advances in remote sensing data processing algorithms, geochemical data analysis, geophysical data analysis, and machine learning algorithms in mineral exploration. It also presents approaches on recent remote sensing and GIS-based mineral prospectivity modeling, which offer a piece of excellent information to professional earth scientists, researchers, mineral exploration communities, and mining companies.

The anticipated goal of this book is to solve high complexity issues in remote sensing data processing, geochemical data analysis, geophysical data analysis, and appropriate applications of GIS techniques for data fusion designed for mineral exploration purposes. This book contains updated and informative knowledge of remote sensing imagery, geochemistry, geophysics, and geospatial techniques that can assist in delineating the signatures and patterns linked to deep-seated, covered, blind, or buried mineral deposits. It includes several color figures for remotely sensed recognition of mineralization-related footprints. Also, this book comprehends several interactive flowcharts for describing the procedures explained therein. Numerous websites are addressed, and their links are provided for data acquisition, extra information, and data resources. The literature is linked to recent scientific publications, conferences, and symposiums in remote sensing and GIS topics in specific subjects and disciplines.

This book is a hands-on guide for getting to know the geospatial toolbox used in mineral exploration. Everyone involved in mineral exploration should have the know-how knowledge of using the geospatial toolbox available. However, there is a knowledge gap in this area, and it is hard for many geologists to grasp these concepts. This book is an effort to bridge this knowledge gap. With the advent of machine learning and deep learning techniques, it is not surprising that these methods are being increasingly applied in many fields. However, applying these concepts in geology, particularly in mineral exploration, has some limitations and challenges. This book outlines these challenges and presents solutions for tackling them. Therefore this book will be needed for those working in image processing and machine learning techniques, aiming to apply their ideas to mineral exploration. Those with jobs in data processing such as department of remote sensing and GIS, department of statistics and data science, and department of geoinformatics, who mostly work for the geo-information technology industry, for example, geospatial companies are also the audience of this book.

I am grateful for technical and management contributions by editors and authors, as well as professional staff at Academic Press. Finally, I must thank my family, Danica and Avina, for they sacrificed more than I did during the long hours devoted to this book. Buoyantly, the result is to some degree worthy of their support.

Amin Beiranvand Pour

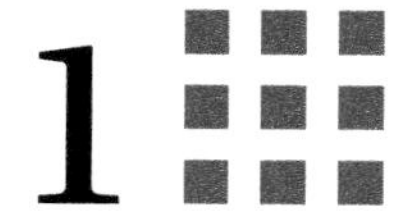

Introduction to mineral exploration

Amin Beiranvand Pour[1], Mohammad Parsa[2], Ahmed M. Eldosouky[3]

[1]INSTITUTE OF OCEANOGRAPHY AND ENVIRONMENT (INOS), UNIVERSITY MALAYSIA TERENGGANU (UMT), KUALA NERUS, TERENGGANU, MALAYSIA [2]MINERAL EXPLORATION RESEARCH CENTRE, HARQUAIL SCHOOL OF EARTH SCIENCES, LAURENTIAN UNIVERSITY, SUDBURY, ON, CANADA [3]GEOLOGY DEPARTMENT, FACULTY OF SCIENCE, SUEZ UNIVERSITY, SUEZ, EGYPT

1.1 Mineral exploration

There are currently over 7 billion people living on the planet Earth, the survival of which is inextricably intertwined with mineral resources. Without iron ore mining and limestone mining, for example, one cannot produce steel and concrete, leaving the world bereft of infrastructures and transportation systems. That being said, mineral resources are an integral part of people's lives. A prerequisite to exploiting these mineral resources is the exploration phase—finding a mineral deposit. This stage requires time, effort, and money, yet exploration does not always lead to success.

Mineral exploration, a multidisciplinary field of science combining computer science, geology, and economics, can be deemed as a series of activities for understanding ore-forming processes and vectoring toward undiscovered mineral deposits. In mineral exploration the knowledge gathered from diverse fields is combined and applied to a scientifically supported search for mineral deposits and their qualitative and quantitative assessments. Mineral exploration is a multifaceted, multistage process, constituting (1) exploration strategy, (2) prospecting, (3) early-stage exploration, and (4) detailed surveys, discussed in the upcoming subsections. The classification, however, is different from the common classifications described elsewhere (e.g., Haldar, 2018) and aims to explain modern-day exploration campaigns.

Aside from the differences in classification schemes, each stage of mineral exploration entails narrowing down the search space for further stages using geospatial modeling and analysis of geoscientific data. Once progressing to a new stage, where the extent of the area being studied decreases, the costs involved in the surveys dramatically rise (Singer and Kouda, 1999).

1.2 Stages

1.2.1 Exploration strategy

In essence, what is defined here as the exploration strategy is to address the question as to whether one should start an exploration campaign from scratch—working from the unknown—or

Geospatial Analysis Applied to Mineral Exploration. DOI: https://doi.org/10.1016/B978-0-323-95608-6.00001-9

build on the previous work—working from the known—conducted by previous campaigns. The former is called greenfield exploration and is defined as exploration in tectonically favorable zones where few or even no already-discovered mineral deposits are presented. The latter, however, is referred to as brownfield exploration and is linked to the premise that some deposit types, such as volcanic-hosted massive sulfide deposits (Galley et al., 2007) and Mississippi Valley-type deposits (Leach et al., 2005), often form in clusters.

1.2.2 Prospecting

Either strategy constitutes reanalyzing the legacy data gathered throughout many years for the area being studied. These data are usually publicly available regional-scale datasets, including geological and tectonic maps and geophysical and geochemical data. The legacy data are integrated into maps representing the favorability of deposits of the type being explored. Based on these maps, exploration firms may acquire exploration rights called claims, granting an exclusive right to search for mineral substances on a given territory.

1.2.3 Early-stage exploration

Once the claims are granted, geological mapping, geochemical sampling, and geophysical surveys in scales varying between 1:25,000 and 1:5000 are often carried out. The results of these surveys are synthesized to define some strategic target zones for detailed exploration surveys.

1.2.4 Detailed surveys

Trenching and drilling are conducted to capture mineralized zones based on the target areas defined in the previous stage. Trenches are dug to sample along continuous zones. These samples are the subject of geochemical and petrological studies. Diamond core drilling allows for various studies, including core logging, in situ geophysical analysis of samples, such as measuring the magnetic susceptibility of samples, geochemical analysis, and petrological studies. However, rotary core drilling merely provides a chance for geochemical analysis of samples.

Once one hits the mineralization, systematic drilling should continue to delineate the boundaries of mineralized zones. This includes infill drilling, drilling in the gaps between previously drilled holes, and drilling some holes to define mineralized zones' shape and spatial pattern. Once the shape and outline of mineralized zones are defined, one can proceed to resource modeling to estimate the average grade and tonnage.

1.3 4D-geographic information system for mineral exploration

Mineral exploration has been defined as the game of probabilities in which success is never guaranteed. Risk and uncertainty are, therefore, intrinsic aspects of mineral exploration projects (Singer and Kouda, 1999). It is now widely held that as the exploration proceeds to the detailed stages, the risk also increases, leading decision-makers to seek strategies for managing and reducing this risk (Haldar, 2018). The absence of grassroots exploration surveys, outlining the fact that a majority of exploration projects are now focused on locating deep-seated, blind, and covered deposits, also jeopardize the success of today's exploration campaigns.

One obvious approach to reducing the risk involved in mineral exploration and quantifying the probability of discovering mineral deposits is conducting different surveys and combining their output using mineral prospectivity modeling (MPM) techniques (Carranza, 2008). Modern-day exploration programs entail an array of geophysical, geochemical, and geological surveys. Therefore these programs are faced with an information overload problem, requiring big data analysis techniques to help distill the information available into predictive models of mineral prospectivity. However, the information provided by individual surveys is often incomplete and targets merely one or a few characteristics of the desired style of mineral deposits. Although a geographic information system (GIS) provides a suite of tools and techniques for visualizing the results of individual surveys and integrating their results into predictive models showing the probability of discovering mineral deposits, it does not mean that all the data compiled and gathered from different exploration surveys should be combined in a GIS. In essence, only signatures that are spatially, temporally, and perhaps genetically linked to the mineralization of the type being explored should be considered for mineral exploration. Therefore diverse data types from different surveys should be categorized, given their geological age and spatial information from mineralization-related signatures.

A 4D-GIS for mineral exploration, a system enabling the analysis, visualization, and integration of 2D- and 3D-based big data concerning their spatial–temporal association with mineralization, is, thus, required for achieving the objectives of modern-day exploration campaigns. Given the previous definition, the main components of this 4D-GIS system are (1) *input datasets*, (2) *user-guided interpretation of the datasets*, and (3) *predictive modeling*.

1.3.1 Input

As outlined previously, diverse surveys are being conducted over individual claims to find the patterns linked to mineral deposits. Irrespective of the scale, the major input feeding the 4D-GIS are remote sensing-based data, exploration geochemistry, exploration geophysics, and geological indicators. These are evaluated in the upcoming subsections.

1.3.1.1 Remote sensing

Remote sensing technology affords a great boost for mineral exploration by collecting data from immense and inaccessible regions using sophisticated sensors mounted on satellite or aircraft. Remote sensing has been used for mineral exploration over the past decades and provides low-cost data and synoptic observation of the ground. The application of remote sensing technology for geological mapping and mineral exploration has been developing since the 1970s. Hydrothermal alteration minerals with diagnostic spectral absorption properties in the visible and near-infrared (VNIR) and the shortwave infrared (SWIR) parts of the electromagnetic spectrum have been identified by multispectral and hyperspectral remote sensing sensors (Rowan et al., 1977; Hunt and Ashley, 1979; Van der Meer et al., 2012a,b). Landsat Thematic Mapper/Enhanced Thematic Mapper + (TM/ETM +) and Landsat-8 data have been used for mapping hydrothermal alteration minerals associated with porphyry copper and gold mineralization. Shortwave infrared bands (bands 5 and 7) of TM/ETM + and (bands 6 and 7) of Landsat-8 are utilized to identify hydroxyl-bearing minerals. Band ratios of 5/7 and 6/7 are sensitive to hydroxyl (OH) minerals, which are found in the alteration zones (Abdelsalam et al., 2000; Pour and Hashim, 2018; Safari et al., 2018). Nonetheless, the broad extent of these bands does not allow discriminating specific alteration zones and minerals by TM/ETM + and Landsat-8 data, which are important for exploring high-economic potential zone for ore mineralization. The spatial resolution of the SPOT imagery is suitable for geological structural mapping. The SPOT data are successfully used for geological structural mapping and lineament extraction (Ahmadirouhani et al., 2017; Hosseini et al., 2019).

The Advanced Spaceborne Thermal Emission and Reflection Radiometer (ASTER) is a high spatial, spectral, and radiometric resolution multispectral remote sensing sensor. It consists of three separate instrument subsystems, including the VNIR, the SWIR, and the thermal infrared. ASTER SWIR spectral properties are unprecedented multispectral data for the discrimination of hydrothermal alteration mineral zones such as phyllic, argillic, and propylitic associated with porphyry copper mineralization (Mars and Rowan, 2006). The ASTER SWIR bands with sufficient spectral resolution have been widely and successfully used in numerous mineral explorations and lithological mapping projects (Abrams and Yamaguchi, 2019). The Advanced Land Imager (ALI) sensor has six unique wavelength channels spanning the VNIR (0.4–1.0 μm). Because of their respective band center positions, ALI is especially useful for discriminating among ferric-iron-bearing minerals for mineral exploration applications (Hubbard et al., 2003).

The Sentinel-2 sensor has 13 bands covering the VNIR (8 bands) and SWIR (2 bands) regions with a spatial resolution of 10, 20, and 60 m. Iron oxide/hydroxides and hydroxyl-bearing alteration minerals can be mapped using band ratios of 4/2 and 11/12 (Van Der Meer et al., 2014; Van Der Werff and Van Der Meer, 2016). WorldView-3 (WV-3) sensor contains the highest spatial, spectral, and radiation resolutions among the multispectral satellite sensors, presently. Iron oxides/hydroxide minerals can be comprehensively mapped and discriminated by VNIR bands of WV-3 (Mars 2018; Pour et al., 2019). Additionally, SWIR bands of WV-3 contain high capability for detailed mapping of Al–OH, Mg–Fe–OH, CO_3, and Si–OH key hydrothermal alteration minerals (Kruse and Perry, 2013; Kruse et al., 2015; Mars 2018).

Hyperion is the first advanced satellite hyperspectral sensor in commission across the spectral coverage from 0.4 to 2.5 and 10-nm spectral resolution. It has 242 spectral channels over a 7.6-km swath width, and 30-m spatial resolution (Liao et al., 2000). The first subset of VNIR bands between 0.4 and 1.3 μm can also be used to highlight iron oxide/hydroxide minerals (Bishop et al., 2011). Hyperion SWIR bands (2.0–2.5 μm) can uniquely identify and map hydroxyl-bearing minerals, sulfates, and carbonates in the hydrothermal alteration assemblages (Kruse et al., 2003; Gersman et al., 2008; Bishop et al., 2011). The new hyperspectral Precursore IperSpettrale della Missione Applicativa (PRISMA) sensor is able to capture images in a continuum of 240 spectral bands ranging between 400 and 2500 nm. Iron oxide/hydroxide minerals can be mapped using VNIR bands, and gypsum and clay minerals are detectable by SWIR bands of PRISMA (Pearlshtien et al., 2021). Generally, PRISMA offers the mineral exploration community and users many applications in the field of environmental monitoring and resources management, such as raw material exploration and mining (Pearlshtien et al., 2021).

The Airborne Visible/Infrared Imaging Spectrometer (AVIRIS) is a hyperspectral airborne instrument flown in a NASA high-altitude aircraft. It was the first optical sensor (1987) to measure the solar reflected spectrum from the VNIR and SWIR (400–2500 nm) using 224 contiguous bands at 10-nm intervals (Green et al., 1998). A variety of alteration minerals such as iron oxide/hydroxide minerals, clay minerals, carbonates, and ammonium minerals were mapped using AVIRIS data (Baugh et al., 1998; Rowan et al., 2000; Kruse et al., 2003). "HyMap" (Hyperspectral Mapper) is an airborne hyperspectral imaging sensor (manufactured by Integrated Spectronics Pty Ltd, Sydney, Australia) operated by HyVista Corporation. HyMap data have 126 spectral bands (at approximately 15-nm intervals) spanning the wavelength interval 0.45–2.5 μm (VNIR—64 bands) and (SWIR—62 bands) and 5-m spatial resolution (Cocks et al., 1998). HyMap data were used for detailed mapping hydrothermal alteration minerals associated with Cu–Au–Mo porphyry deposits, carbonatite complex, and metamorphic–hydrothermal U–REE deposit (Bedini, 2009, 2011, 2012; Huo et al., 2014; Salles et al., 2017).

Recently, unmanned aerial vehicles (UAVs) or drones have been utilized in the mining industry for various applications from mineral exploration to mine reclamation. One of the most significant benefits of using UAVs is the low operational cost and ease of use (Park and Choi, 2020). UAVs can be equipped with optical devices, cameras covering different ranges of the electromagnetic spectrum [multispectral, hyperspectral, short/mid-wave range cameras (e.g., thermal)], and geophysical instrumentation (magnetometer) for mineral exploration purposes such as aerial mapping, resource modeling, and geophysical surveys (Kirsch et al., 2018; Walter et al., 2020; Heincke et al., 2019; Jackisch et al., 2019). Drone-based hyperspectral imaging has been utilized to map rare-earth-element-rich minerals in Namibia. Furthermore, drone-based magnetic surveys were deployed to identify subsurface ore potential at a fraction of the cost of traditional surveys in Greenland (Jackisch et al., 2019). Integration of drone-based hyperspectral and magnetic data was used for mapping Otanmäki Fe–Ti–V deposits in central Finland (Jackisch et al., 2019).

Synthetic aperture radar (SAR) sensors transmit electromagnetic radiation at specified wavelengths and measure the reflected energy. Radar transmits and detects radiation between 2 and 100 cm, typically at 2.5–3.8 cm (X-band), 4.0–7.5 cm (C-band), and 15.0–30.0 cm (L-band) (Woodhouse, 2006). SAR imagery such as RADARSAT, JERS-1 SAR, ERS SAR, PALSAR, and Sentinel-1 data have been extensively used for geological mapping, especially structural analysis related to ore mineralization (Singhroy, 2001; Paganelli et al., 2003; Pour et al., 2013, 2014, 2016, 2018; Zoheir et al., 2019). More explanations about the multispectral, hyperspectral, UAVs' and SAR remote sensing sensors, data acquisition, preprocessing techniques, image processing algorithms for alteration mineral detection, lithological and structural mapping, and case studies will be comprehensively discussed in Chapter 2.

1.3.1.2 Exploration geochemistry

Exploration geochemistry entails conducting a sampling survey over an area, chemically analyzing the samples, and interpreting the results. Each of the stages mentioned earlier exerts control over the success of an exploration campaign. The choice of sampling media, be they soil, till, rock chip, water, stream sediment, and the like, depends on the geomorphological and geological setting of an area of interest and substantially controls the methods of interpretation. In addition, one must consider some factors while designing the sampling scheme of a geochemical exploration project; these are the topographical condition, the budget, and accessibility. There are many methods available for chemically analyzing the geochemical samples. Generally, samples can be analyzed for unielement or multielements. All these methods will be discussed in Chapter 4.

With the dwindling number of exposed mineral deposits, especially in developed countries, the exploration industry is faced with the challenge of discovering deep-seated, blind, or covered deposits, which are, granted, marked by weak geochemical signatures, most of the time (Beus and Grigorian, 1977). Exploration geochemistry can play a pivotal role in delineating weak geochemical anomalies that might represent undiscovered mineral deposits. Chapter 4 discusses three breakthroughs of exploration geochemistry, considering the compositional nature of geochemical data (Aitchison, 1982), fractal/multifractal-based delineation of geochemical anomalies (Zuo and Wang, 2016), and artificial-intelligence-based pattern recognition applied to exploration geochemistry (Zuo et al., 2019), and describes how these major steps should be adopted to interpret the results of a geochemical survey for mineral exploration.

1.3.1.3 Exploration geophysics

The application of geophysical methods to the exploration of mineral deposits is a common practice since the early 20th century (Best, 2015). These methods are based on contrasts in the physical properties between mineral deposits and the surrounding rock. These physical properties are velocity, resistivity, density, magnetic susceptibility, chargeability, gamma radiation, and dielectric constant. In 1920 Schlumberger applied the electrical method in the iron-bearing basin of May-Saint-Andié (Binley and Slater, 2020). Many resistivity methods have been established throughout the first half of the 20th century (Binley and Slater, 2020). The examples of early successes of these methods for exploring minerals in North America and Europe were documented by Nostrand (1966). Since the early 1920s, electromagnetic methods have been developed and applied in exploring mineral deposits. Many ore deposits (e.g., Ernest Henry copper–gold, copper deposit in Tibet; iron ore deposit in Qiumu mine,

Heath Steele zinc–lead–copper–silver deposit, and Damingshan gold deposit) have been discovered by applying these methods (Sheard et al., 2005; Guo et al., 2020).

The earliest study on the use of seismic data in mineral exploration is reported by Schmidt (1959) for detecting the iron ores of the Siegerland district (Germany). The seismic methods have also been used to locate other deposits such as massive sulfide, uranium, gold, Ni–sulfide, Ni–Cu–PGE, and nickel deposits (Salisbury and Snyder, 2007; Malehmir et al., 2012). The electromagnetic method was independently developed by Rikitake (1950), Tikhonov (1950), and Cagniard (1953), and its effectiveness was demonstrated on a variety of mineral systems in the world (Zhdanov, 2010). Aside from the earlier methods, the potential field methods are also widely used techniques for mineral exploration. The earliest published reports on using these methods in mineral exploration are included in the report on gravity interpretation of Arthur (1940) for exploring hematite at Steep Rock Lake and the report on the use of magnetic data for mapping an iron–copper–sulfide ore body in the Sudbury district (Canada) in 1930 by Eve and Keys (see Hinze et al., 2013). The effectiveness of these methods in mineral exploration is well documented by Hinze (1960), Nabighian et al. (2005a,b), and Zhang et al. (2020). The role of integration between geophysical and remote sensing datasets for mapping hydrothermal deposits is well described by Eldosouky et al. (2017, 2020, 2021), Elkhateeb and Eldosouky (2016), and Elkhateeb et al. (2021). The use of geophysical data in mineral exploration will be explained in more detail in Chapter 5.

1.3.1.4 Geological indicators

Mineral deposit types are groups of minerals having similar geologic characteristics, geologic environments of occurrence, and geologic processes of formation (Guilbert and Park, 1986; Cox and Singer, 1986; Bliss, 1992). The distribution of mineral deposits is determined by the geological processes that formed them. Mineral deposits are, therefore, generally clustered in geological provinces (mineral provinces or mineral districts) with some provinces being strongly endowed with particular mineral commodities (Jaireth and Huston, 2010). Three main groups of metallic mineral deposit types can be recognized based on their mode of formation, including magmatic ore deposits, hydrothermal ore deposits, and sedimentary ore deposits (Plumlee, 1999).

Descriptions of an ore mineral deposit often refer to host rocks and geologic structures or features that are important to the deposit. Furthermore, hydrothermal activity and ore-bearing fluids are linked to most economic mineral deposits. Therefore identification of host-rock mineralization, hydrothermal alteration zones, and geological structures is one of the most important indicators for mineral exploration during fieldwork. The use of Global Positioning System (GPS) technology during field surveys helps to accurately position the sampling points for high-resolution mineral exploration mapping. Mineralogy and petrography analysis include macroscopic-to-microscopic investigations of the rock samples. Once macroscopic details about the rock samples have been determined by visually inspecting, a petrographic thin section is typically made to characterize the microscopic features. Thin sections are great for identifying the minerals present, porosity, alteration, microstructures, and provenance.

To acquire quantitative bulk rock compositions X-ray diffraction (XRD) is a common technique. There are various XRD machines and techniques, but typically a rock sample is crushed into a fine powder, which is packed and mounted onto a stage that is analyzed by X-rays. The X-ray detector captures this information which gets plotted onto a diffractogram, where characteristic peaks can be identified as specific minerals. The XRD is used to identify the exact mineral assemblages of a rock. Analytical Spectral Devices' (ASD') spectrometers and spectroradiometers are used to provide spectral characteristics of minerals for identifying mineral zonation and information for vectoring at the regional and deposit scale. Further explanations for mineral deposits and occurrence, geological mapping, laboratory analysis, and accuracy assessment techniques will be explained in Chapter 6.

1.3.2 User-guided interpretation of datasets

Despite the diversity of data available for exploration, only information directly or indirectly linked to mineralization should be extracted from these data for the purpose of mineral exploration. Therefore knowledge-guided interpretation is required for the recognition of ore-related patterns embedded in these datasets. As to how the patterns associated with mineralization are extracted is a challenge. Traditionally, mineralization-related patterns are recognized using the features described in descriptive deposit models (Cox and Singer, 1986). However, adopting this approach may hinder mineral exploration given the fact that descriptive deposit models only describe local-scale features. Given the complexity of local-scale geological processes, hardly a deposit model can be generalized, and many of the features described in descriptive deposit models might be absent in different geological settings. That being said, a growing body of research (e.g., Kreuzer et al., 2008; Hagemann et al., 2016; Ford et al., 2019; Parsa and Pour, 2021; Parsa and Maghsoudi, 2021; Parsa and Carranza, 2021) have adopted a scale-independent, process-oriented, probabilistic framework, referred to as the mineral systems framework (Knox-Robinson and Wyborn, 1997), to distill the data compiled from different surveys into a suite of 2D or 3D exploration targeting criteria. These exploration targeting criteria either directly or indirectly (i.e., by proxy) represent the main components of a mineral system: source, fluid pathways, trapping, deposition, and preservation (Mccuaig et al., 2010). Considering the probabilistic nature of mineral exploration, the ability to be used in various scales, and that this system relies on ore-forming processes rather than local geological features (Mccuaig et al., 2010), makes the mineral systems framework the ipso facto method of choice for today's mineral exploration campaigns.

This method starts by breaking the main ore-forming process into a set of constituent processes unique to individual styles of mineralization. The user should exploit the potential of data available to translate the constituent processes into a suite of 2D maps or 3D models that represent the constituent processes directly or by proxy (Mccuaig et al., 2010). An example of using the mineral system approach is illustrated in Table 1.1. In this table, fundamental processes leading to the deposition of Mississippi Valley-type Zn–Pb systems (Leach et al., 2010), namely, the transportation of ore-bearing fluids and the trapping

Table 1.1 Mineral systems–based translation of ore-forming processes of Mississippi Valley-Type Zn–Pb (MVT) systems into a suite of 2D exploration targeting criteria.

Fundamental processes	Constituent processes	Exploration targeting criteria
Transport	Faults provide pathways for the ascent and circulation of ore-bearing hydrothermal fluids	Maps of fault density can be used as an indicator of the transportation
The trapping mechanism	Carbonate platforms, be they limestones or dolostones, are porous media allowing the migration of ore-bearing fluids. Lithological transitions between the carbonate platforms and nonporous media, such as shales, deter the fluids from moving to neighboring lithologies, providing conditions necessary for ore deposition.	Maps showing the intersections of carbonate platforms and shales can be deemed as proxies describing the trapping mechanism
	Owing to the inherent porosity of the carbonate platforms decreases the pressure of ore-bearing fluids, which is another mechanism for trapping and deposition of fluids	Maps showing the extent of carbonate platforms are indicators directly showing the trapping mechanism

Source: After Parsa, M., 2021. A data augmentation approach to XGboost-based mineral potential mapping: an example of carbonate-hosted Zn Pb mineral systems of Western Iran. J. Geochem. Explor. 228, 106811. https://doi.org/10.1016/j.gexplo.2021.106811.

mechanism, have been described by their constituent processes, followed by the suite of exploration targeting criteria that may be employed in the predictive modeling stage.

1.3.3 Predictive modeling

The exploration targeting criteria defined are integrated into predictive models showing the favorability of discovering new mineral deposits in different locations of an area of interest. Predictive modeling constitutes several consecutive stages: (1) opting for a proper mathematical framework for generating predictive models, (2) interpreting predictive models, and finally (3) evaluating the predictive model's performance, thoroughly discussed in Chapter 7.

There are two general types of mathematical frameworks employed in MPM—knowledge- and data-driven (Carranza, 2008). The former entails a subjective weighting and synthesis of exploration targeting criteria; this strategy is appropriate for greenfield exploration with a few already-discovered mineral deposits. On the contrary, the latter suits data-rich terrains with information available on a decent number of already-discovered mineral deposits. A diverse array of mathematical frameworks, including multicriteria decision-making techniques (e.g., Feizi et al., 2017), fuzzy logic (e.g., Porwal et al., 2003; Sadr et al., 2014a; Parsa and Pour, 2021), and a whole array of other techniques (e.g., Bonham-Carter, 1994; Sadr et al., 2014b), have been exploited as knowledge-driven techniques for MPM. Likewise, different approaches have been adopted for data-driven MPM, ranging from the traditional use of weights of evidence (Bonham-Carter, 1994) approach to modern-day use of machine- and deep-learning-based algorithms (e.g., Zuo and Carranza, 2011; Parsa et al., 2018; Zuo et al., 2019; Parsa, 2021).

A predictive model is usually a stretched or continuous 2D/3D model, constituting cells or blocks with diverse predictive values. These maps are generally devoid of interpretation, requiring additional techniques for demarcating exploration targets. Together with the methods used for verifying the results of predictive models, these techniques will be discussed in Chapter 7 (Ericsson, 2012; Wright, 1989; Stevens, 2010).

References

Abdelsalam, M.G., Stern, R.J., Berhane, W.G., 2000. Mapping gossans in arid regions with Landsat TM and SIR-C images: the Beddaho Alteration Zone in northern Eritrea. J. Afr. Earth Sci. 30 (4), 903–916. Available from: https://doi.org/10.1016/S0899-5362(00)00059-2.

Abrams, M., Yamaguchi, Y., 2019. Twenty years of ASTER contributions to lithologic mapping and mineral exploration. Remote Sens. 1394. Available from: https://doi.org/10.3390/rs11111394.

Ahmadirouhani, R., Rahimi, B., Karimpour, M.H., Malekzadeh Shafaroudi, A., Afshar Najafi, S., Pour, A.B., 2017. Fracture mapping of lineaments and recognizing their tectonic significance using SPOT-5 satellite data: a case study from the Bajestan area, Lut Block, east of Iran. J. Afr. Earth Sci. 134, 600–612. Available from: https://doi.org/10.1016/j.jafrearsci.2017.07.027; http://www.sciencedirect.com/science/journal/1464343X.

Aitchison, J., 1982. The statistical analysis of compositional data. J. R. Stat. Soc.: Ser. B (Methodol.) 44 (2), 139–160. Available from: https://doi.org/10.1111/j.2517-6161.1982.tb01195.x.

Arthur, A.B., 1940. Exploration for hematite at Steep Rock Lake (Ontario). Eng. J. 23.

Baugh, W.M., Kruse, F.A., Atkinson, W.W., 1998. Quantitative geochemical mapping of ammonium minerals in the southern Cedar Mountains, Nevada, using the Airborne Visible/Infrared Imaging Spectrometer (AVIRIS). Remote Sens. Environ. 65 (3), 292–308. Available from: https://doi.org/10.1016/S0034-4257(98)00039-X.

Bedini, E., 2009. Mapping lithology of the Sarfartoq carbonatite complex, southern West Greenland, using HyMap imaging spectrometer data. Remote. Sens. Environ. 113 (6), 1208–1219. Available from: https://doi.org/10.1016/j.rse.2009.02.007.

Bedini, E., 2011. Mineral mapping in the Kap Simpson complex, central East Greenland, using HyMap and ASTER remote sensing data. Adv. Space Res. 47, 60–73. Available from: https://doi.org/10.1016/j.asr.2010.08.021.

Bedini, E., 2012. Mapping alteration minerals at Malmbjerg molybdenum deposit, central East Greenland, by Kohonen self-organizing maps and matched filter analysis of HyMap data. Int. J. Remote. Sens. 33 (4), 939–961. Available from: https://doi.org/10.1080/01431161.2010.542202; https://www.tandfonline.com/loi/tres20.

Best, M.E., 2015. Mineral Resources Treatise on Geophysics, second ed. Elsevier Inc, Canada, pp. 525–556. Available from: http://www.sciencedirect.com/science/book/9780444538031; https://doi.org/10.1016/B978-0-444-53802-4.00200-1.

Beus, A.A., Grigorian, S.V., 1977. Geochemical exploration methods. Miner. Depos. 287.

Binley, A., Slater, L., 2020. Resistivity and Induced Polarization Theory and Applications to the Near-Surface Earth. Cambridge University Press

Bishop, C.A., Liu, J.G., Mason, P.J., 2011. Hyperspectral remote sensing for mineral exploration in Pulang, Yunnan province, China. Int. J. Remote. Sens. 32 (9), 2409–2426. Available from: https://doi.org/10.1080/01431161003698336; https://www.tandfonline.com/loi/tres20.

Bliss, J.D., 1992. Developments in mineral deposit modeling. Geol. Surv. Bull. 168.

Bonham-Carter, G.F., 1994. Geographic Information Systems for Geoscientists: Modelling with GIS Pergamon, Oxford.

Cagniard, L., 1953. Basic theory of the Magneto-Telluric method of geophysical prospecting. Geophysics 605–635. Available from: https://doi.org/10.1190/1.1437915.

Carranza, E.J.M., 2008. Geochemical Anomaly and Mineral Prospectivity Mapping in GIS. Elsevier

Cocks, T., Jenssen, R., Stewart, W.I., Shields, T., 1998. The HyMap airborne hyperspectral sensor: The system, calibration and performance. In: Schaepman, M., Schläpfer, D., Itten, K.I. (Eds.), Proceedings of the 1st EARSeL Workshop on Imaging Spectroscopy, 6–8 October, Zurich. EARSeL, Paris, pp. 37–43.

Cox, D.P., Singer, D.A., 1986. Mineral deposit models. U.S. Geol. Surv. Bull.

Eldosouky, A.M., Abdelkareem, M., Elkhateeb, S.O., 2017. Integration of remote sensing and aeromagnetic data for mapping structural features and hydrothermal alteration zones in Wadi Allaqi area, South Eastern Desert of Egypt. J. Afr. Earth Sci. 130, 28–37. Available from: https://doi.org/10.1016/j.jafrearsci.2017.03.006; http://www.sciencedirect.com/science/journal/1464343X.

Eldosouky, A.M., Sehsah, H., Elkhateeb, S.O., Pour, A.B., 2020. Integrating aeromagnetic data and Landsat-8 imagery for detection of post-accretionary shear zones controlling hydrothermal alterations: the Allaqi-Heiani Suture zone, South Eastern Desert, Egypt. Adv. Space Res. 65 (3), 1008–1024. Available from: https://doi.org/10.1016/j.asr.2019.10.030; http://www.journals.elsevier.com/advances-in-space-research/.

Eldosouky, A.M., El-Qassas, R.A.Y., Pour, A.B., Mohamed, H., Sekandari, M., 2021. Integration of ASTER satellite imagery and 3D inversion of aeromagnetic data for deep mineral exploration. Adv. Space Res. 68 (9), 3641–3662. Available from: https://doi.org/10.1016/j.asr.2021.07.016; http://www.journals.elsevier.com/advances-in-space-research/.

Elkhateeb, S.O., Eldosouky, A.M., 2016. Detection of porphyry intrusions using analytic signal (AS), Euler Deconvolution, and Center for Exploration Targeting (CET) Technique Porphyry Analysis at Wadi Allaqi Area, South Eastern Desert, Egypt. Int. J. Sci. Eng. Res. 7 (6), 471–477.

Elkhateeb, S.O., Eldosouky, A.M., Khalifa, M.O., Aboalhassan, M., 2021. Probability of mineral occurrence in the Southeast of Aswan area, Egypt, from the analysis of aeromagnetic data. Arab. J. Geosci. 14 (15), Available from: https://doi.org/10.1007/s12517-021-07997-1; http://www.springer.com/geosciences/journal/12517?cm_mmc = AD-_-enews-_-PSE1892-_-0.

Ericsson, M. Mining technology-trends and development. POLINARES Working Paper Number 29, 2012.

Feizi, F., Karbalaei-Ramezanali, A., Tusi, H., 2017. Mineral potential mapping via TOPSIS with hybrid AHP–Shannon entropy weighting of evidence: a case study for Porphyry-Cu, Farmahin Area, Markazi Province, Iran. Nat. Resour. Res. 26 (4), 553–570. Available from: https://doi.org/10.1007/s11053-017-9338-3; http://www.kluweronline.com/issn/1520-7439.

Ford, A., Peters, K.J., Partington, G.A., Blevin, P.L., Downes, P.M., Fitzherbert, J.A., et al., 2019. Translating expressions of intrusion-related mineral systems into mappable spatial proxies for mineral potential mapping: case studies from the Southern New England Orogen, Australia. Ore Geol. Rev. 111, 102943. Available from: https://doi.org/10.1016/j.oregeorev.2019.102943.

Galley, A.G., Hannington, M.D., Jonasson, I.R., 2007. Volcanogenic massive sulphide deposits. In:. Geological Association of Canada, Mineral Deposits Division. Springer, pp. 141–161.

Gersman, R., Ben-Dor, E., Beyth, M., Avigad, D., Abraha, M., Kibreab, A., 2008. Mapping of hydrothermally altered rocks by the EO-1 Hyperion sensor, Northern Danakil Depression, Eritrea. Int. J. Remote. Sens. 29 (13), 3911–3936. Available from: https://doi.org/10.1080/01431160701874587; https://www.tandfonline.com/loi/tres20.

Green, R.O., Eastwood, M.L., Sarture, C.M., Chrien, T.G., Aronsson, M., Chippendale, B.J., et al., 1998. Imaging spectroscopy and the Airborne Visible/Infrared Imaging Spectrometer (AVIRIS). Remote Sens. Environ. 65 (3), 227–248. Available from: https://doi.org/10.1016/S0034-4257(98)00064-9.

Guilbert, J., Park, C., 1986. Geol. Ore Depos. 985.

Guo, Z., Xue, G., Liu, J., Wu, X., 2020. Electromagnetic methods for mineral exploration in China: a review. Ore Geol. Rev. 118, 103357. Available from: https://doi.org/10.1016/j.oregeorev.2020.103357.

Hagemann, S.G., Lisitsin, V.A., Huston, D.L., 2016. Mineral system analysis: Quo vadis. Ore Geol. Rev. 76, 504–522. Available from: https://doi.org/10.1016/j.oregeorev.2015.12.012; http://www.sciencedirect.com/science/journal/01691368.

Haldar, S.K., 2018. Mineral Exploration: Principles and Applications. Elsevier

Heincke, B., Jackisch, R., Saartenoja, A., Salmirinne, H., Rapp, S., Zimmermann, R., et al., 2019. Developing multi-sensor drones for geological mapping and mineral exploration: setup and first results from the MULSEDRO project. Geol. Surv. Den. Greenl. Bulletin. 43, Available from: https://doi.org/10.34194/GEUSB-201943-03-02; https://geusbulletin.org/index.php/geusb.

Heller Pearlshtien, D., Pignatti, S., Greisman-Ran, U., Ben-Dor, E., 2021. PRISMA sensor evaluation: a case study of mineral mapping performance over Makhtesh Ramon, Israel. Int. J. Remote. Sens. 42 (15), 5882–5914. Available from: https://doi.org/10.1080/01431161.2021.1931541.

Hinze, W.J., 1960. Application of the gravity method to iron ore exploration. Econ. Geol. 55 (3), 465–484. Available from: https://doi.org/10.2113/gsecongeo.55.3.465; http://economicgeology.org/content/55/3/465.full.pdf + html. United States.

Hinze, W.J., Frese, R.R.B., Saad, A.H., 2013. Gravity Magnetic Exploration.

Hosseini, S., Lashkaripour, G.R., Moghadas, N.H., Ghafoori, M., Pour, A.B., 2019. Lineament mapping and fractal analysis using SPOT-ASTER satellite imagery for evaluating the severity of slope weathering process. Adv. Space Res 63 (2), 871–885. Available from: https://doi.org/10.1016/j.asr.2018.10.005; http://www.journals.elsevier.com/advances-in-space-research/.

Hubbard, B.E., Crowley, J.K., Zimbelman, D.R., 2003. Comparative alteration mineral mapping using visible to shortwave infrared (0.4–2.4 μm) Hyperion, ALI, and ASTER imagery. IEEE Trans. Geosci. Remote Sens. 41 (6), 1401–1410. Available from: https://doi.org/10.1109/TGRS.2003.812906.

Hunt, G.R., Ashley, R.P., 1979. Spectra of altered rocks in the visible and near infrared. Econ. Geol. 74 (7), 1613–1629. Available from: https://doi.org/10.2113/gsecongeo.74.7.1613.

Huo, H., Ni, Z., Jiang, X., Zhou, P., Liu, L., 2014. Mineral mapping and ore prospecting with HyMap data over Eastern Tien Shan, Xinjiang Uyghur autonomous region. Remote. Sens. 6 (12), 11829–11851. Available from: https://doi.org/10.3390/rs61211829.

Jackisch, R., Madriz, Y., Zimmermann, R., Pirttijärvi, M., Saartenoja, A., Heincke, B.H., et al., 2019. Drone-borne hyperspectral and magnetic data integration: Otanmäki Fe-Ti-V deposit in Finland. Remote Sens. 11 (18), 2084. Available from: https://doi.org/10.3390/rs11182084.

Jaireth, S., Huston, D., 2010. Metal endowment of cratons, terranes and districts: insights from a quantitative analysis of regions with giant and super-giant deposits. Ore Geol. Rev. 38 (3), 288–303. Available from: https://doi.org/10.1016/j.oregeorev.2010.05.005.

Kirsch, M., Lorenz, S., Zimmermann, R., Tusa, L., Möckel, R., Hödl, P., et al., 2018. Integration of terrestrial and drone-borne hyperspectral and photogrammetric sensing methods for exploration mapping and mining monitoring. Remote Sens. 10 (9), 1366. Available from: https://doi.org/10.3390/rs10091366.

Knox-Robinson, C.M., Wyborn, L.A.I., 1997. Towards a holistic exploration strategy: using geographic information systems as a tool to enhance exploration. Aust. J. Earth Sci. 44 (4), 453–463. Available from: https://doi.org/10.1080/08120099708728326.

Kreuzer, O.K., Etheridge, M.A., Guj, P., McMahon, M.E., Holden, D.J., 2008. Linking mineral deposit models to quantitative risk analysis and decision-making in exploration. Econ. Geol. 103 (4), 829–850. Available from: https://doi.org/10.2113/gsecongeo.103.4.829.

Kruse, F.A., Perry, S.L., 2013. Mineral mapping using simulated worldview-3 short-wave-infrared imagery. Remote. Sens. 5 (6), 2688–2703. Available from: https://doi.org/10.3390/rs5062688; http://www.mdpi.com/2072-4292/5/6/2688/pdf. United States.

Kruse, F.A., Boardman, J.W., Huntington, J.F., 2003. Comparison of airborne hyperspectral data and EO-1 Hyperion for mineral mapping. IEEE Trans. Geosci. Remote. Sens. 41 (6), 1388–1400. Available from: https://doi.org/10.1109/TGRS.2003.812908.

Kruse, F.A., Baugh, W.M., Perry, S.L., 2015. Validation of DigitalGlobe WorldView-3 Earth imaging satellite shortwave infrared bands for mineral mapping. J. Appl. Remote. Sens. 9 (1), Available from: https://doi.org/10.1117/1.JRS.9.096044; http://www.spie.org/x3636.xml.

Leach, D.L., Sangster, D.F., Kelley, K.D., Large, R.R., Garven, G., Allen, C.R., et al., 2005. Sediment-hosted lead-zinc deposits: a global perspective. Econ. Geol. 100, 561–607.

Leach, D.L., Taylor, R.D., Fey, D.L., Diehl, S.F., Saltus, R.W., 2010. A deposit model for Mississippi Valley-type lead-zinc ores. Chapter A of Mineral Deposit Models for Resource Assessment. USGS.

Liao, L., Jarecke, P., Gleichauf, D., Hedman, T., 2000. Performance characterization of the Hyperion imaging spectrometer instrument. In: Proceedings of SPIE – The International Society for Optical Engineering, United States, 4135, 264–275. Available from: https://doi.org/10.1117/12.494253, 0277786X.

Malehmir, A., Durrheim, R., Bellefleur, G., Urosevic, M., Juhlin, C., White, D.J., et al., 2012. Campbell, Seismic methods in mineral exploration and mine planning: a general overview of past and present case histories and a look into the future. Geophysics 77 (5). Available from: https://doi.org/10.1190/geo2012-0028.1, WC173-WC190.

Mars, J.C., 2018. Mineral and lithologic mapping capability of worldview 3 data at Mountain Pass, California, using true- and false-color composite images, band ratios, and logical operator algorithms. Econ. Geol. 113 (7), 1587–1601. Available from: https://doi.org/10.5382/econgeo.2018.4604; https://watermark.silverchair.com.

Mars, J.C., Rowan, L.C., 2006. Regional mapping of phyllic- and argillic-altered rocks in the Zagros magmatic arc, Iran, using Advanced Spaceborne Thermal Emission and Reflection Radiometer (ASTER) data and logical operator algorithms. Geosphere 2 (3), 161–186. Available from: https://doi.org/10.1130/GES00044.1.

Mccuaig, T.C., Beresford, S., Hronsky, J., 2010. Translating the mineral systems approach into an effective exploration targeting system. Ore Geol. Rev. 38 (3), 128–138. Available from: https://doi.org/10.1016/j.oregeorev.2010.05.008.

Nabighian, M.N., Ander, M.E., Grauch, V.J.S., Hansen, R.O., LaFehr, T.R., Li, Y., et al., 2005a. Historical development of the gravity method in exploration. Geophysics 70 (6), Available from: https://doi.org/10.1190/1.2133785; http://library.seg.org/loi/gpysa7.

Nabighian, M.N., Grauch, V.J.S., Hansen, R.O., LaFehr, T.R., Li, Y., Peirce, J.W., et al., 2005b. The historical development of the magnetic method in exploration. Geophysics 70 (6), Available from: https://doi.org/10.1190/1.2133784; http://library.seg.org/loi/gpysa7.

Nostrand, K.L.C., 1966. Interpretation of Resistivity Data. United States Government Printing Office, Washington.

Paganelli, F., Grunsky, E., Richards, J., Pryde, R., 2003. Use of RADARSAT-1 principal component imagery for structural mapping: a case study in the Buffalo Head Hills area, northern central Alberta. Canada. Can. J. Remote. Sens. 29, 111–140. Available from: https://doi.org/10.5589/m02-084.

Park, S., Choi, Y., 2020. Applications of unmanned aerial vehicles in mining from exploration to reclamation: a review. Minerals 10 (8), 663. Available from: https://doi.org/10.3390/min10080663.

Parsa, M., 2021. A data augmentation approach to XGboost-based mineral potential mapping: an example of carbonate-hosted Zn Pb mineral systems of Western Iran. J. Geochem. Explor 228, 106811. Available from: https://doi.org/10.1016/j.gexplo.2021.106811.

Parsa, M., Carranza, E.J.M., 2021. Modulating the impacts of stochastic uncertainties linked to deposit locations in data-driven predictive mapping of mineral prospectivity. Nat. Resour. Res. 30 (5), 3081–3097. Available from: https://doi.org/10.1007/s11053-021-09891-9; https://link.springer.com/journal/11053.

Parsa, M., Maghsoudi, A., 2021. Assessing the effects of mineral systems-derived exploration targeting criteria for random Forests-based predictive mapping of mineral prospectivity in Ahar-Arasbaran area, Iran. Ore Geol. Rev. 138, 104399. Available from: https://doi.org/10.1016/j.oregeorev.2021.104399.

Parsa, M., Pour, A.B., 2021. A simulation-based framework for modulating the effects of subjectivity in greenfield Mineral Prospectivity Mapping with geochemical and geological data. J. Geochem. Explor. 229, 106838. Available from: https://doi.org/10.1016/j.gexplo.2021.106838.

Parsa, M., Maghsoudi, A., Yousefi, M., 2018. Spatial analyses of exploration evidence data to model skarn-type copper prospectivity in the Varzaghan district, NW Iran. Ore Geol. Rev. 92, 97–112. Available from: https://doi.org/10.1016/j.oregeorev.2017.11.013; http://www.sciencedirect.com/science/journal/01691368.

Plumlee, G., 1999. The environmental geology of mineral deposits. Rev. Economic Geol. 6, 71–116.

Porwal, A., Carranza, E.J.M., Hale, M., 2003. Knowledge-driven and data-driven fuzzy models for predictive mineral potential mapping. Nat. Resour. Res. 12 (1), 1–25. Available from: https://doi.org/10.1023/A:1022693220894.

Pour, A.B., Hashim, M., 2018. Hydrothermal alteration mapping from Landsat-8 data, Sar Cheshmeh copper mining district, south-eastern Islamic Republic of Iran. J. Taibah Univ. Sci. 155–166. Available from: https://doi.org/10.1016/j.jtusci.2014.11.008.

Pour, A.B., Hashim, M., van Genderen, J., 2013. Detection of hydrothermal alteration zones in a tropical region using satellite remote sensing data: Bau goldfield, Sarawak, Malaysia. Ore Geol. Rev. 54, 181–196. Available from: https://doi.org/10.1016/j.oregeorev.2013.03.010.

Pour, A.B., Hashim, M., Marghany, M., 2014. Exploration of gold mineralization in a tropical region using Earth Observing-1 (EO1) and JERS-1 SAR data: a case study from Bau gold field, Sarawak, Malaysia. Arab. J. Geosci. 7 (6), 2393–2406. Available from: https://doi.org/10.1007/s12517-013-0969-3; http://www.springer.com/geosciences/journal/12517?cm_mmc = AD-_-enews-_-PSE1892-_ – 0.

Pour, A.B., Hashim, M., Makoundi, C., Zaw, K., 2016. Structural mapping of the Bentong-Raub suture zone using PALSAR remote sensing data, peninsular Malaysia: implications for sediment-hosted/orogenic gold mineral systems exploration. Resour. Geol 66 (4), 368–385. Available from: https://doi.org/10.1111/rge.12105; http://onlinelibrary.wiley.com/journal/10.1111/(ISSN)1751-3928.

Pour, A.B., Park, T.Y.S., Park, Y., Hong, J.K., Zoheir, B., Pradhan, B., et al., 2018. Application of multi-sensor satellite data for exploration of Zn-Pb sulfide mineralization in the Franklinian Basin, North Greenland. Remote Sens. 10 (8), Available from: https://doi.org/10.3390/rs10081186; https://res.mdpi.com/remotesensing/remotesensing-10-01186/article_deploy/remotesensing-10-01186.pdf?filename = &attachment = 1.

Pour, A.B., Park, T.Y.S., Park, Y., Hong, J.K., Muslim, A.M., Läufer, A., et al., 2019. Landsat-8, advanced spaceborne thermal emission and reflection radiometer, and WorldView-3 multispectral satellite imagery for prospecting copper-gold mineralization in the northeastern Inglefield Mobile Belt (IMB), northwest Greenland. Remote Sens. 11 (20), Available from: https://doi.org/10.3390/rs11202430; https://res.mdpi.com/d_attachment/remotesensing/remotesensing-11-02430/article_deploy/remotesensing-11-02430-v2.pdf.

Rikitake, T., 1950. Electromagnetic induction within the Earth and its relation to the electrical state of the Earth's interior. Part II. Tokyo Univ. Bull. Earthq. Res. Inst. 28, 263–283.

Rowan, L.C., Goetz, A.F.H., Ashley, R.P., 1977. Discrimination of hydrothermally altered and unaltered rocks in visible and near infrared multispectral images. Geophysics 42 (3), 522–535. Available from: https://doi.org/10.1190/1.1440723; http://library.seg.org/loi/gpysa7.

Rowan, L.C., Crowley, J.K., Schmidt, R.G., Ager, C.M., Mars, J.C., 2000. Mapping hydrothermally altered rocks by analyzing hyperspectral image (AVIRIS) data of forested areas in the Southeastern United States. J. Geochem. Explor. 68 (3), 145–166. Available from: https://doi.org/10.1016/S0375-6742(99)00081-3.

Sadr, M.P., Maghsoudi, A., Saljoughi, B.S., 2014a. Landslide susceptibility mapping of Komroud sub-basin using fuzzy logic approach. Geodynam. Res. Int. Bull. 2, XVI-XXVIII.

Sadr, M.P., Hassani, H., Maghsoudi, A., 2014b. Slope instability assessment using a weighted overlay mapping method, A case study of Khorramabad-Doroud railway track, W Iran. J. Tethys 2 (3), 254–271.

Safari, M., Maghsoudi, A., Pour, A.B., 2018. Application of Landsat-8 and ASTER satellite remote sensing data for porphyry copper exploration: a case study from Shahr-e-Babak, Kerman, South Iran. Geocarto Int. 33 (11), 1186–1201. Available from: https://doi.org/10.1080/10106049.2017.1334834; http://www.tandfonline.com/toc/tgei20/current.

Salisbury, M.H., Snyder, d, 2007. Mineral deposits of Canada: a synthesis of major deposit types, district metallogeny, the evolution of geological provinces, and exploration methods: geological Association of Canada. Miner. Depos. Division. Spec. Publ. 5, 971–982.

Salles, R.D.R., de Souza Filho, C.R., Cudahy, T., Vicente, L.E., Monteiro, L.V.S., 2017. Hyperspectral remote sensing applied to uranium exploration: a case study at the Mary Kathleen metamorphic-hydrothermal U-REE deposit, NW, Queensland, Australia. J. Geochem. Explor. 179, 36–50. Available from: https://doi.org/10.1016/j.gexplo.2016.07.002; http://www.sciencedirect.com/science/journal/03756742.

Schmidt, G., 1959. Results of underground-seismic reflection investigations in the siderite district of the Siegerland*. Geophys. Prospecting 7 (3), 287–290. Available from: https://doi.org/10.1111/j.1365-2478.1959.tb01470.x.

Sheard, S.N., Ritchie, T.J., Christopherson, K.R., Brand, E., 2005. Mining, environmental, petroleum, and engineering industry applications of electromagnetic techniques in geophysics. Surv. Geophys. 26 (5), 653–669. Available from: https://doi.org/10.1007/s10712-005-1760-0.

Singer, D.A., Kouda, R., 1999. Examining risk in mineral exploration. Nat. Resour. Res. 8 (2), 111–122. Available from: https://doi.org/10.1023/A:1021838618750; http://www.kluweronline.com/issn/1520-7439.

Singhroy, V.H., 2001. Canada Geological applications of RADARSAT-1: a review. In: International Geoscience and Remote Sensing Symposium (IGARSS) 1, 468–470.

Stevens, R., 2010. Mineral Exploration and Mining EssentialsPakawau Geomanagement Inc., Port Coquitlam, p. 322.

Tikhonov, A.N., 1950. On the determination of electrical characteristics of deep layers of the Earth's crust (in Russian). Doklady Akademii Nauk, SSSR 73, 295–297.

Walter, C., Braun, A., Fotopoulos, G., 2020. High-resolution unmanned aerial vehicle aeromagnetic surveys for mineral exploration targets. Geophys. Prospect. 68 (1), 334–349. Available from: https://doi.org/10.1111/1365-2478.12914; http://onlinelibrary.wiley.com/journal/10.1111/(ISSN)1365-2478.

Woodhouse, I.H., 2006. Introduction to Microwave Remote Sensing. CRC Press, Taylor & Francis Group, Boca Raton

Wright, D.F., 1989. Weights of evidence modelling: a new approach to mapping mineral potential. Stat. Appl. Earth Sci. 89 (9), 171–183.

Van Der Meer, F.D., Van Der Werff, H.M.A., Van Ruitenbeek, F.J.A., 2014. Potential of ESA's sentinel-2 for geological applications. Remote Sens. Environ. 148, 124–133. Available from: https://doi.org/10.1016/j.rse.2014.03.022.

van der Meer, F.D., van der Werff, H.M.A., van Ruitenbeek, F.J.A., Hecker, C.A., Bakker, W.H., Meijde, N., et al., 2012a. Multi- and Hyperspectral Geologic Remote Sensing: A Review.

van der Meer, F.D., van der Werff, H.M.A., van Ruitenbeek, F.J.A., 2012b. Potential of ESA's Sentinel-2 for geological applications. Remote Sens. Environ. 148, 112–128.

Van Der Werff, H., Van Der Meer, F., 2016. Sentinel-2A MSI and Landsat 8 OLI Provide Data Continuity for Geological Remote Sensing. 8.

Zhang, J., Zeng, Z., Zhao, X., Li, J., Zhou, Y., Gong, M., 2020. Deep mineral exploration of the Jinchuan Cu–Ni sulfide deposit based on aeromagnetic, gravity, and CSAMT methods. Minerals 10 (2). Available from: https://doi.org/10.3390/min10020168; https://www.mdpi.com/2075-163X/10/2/168/pdf.

Zhdanov, M.S., 2010. Electromagnetic geophysics: notes from the past and the road ahead. Geophysics 75 (5), 75A49–75A66. Available from: https://doi.org/10.1190/1.3483901.

Zoheir, B., El-Wahed, M.A., Pour, A.B., Abdelnasser, A., 2019. Orogenic gold in transpression and transtension zones: field and remote sensing studies of the Barramiya–Mueilha sector, Egypt. Remote. Sens. 11 (18), Available from: https://doi.org/10.3390/rs11182122; https://res.mdpi.com/d_attachment/remotesensing/remotesensing-11-02122/article_deploy/remotesensing-11-02122.pdf.

Zuo, R., Carranza, E.J.M., 2011. Support vector machine: a tool for mapping mineral prospectivity. Comput. Geosci. 37 (12), 1967–1975. Available from: https://doi.org/10.1016/j.cageo.2010.09.014.

Zuo, R., Wang, J., 2016. Fractal/multifractal modeling of geochemical data: a review. J. Geochem. Explor. 164, 33–41. Available from: https://doi.org/10.1016/j.gexplo.2015.04.010; http://www.sciencedirect.com/science/journal/03756742.

Zuo, R., Xiong, Y., Wang, J., Carranza, E.J.M., 2019. Deep learning and its application in geochemical mapping. Earth-Sci. Rev. 192, 1–14. Available from: https://doi.org/10.1016/j.earscirev.2019.02.023; http://www.sciencedirect.com/science/journal/00128252.

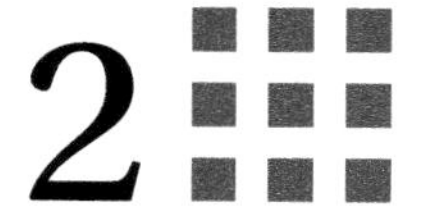

Remote sensing for mineral exploration

Amin Beiranvand Pour[1], Hojjatollah Ranjbar[2], Milad Sekandari[2], Mohamed Abd El-Wahed[3], Mohammad Shawkat Hossain[1], Mazlan Hashim[4], Mastoureh Yousefi[5], Basem Zoheir[6,7], Jonas Didero Takodjou Wambo[8], Aidy M. Muslim[1]

[1]INSTITUTE OF OCEANOGRAPHY AND ENVIRONMENT (INOS), UNIVERSITY MALAYSIA TERENGGANU (UMT), KUALA NERUS, TERENGGANU, MALAYSIA [2]MINING ENGINEERING DEPARTMENT, SHAHID BAHONAR UNIVERSITY OF KERMAN, KERMAN, IRAN [3]GEOLOGY DEPARTMENT, FACULTY OF SCIENCE, TANTA UNIVERSITY, TANTA, EGYPT [4]GEOSCIENCE AND DIGITAL EARTH CENTRE (INSTEG), RESEARCH INSTITUTE FOR SUSTAINABLE ENVIRONMENT, UNIVERSITI TEKNOLOGI MALAYSIA, JOHOR BAHRU, MALAYSIA [5]DEPARTMENT OF MINING ENGINEERING, ISFAHAN UNIVERSITY OF TECHNOLOGY, ISFAHAN, IRAN [6]GEOLOGY DEPARTMENT, FACULTY OF SCIENCE, BENHA UNIVERSITY, BENHA, EGYPT [7]INSTITUTE OF GEOSCIENCES, UNIVERSITY OF KIEL, KIEL, GERMANY [8]DIVISION OF GEOLOGICAL AND PLANETARY SCIENCES, CALIFORNIA INSTITUTE OF TECHNOLOGY, PASADENA, CA, UNITED STATES

2.1 Introduction

Ore minerals particularly contribute to the economic evolution of a country and provide raw materials for the appropriate functioning of modern societies. Remote sensing technology is a greatly applicable tool for regional geological mapping and ore mineral exploration at a low cost. The application of remote sensing techniques in the initial stages of ore mineral exploration has a significant impact on reducing exploration and exploitation prices. In recent decades, hydrothermal alteration mineral detection, lithological and structural mapping are the most conspicuous applications of multispectral, hyperspectral, synthetic aperture radar (SAR), and unmanned aerial vehicle (UAV) remote sensing data for mineral exploration. In the metallogenic provinces and frontier areas around the world, several ore mineralizations such as orogenic gold, porphyry copper, massive sulfide, epithermal gold, podiform chromite, uranium, magnetite, and iron oxide copper–gold (IOCG) deposits have been auspiciously explored using remote sensing data. The Landsat data series, the Satellite Pour l'Observation de la Terre-1 (SPOT) series, the Advanced Spaceborne Thermal Emission

Geospatial Analysis Applied to Mineral Exploration. DOI: https://doi.org/10.1016/B978-0-323-95608-6.00002-0

and Reflection Radiometer (ASTER), the Advanced Land Imager (ALI), Sentinel series, Worldview series, Hyperion, HyMap, the Airborne Visible/IR Image Spectrometer (AVIRIS), a variety of UAVs, and SAR remote sensing data support cost-effective techniques for mineral exploration. Advanced image-processing algorithms based on state-of-the-art data extraction techniques have been executed and developed for identifying key alteration minerals associated with a variety of ore deposits and geological structural features. Additionally, fieldwork, analytical spectral device spectroscopy, and X-ray diffraction (XRD) analysis in conjunction with remote sensing data can provide extensive information about hydrothermal alteration zones and structural features related to ore mineralization.

2.2 Spectral features of hydrothermal alteration minerals and lithologies

The capability of discriminating between hydrothermally altered and unaltered rocks is one of the significant factors for mineral exploration programs. In the region of solar-reflected light (0.325–2.5 μm), many minerals show diagnostic absorption features due to vibrational overtones, electronic transition, charge transfer, and conduction processes (Hunt, 1977; Hunt and Ashley, 1979; Clark et al., 1990; Cloutis, 1996). Hydrothermally altered rocks are frequently indicated by iron oxide/hydroxides, clay, carbonate, and sulfate mineral assemblages, which produce diagnostic absorption signatures throughout the visible and near-infrared (VNIR) and shortwave-infrared (SWIR) regions. Quartz is one of the important rock-forming minerals that does not have diagnostic spectral absorption features in the VNIR and SWIR regions but displays strong fundamental molecular absorption features in the thermal infrared (TIR) region (8–12 μm) (Ninomiya et al., 2005; Hunt and Salisbury, 1976).

2.2.1 Iron oxide minerals

Iron oxides are one of the significant mineral groups that are associated with hydrothermally altered rocks of many ore deposits (Sabins, 1999). Iron oxide/hydroxide minerals are formed during supergene alteration and represent a yellowish or reddish tint to the altered rocks, which are generally called gossan (Abdelsalam et al., 2000). In the VNIR part of the electromagnetic spectrum, the ion transition energy distribution depends on the cation and the mineral structure, which influence spectral characteristics (Hunt and Ashley, 1979). Electronic processes produce absorption features in the VNIR radiation (0.4–1.1 μm) due to the presence of transition elements, such as $Fe^{2+}$and Fe^{3+}, and often substituted by Mn, Cr, and Ni in the crystal structure of the minerals (Hunt, 1977; Hunt and Ashley, 1979; Clark et al., 1993a,b). The concentration of transition metal cations, such as Fe^{3+} and Fe^{2+}, can affect the intensities of absorption features (Sherman and Waite, 1985). Fe^{3+} produces absorption features near 0.45–0.90 μm, while broad absorption features near 0.90–1.2 μm are related to Fe^{2+} (Morris et al., 1985). Specifically, the absorption features related to Fe^{3+} are typically at 0.49, 0.70, and 0.87 μm, while Fe^{2+} shows absorption properties at 0.51, 0.55, and 1.20 μm (Hunt, 1977; Clark, 1999).

Charge-transfer absorption features between 0.48 and 0.72 μm and crystal-field absorption properties between 0.63 and 0.72 μm are documented for iron oxide/hydroxide minerals such as hematite, limonite, goethite, and jarosite (Sherman and Waite, 1985; Morris et al., 1985). Accordingly, iron oxide/hydroxide minerals such as goethite, limonite, jarosite, and hematite tend to have spectral absorption features in the visible to middle infrared from 0.4 to 1.1 μm (absorption features of Fe^{3+} about 0.45–0.90 μm and Fe^{2+} about 0.90–1.2 μm) of the electromagnetic spectrum (Hunt, 1977; Hunt and Ashley, 1979; Hunt and Salisbury, 1974). Goethite, hematite, and limonite have strong Fe^{3+} absorption features at 0.97–0.83 and 0.48 μm. Jarosite has Fe–O–H absorption features at 0.94 and 2.27 μm (Hunt et al., 1971a,b; Bishop and Murad, 2005) (Fig. 2.1).

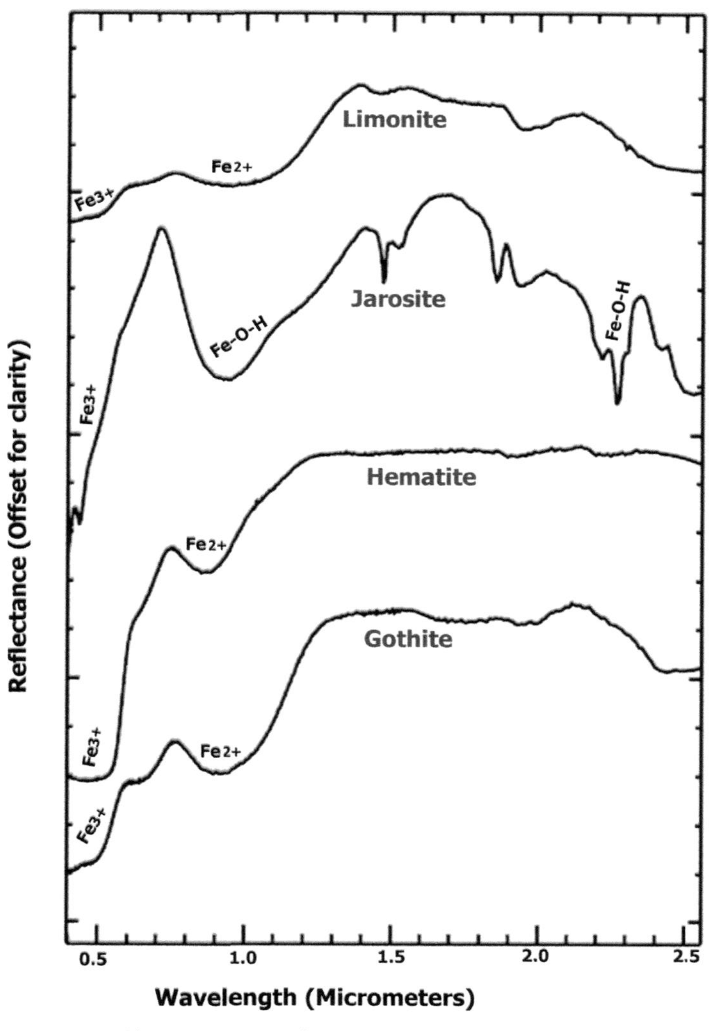

FIGURE 2.1 Laboratory spectra of limonite, jarosite, hematite, and goethite.

2.2.2 Hydroxyl-bearing minerals and carbonates

SWIR radiation is the best spectral part of the electromagnetic spectrum for identifying various assemblages of alteration minerals and zones associated with hydrothermal ore mineralizations. Generally, the spatial distribution of hydrothermal alteration minerals having OH and CO_3 groups can be identified and mapped in the SWIR region (Huntington, 1996). It is due to the fact that hydroxyl-bearing minerals, including clay and sulfate groups, as well as carbonate minerals represent diagnostic spectral absorption features due to vibrational processes of fundamental absorptions of Al–O–H, Mg–O–H, Si–O–H, and CO_3 groups in the SWIR region (Hunt and Ashley, 1979; Gaffey, 1986; Crowley, 1991; Clark et al., 1993a,b; Cloutis, 1996). Phyllosilicates, including Al–Si–(OH) and Mg–Si–(OH) bearing minerals (i.e., kaolinite, montmorillonite, muscovite, illite, talc, and chlorite), and sorosilicate group, including Ca–Al–Si–(OH) bearing minerals such as epidote group, and OH-bearing sulfates, including alunite and gypsum, and also carbonates can be identified because of their spectral characteristics in the SWIR region (Hunt, 1977; Hunt and Ashley, 1979; James et al., 1988; Clark et al., 1990; Cloutis et al., 2006; Bishop et al., 2008). Accordingly, the remote sensing SWIR data can identify hydrothermal alteration mineral assemblages, including (1) mineralogy generated by the passage of low PH fluids (alunite and pyrophyllite); (2) Al–Si–(OH) and Mg–Si–(OH)-bearing minerals, including kaolinite and mica and chlorite groups; and (3) Ca–Al–Si–(OH)-bearing minerals such as epidote group, as well as carbonate group (e.g., calcite and dolomite).

Sericitically altered rocks typically contain sericite (a fine-grained form of muscovite) that has a distinct Al–OH absorption feature at 2.2 μm and a less intense absorption feature at 2.35 μm (Fig. 2.2) (Abrams and Brown, 1984; Spatz and Wilson, 1995). Kaolinite and alunite are typical constituents of advanced argillic alteration that exhibit Al–OH 2.165- and 2.2-μm absorption features (Fig. 2.2) (Hunt, 1977; Hunt and Ashley, 1979; Rowan et al., 2003). Although less common than alunite or kaolinite, advanced argillic-altered rocks can also contain pyrophyllite that has an intense 2.165-μm Al–O–H absorption feature. Propylitically altered rocks typically contain variable amounts of chlorite, epidote, and calcite, which exhibit Fe, Mg–O–H, and CO_3 about 2.31–2.33-μm absorption features (Fig. 2.2) (Rowan and Mars, 2003). VNIR–SWIR spectra of epidote and chlorite also exhibit broad, prominent Fe^{2+} absorption spanning the 0.62–1.65-μm region (Fig. 2.2). The absorption features near 1.40 and 1.90 μm can be attributed to OH stretches occurring at about 1.4 μm and the combination of the H–O–H bend with OH stretches near 1.90 μm (Fig. 2.2). Hence, two overall absorption features at about 1.40 and 1.90 μm present in the alteration minerals due to the OH and H_2O vibrational bands, respectively (Bishop et al., 2008).

2.2.3 Silicate minerals and lithologies

Mafic-to-ultramafic and felsic (granitic) lithologies are potential geological units to host or accompany magmatic ore mineralization. The TIR data can be used for detecting silica (quartz) to explore magmatic ore deposits (i.e., Ni–PGE sulfide deposits in mafic-to-ultramafic

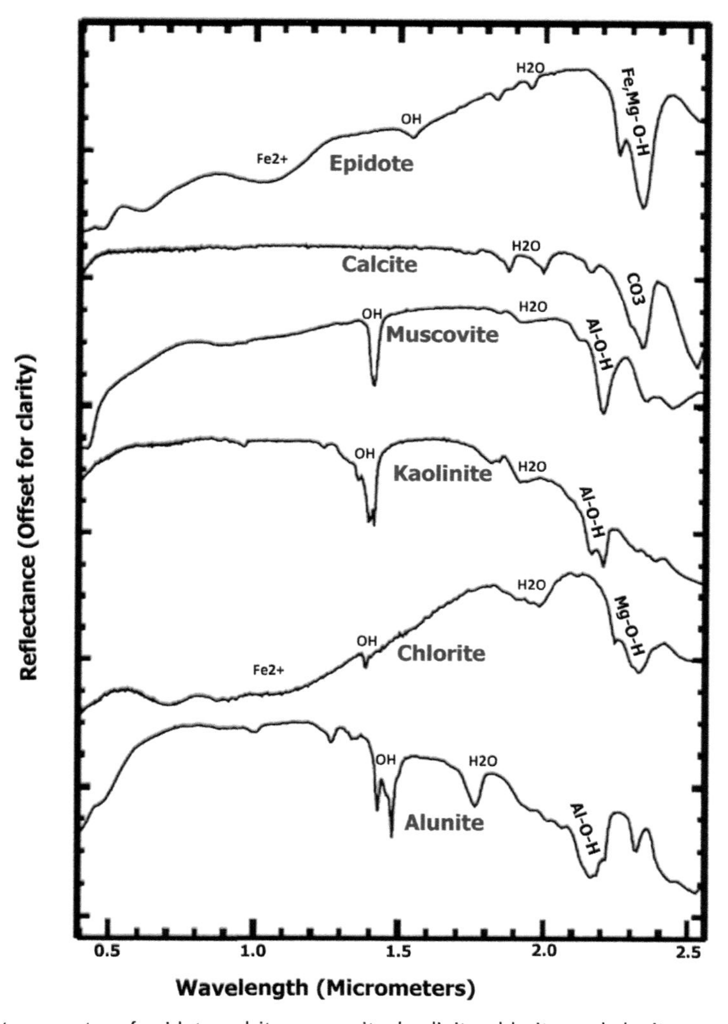

FIGURE 2.2 Laboratory spectra of epidote, calcite, muscovite, kaolinite, chlorite, and alunite.

units and nonferrous-metals, rare-metals, and radioactive-element ore deposits in felsic units), which are not associated with extensive alteration minerals and zones (Lyon, 1972; Hunt and Salisbury, 1976; Salisbury and D'Aria, 1992; Ninomiya, 1995; Ninomiya et al., 2005; Aboelkhair et al., 2010; Yajima and Yamaguchi, 2013; Ding et al., 2015). In the TIR region (8–14 μm), spectral emissivity features due to Si–O–Si-stretching vibrations are useful for the spectral discrimination of silicate lithological units (Ninomiya et al., 2005; Rockwell and Hofstra, 2008).

Alkali feldspar-bearing granites are characterized by high silica content, although in granodiorite and tonalite, the abundance of plagioclase is known to increase with decreasing quartz content. Spectra of granitoid rocks and silicates characteristically show a broad emission minimum (reststrahlen band) in the 8.5–14-μm interval (Si–O-stretching region), and a well-defined maximum (Christiansen frequency peak) in the 7–9-μm interval (Logan et al., 1973; Sabine et al., 1994; Hook et al., 1999). The subtle differences in the bonding of silicates are responsible for the shifting of the emissivity minima in the TIR spectra of silicate-bearing rocks (Fig. 2.3). The broad spectral emissivity minimum shifts to a longer wavelength for silicate rocks as the SiO_2 content decreases (the rock-type changes from felsic to mafic) (Fig. 2.3) (Salisbury and Walter, 1989; Salisbury and D'Aria, 1992; Ninomiya, 1995, 2003b).

Quartz can also be found in jasperoid, quartz–alunite, and quartz–sericite–pyrite alteration, hot spring silica sinter terraces, and several diatomite and perlite mines and prospects. Hydrothermal silica minerals typically consist of quartz, opal, and chalcedony. TIR emissivity spectra illustrate that quartz and opal contain a prominent reststrahlen feature in the 9.1-μm region (Fig. 2.3) (Rockwell and Hofstra, 2008). Numerous bodies of hydrothermal quartz were found in or near Carlin-type gold deposits, distal disseminated Au–Ag deposits, high- and low-sulfidation epithermal Au–Ag deposits, and geothermal areas. In recent years,

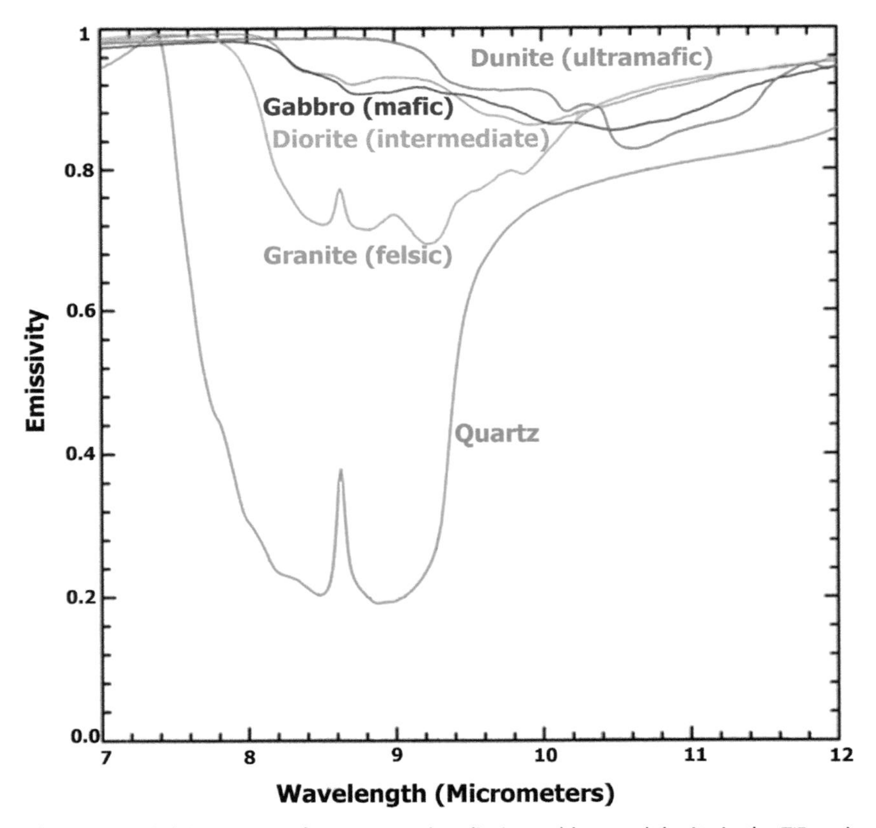

FIGURE 2.3 Laboratory emissivity spectra of quartz, granite, diorite, gabbro, and dunite in the TIR region. *TIR*, Thermal infrared.

several investigations documented the high capability of TIR remote sensing data for lithological discrimination and mapping around the world (Ninomiya et al., 2005; Scheidt et al., 2008; Aboelkhair et al., 2010; Pour and Hashim, 2012a,b; Yajima and Yamaguchi, 2013; Ramakrishnan et al., 2013; Ding et al., 2014, 2015; Guha and K., 2016; Guha and Kumar, 2016; Pour et al., 2018a; Ninomiya and Fu, 2019; Pour et al., 2021a,b).

2.3 Multispectral sensors

Multispectral remote sensing involves the acquisition of VNIR, SWIR, and TIR images in several broad wavelength bands. Various materials reflect, absorb, and emit inversely at different wavelengths. Essentially, it is feasible to distinguish between materials by their spectral reflectance and emissivity signatures as detected in multispectral remotely sensed images. Multispectral imagery generally refers to 3–10 (or more) bands that are represented in pixels. Each band is acquired using a remote sensing radiometer. Multispectral imaging concentrates on several preselected wavebands based on the application at hand. In this section, the technical characteristics of Landsat data, the Satellite Pour l'Observation de la Terra-1 (SPOT) series, the ASTER, the ALI, Sentinel Sentinel-2 MSI data, and Worldview data are briefly described.

2.3.1 Landsat data

The Landsat satellites era that began in 1972 will become a nearly 50-year global land record with the successful launch and operation of the Landsat Data Continuity Mission (LDCM) (Roy et al., 2014). The first generation (Landsats 1–3) operated from 1972 to 1985 and is essentially replaced by the second generation (Landsats 4, 5, and 7), which began in 1982 and continues to the present (Sabins, 1999; Roy et al., 2014). Landsat 7 (Enhanced Thematic Mapper Plus (ETM +)) was launched on April 15, 1999. It detects spectrally filtered radiation in the VNIR, SWIR, TIR, and panchromatic bands with a spatial resolution of 30, 60, and 15 m, respectively (Wulder et al., 2008; Irons et al., 2012). Landsat 8 was launched on February 4, 2013 and joins Landsat 7 in orbit, providing an increased coverage of the Earth's surface. It has two sensors, namely, the Operational Land Imager (OLI) and the Thermal Infrared Sensor (TIRS). These two instruments collect image data for nine VNIR and SWIR bands and two TIR bands (Roy et al., 2014). Landsat 9 was launched on September 27, 2021. It carries approximately identical sensors as Landsat 8, in particular the OLI-2, and the TIRS-2. The OLI-2 data contain 14-bit quantization. However, the quality of split-window thermal data usability for the TIRS-2 has been increased and improved. Fig. 2.4 shows spectral bandpasses for all Landsat sensors.

Landsat Thematic Mapper/Enhanced Thematic Mapper + (TM/ETM +) and Landsat 8 data have been used for mapping lithological units and hydrothermal alteration minerals associated with a variety of ore mineralizations (Pour et al., 2018a,b, 2019a,b,c; Sekandari et al., 2020; Traore et al., 2020; Takodjou Wambo et al., 2020; Moradpour et al., 2022). Bands 1, 3, 5, and 7 of TM/ETM + and bands 2, 4, 6, and 7 of Landsat 8 are specifically used to map iron oxide/hydroxides and hydroxyl-bearing minerals, correspondingly. For instance, in mapping hydrothermal

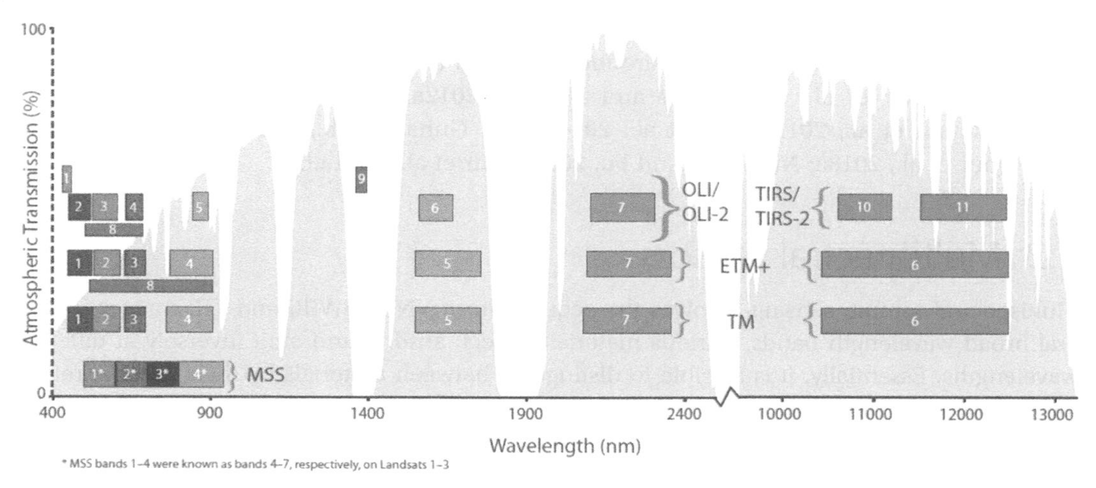

FIGURE 2.4 The bandpass wavelengths for the Landsat 1–9 sensors are shown.

alteration zones, the band ratios of 3/1 and 5/7 (from TM/ETM +) and the band ratios of 4/2 and 6/7 (from Landsat 8) can exclusively be implemented to identify gossan (iron oxide/hydroxides) and hydroxyl (OH) minerals, respectively (Abdelsalam et al., 2000; Pour and Hashim, 2015a,b,c; Safari et al., 2018; Takodjou Wambo et al., 2020; Bolouki et al., 2020; Moradpour et al., 2022). Nonetheless, the broad extent of these bands does not allow the discrimination of specific alteration zones (e.g., argillic and phyllic alteration zones) and mineral assemblages using the TM/ETM + and Landsat 8 bands. Distinguishing argillic, advanced argillic, phyllic and propylitic alteration zones is crucial for prospecting high-potential zone of porphyry copper mineralization and massive sulfides and many other hydrothermal ore mineralizations (Ranjbar et al., 2004; Honarmand et al., 2012; Pour and Hashim, 2011a,b, 2012a,b; Shahriari et al., 2013; Rajendran and Nasir, 2017; Yousefi et al., 2021, 2022). The performance characteristics of the ETM + and Landsat 8 and 9 sensors are shown in Table 2.1.

2.3.2 SPOT data

The Satellite Pour I'Observation de la Terra-1 (SPOT-l) was successfully launched into orbit by the French Space Agency, Centre National d'études Spatiales (CNES) in 1986, which delivered high-quality imageries from space. SPOT-2 was launched in 1990, followed by SPOT-3 in 1993, SPOT-4 in 1998, SPOT-5 in 2002, SPOT-6 in 2013, and SPOT-7 in 2014. It is anticipated that SPOT-6 and SPOT-7 can deliver data continuity by 2024 (Dagras et al., 1995). SPOT-5 presents enhanced facilities and cost-effective imaging solutions for geological applications, for example, oil and gas exploration, natural disaster management, urban and rural planning, and medium-scale (e.g., 1:25,000 and 1:10,000) geological–structural mapping (Huurneman et al., 2009; Hosseini et al., 2019). The spatial resolution of the SPOT-5 imagery is 3 m in the panchromatic band and 10 m in the multispectral mode, which is suitable for geological–structural mapping. Table 2.2 shows the technical characteristics of SPOT-5 data.

Table 2.1 The performance characteristics of the ETM$^+$ and Landsat-8 and 9 sensors.

Sensors	Subsystem	Band number	Spectral range (μm)	Ground resolution (m)	Swath width (km)
ETM +	VNIR	Pan	0.520–0.900	14.25	185
		1	0.450–0.515	28.50	
		2	0.525–0.605		
		3	0.633–0.690		
		4	0.780–0.900		
	SWIR	5	1.550–1.750		
		7	2.090–2.350		
	TIR	6	10.45–12.50		
Landsat-8/9	VNIR	1	0.433–0.453	30	185
		2	0.450–0.515		
		3	0.525–0.600		
		4	0.630–0.680		
		5	0.845–0.885		
	SWIR	6	1.560–1.660		
		7	2.100–2.300		
		Pan	0.500–0.680	15	
		9	1.360–1.390		
	TIR	10	10.30–11.30	100	
		11	11.50–12.50		

SWIR, Ahortwave infrared; *TIR*, thermal infrared; *VNIR*, visible and near-infrared.

Table 2.2 The technical characteristics of SPOT-5 satellite data.

Subsystem	Band number	Spectral range (nm)	Spatial resolution (m)	Location accuracy
Spectral bands	Band 1 (green)	500–590	10	<30 m (1σ)
	Band 2 (red)	610–680	10	
	Band 3 (near IR)	780–890	10	
	Band 4 (SWIR)	1.580–1.750	20	
	Panchromatic band	480–710	2.5 or 5	
Sensors	HRG			
	HRS			
Swath width	60 km × 60 km to 80 km at nadir			
Orbital altitude	822 km			
Orbital inclination	98.7 degree, sun-synchronous			
Speed	7.4 km/s (26,640 km/h)			
Equator-crossing time	10:30 a.m. (descending node)			
Orbit time	101.4 min			
Revisit time	2–3 days, depending on latitude			
Digitization	8 bits			
Launch date	May 3, 2002			
Launch vehicle	Ariane 4			
Launch location	Guiana Space Centre, Kourou, French Guyana			
Footprint	60 km × 60 km			

HRG, High-resolution geometric; *HRS*, high-resolution stereoscopic; *SWIR*, shortwave infrared.

SPOT-6 and SPOT-7 image product resolutions include panchromatic (1.5 m), color merge (1.5 m), and multispectral (6 m). Spectral bands of SPOT-6 and SPOT-7 are panchromatic (450–745 nm), the blue band (450–525 nm), the green band (530–590 nm), red band (625–695 nm), and near-infrared band (760–890 nm) with 60 km × 60 km swath width.

SPOT-5 contains the unique acquisition ability of the onboard high-resolution stereoscopic (HRS) viewing instrument, which encompasses vast areas in a single pass. Stereoscopic satellite imagery is essential for 3D terrain modeling. A 2.5-m color image can be acquired by fusing two separate images, specifically the image with a panchromatic mode at 2.5 m resolution and the images with a three-band multispectral mode at 10 m resolution. Because the 2.5 m images are produced by fusing two 5-m images, one of the High-Resolution Geometric (HRG) instruments must acquire three images concurrently to generate a 2.5-m color image. Consequently, the images acquired are like a three-band color image with a resolution of 2.5 m and panchromatic viewing geometry (Huurneman et al., 2009). The SPOT-5 data were successfully processed for geological–structural mapping and lineament extraction for Cu–Au prospecting in the Lut Block, east of Iran (Ahmadirouhani et al., 2017, 2018).

2.3.3 Advanced Spaceborne Thermal Emission and Reflection Radiometer data

The ASTER is a high spatial, spectral, and radiometric resolution multispectral remote sensing sensor. It was launched on NASA's Earth Observing System AM-1 (EOS AM-1) polar-orbiting spacecraft in December 1999. EOS AM-1 spacecraft operates in a near-polar, sun-synchronous circular orbit at 705 km altitude. The recurrent cycle is 16 days, with an additional 4-day repeat coverage due to its off-nadir pointing capabilities. ASTER is a cooperative effort between the Japanese Ministry of Economic Trade and Industry (METI) and the National Aeronautics and Space Administration (NASA). It consists of three separate instrument subsystems, which provide observation in three different spectral parts of the electromagnetic spectrum, including the VNIR, SWIR, and TIR (Fig. 2.5) (Abrams et al., 2004).

The VNIR subsystem has three recording channels between 0.52 and 0.86 μm and an additional backward-looking band for the stereo construction of Digital Elevation Models (DEMs) with a spatial resolution of up to 15 m. The SWIR subsystem has six recording channels from 1.6 to 2.43 μm, at a spatial resolution of 30 m, while the TIR subsystem has five recording channels, covering the 8.125–11.65-μm wavelength region with a spatial resolution of 90 m. ASTER swath width is 60 km (each individual scene is cut to a 60 × 60-km^2 area), which makes it useful for regional mapping, though its off-nadir pointing capability extends its total possible field of view to up to 232 km. ASTER can acquire approximately 600 scenes daily but is generally targeted and tasked without continuous operation unlike other multispectral sensors such as Landsat (Abrams, 2000; Abrams et al., 2004; Abrams and Hook, 1995; Yamaguchi et al., 2001; Abrams and Yamaguchi, 2019).

The performance characteristics of ASTER data are shown in Table 2.3. ASTER provides data useful for studying the interaction among the geosphere, hydrosphere, cryosphere, lithosphere, and atmosphere of the Earth from an Earth system science perspective (Fig. 2.6).

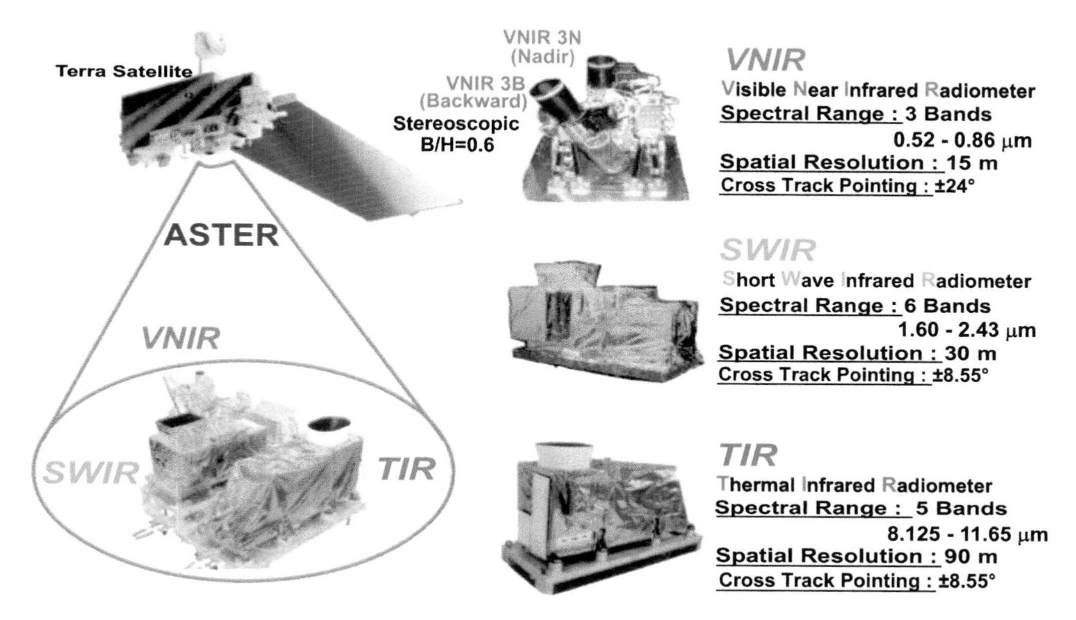

FIGURE 2.5 Stereoscopic capability and characteristics of three separate subsystems of an ASTER sensor. *ASTER*, Advanced Spaceborne Thermal Emission and Reflection Radiometer.

To be more specific, a wide range of science investigations and applications include (1) geology and soil studies; (2) land surface climatology studies; (3) vegetation and ecosystem dynamic studies; (4) volcano monitoring; (5) other hazard monitoring; (6) carbon cycle and marine ecosystem studies; (7) hydrology and water resource applications; (8) aerosol and cloud studies; (9) seasonal evapotranspiration measurement; and (10) land surface and land cover change analyses (Fujisada, 1995; Gillespie et al., 2005).

ASTER standard data products are available "on-demand" from the Earth Remote Sensing Data Analysis Center (ERSDAC; Japan) and the EROS Data Center (EDC; USA). Basically, all the ASTER captured data are processed to generate Level-1A data product, which consists of unprocessed raw image data and coefficients for radiometric correction. The Level-1B (radiance-at-sensor) data product is a resampled image data generated from the Level-1A data by applying the radiometric correction coefficients (Abrams et al., 2004). Level-2 data products of measured physical parameters include surface radiance data with nominal atmospheric corrections (Level-2B01), surface reflectance data containing atmospherically corrected VNIR-SWIR data (Level-2B07 or AST-07), surface emissivity data with MODTRAN atmospheric correction, and a temperature-emissivity separation (TES) algorithm (Level-2B04), which are all generated based on user request (Gillespie et al., 1998; Thome et al., 1998; Yamaguchi et al., 1999).

Level-4B data product is also generated under the user request from the along-track stereo observation in the near-infrared channel (band 3N and 3B) to construct DEMs. Level-3A is a geometrically well-corrected orthorectified ASTER standard data product with

Table 2.3 The technical characteristics of Advanced Spaceborne Thermal Emission and Reflection Radiometer data (Yamaguchi et al., 1999; Fujisada, 1995).

Subsystem	Band number	Spectral range (μm)	Radiometric resolution	Absolute accuracy (σ)	Spatial resolution
VNIR	1	0.52–0.60	NE $\Delta\rho \leq 0.5\%$	$\leq$4%	15 m
	2	0.63–0.69			
	3N	0.78–0.86			
	3B	0.78–0.86			
SWIR	4	1.600–1.700	NE $\Delta\rho \leq 0.5\%$		
	5	2.145–2.185	NE $\Delta\rho \leq 1.3\%$		
	6	2.185–2.225	NE $\Delta\rho \leq 1.3\%$	$\leq$4%	30 m
	7	2.235–2.285	NE $\Delta\rho \leq 1.3\%$		
	8	2.295–2.365	NE $\Delta\rho \leq 1.0\%$		
	9	2.360–2.430	NE $\Delta\rho \leq 1.3\%$		
TIR	10	8.125–8.475		$\leq$3K (200–240K)	
	11	8.475–8.825		$\leq$2K (240–270K)	90 m
	12	8.925–9.275	NE $\Delta T \leq 0.3$K	$\leq$1K (270–340K)	
	13	10.25–10.95		$\leq$2K (340–370K)	
	14	10.95–11.65			
Signal quantization levels					
Stereo base-to-height ratio	0.6 (along-track)				
Swath width	60 km				
Total coverage in cross-track direction by pointing	232 km				
Coverage interval	16 days				
Altitude	705 km				
MTF at Nyquist frequency	0.25 (cross-track)				
	0.20 (along-track)				
Band-to-band registration	Intratelescope: 0.2 pixels				
	Intratelescope: 0.3 pixels				
Peak power	726 W				
Mass	406 kg				
Peak data rate	89.2 Mbps				

Band number 3N refers to the nadir-pointing view, whereas 3B designates the backward-pointing view. *SWIR*, Shortwave infrared; *TIR*, thermal infrared; *VNIR*, visible and near-infrared.

ASTER-driven DEM, which is radiometrically equivalent to Level-1B radiance-at-sensor data (Abrams, 2000; Yamaguchi et al., 2001; Ninomiya, 2003a,b; Ninomiya et al., 2005). During scene acquisition of ASTER data, there is an optical "crosstalk" effect caused by a stray of light from the band 4 detector into adjacent band 5 and 9 detectors on the SWIR subsystem. Such deviations from correct reflectance result in false absorption features and a distortion of diagnostic signatures that results in spectroscopic misidentification of minerals (Mars and Rowan, 2010). Fortunately, ASTER Cross-Talk correction software is available from http://www.gds.aster.ersdac.or.jp (Kanlinowski and Oliver, 2004; Iwasaki and Tonooka, 2005; Hewson et al., 2005; Mars and Rowan, 2006, 2010).

Recently, two new crosstalk-corrected ASTER SWIR reflectance products had been released, which include (1) AST-07XT SWIR reflectance data product available "on-demand" from ERSDAC and EDC (Iwasaki and Tonooka, 2005), and (2) RefL1b SWIR reflectance data product generated and described by Mars and Rowan (2010). The AST-07XT SWIR surface

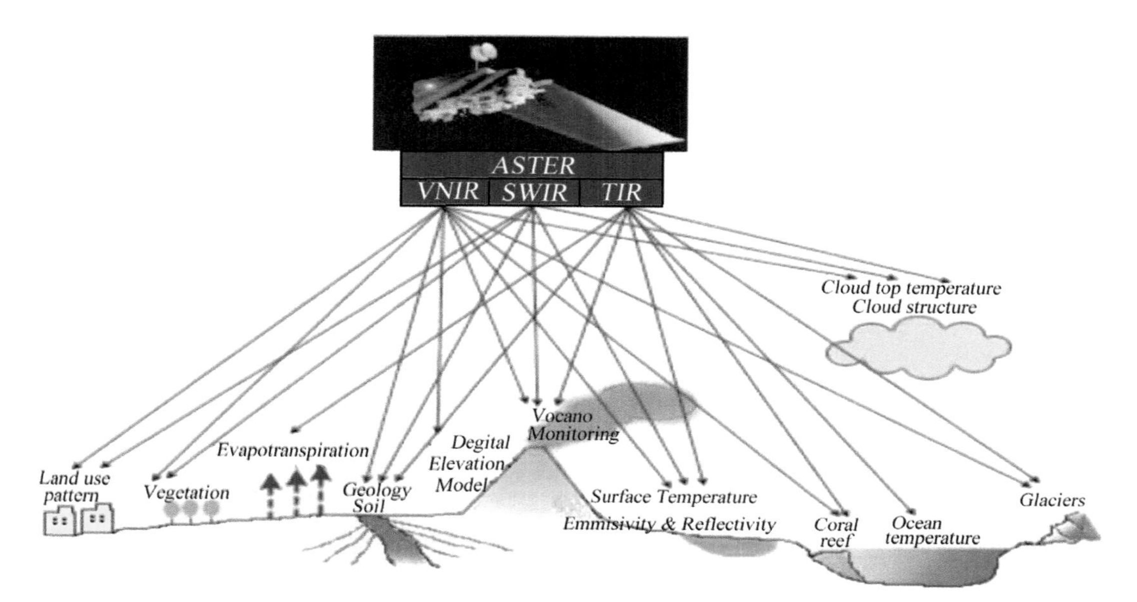

FIGURE 2.6 Applications of ASTER data in a wide range of scientific research studies. *ASTER*, Advanced Spaceborne Thermal Emission and Reflection Radiometer.

reflectance data product is like AST-07 surface reflectance data in that it consists of the same VNIR and SWIR bands. However, the crosstalk correction algorithm and atmospheric correction (nonconcurrently acquired MODIS water vapor data) have been preapplied to the data (Iwasaki and Tonooka, 2005; Biggar et al., 2005; Mars and Rowan, 2010). RefL1b and AST-07XT reflectance datasets both attempt to correct the SWIR anomaly. However, the differences between these datasets are caused by the addition of the radiometric correction factors and the use of concurrently acquired water vapor data for atmospheric correction in the case of the RefL1b data (Mars and Rowan, 2010). Spectral analyst results for AST-07XT and RefL1b indicated that these new ASTER products can be used for regional mineral mapping without the use of additional spectral data from the site for calibration. Especially, ASTER mineral mapping projects will be more feasible using RefL1b data (Mars and Rowan, 2010). The use of ASTER data in mineral exploration and lithological mapping has increased in recent years due to (1) spectral characteristics of the unique integral bands of ASTER, which are highly sensitive to hydrothermal alteration minerals, especially in SWIR radiation region; (2) the possibility of applying several image processing techniques; (3) "on-demand" data product availability with low cost ($\sim$\$60 US or Yen equivalent) and free data products; and (4) broad 60 km $\times$ 60 km scene coverage useful for regional scale mapping (Abrams and Yamaguchi, 2019).

Fig. 2.7 shows the location of ASTER spectral bands in the wavelength of the electromagnetic spectrum and the comparison of spectral bands between Landsat 7 TM/ETM+ and ASTER. ASTER VNIR bands are capable of mapping iron oxide/hydroxide (gossan) associated

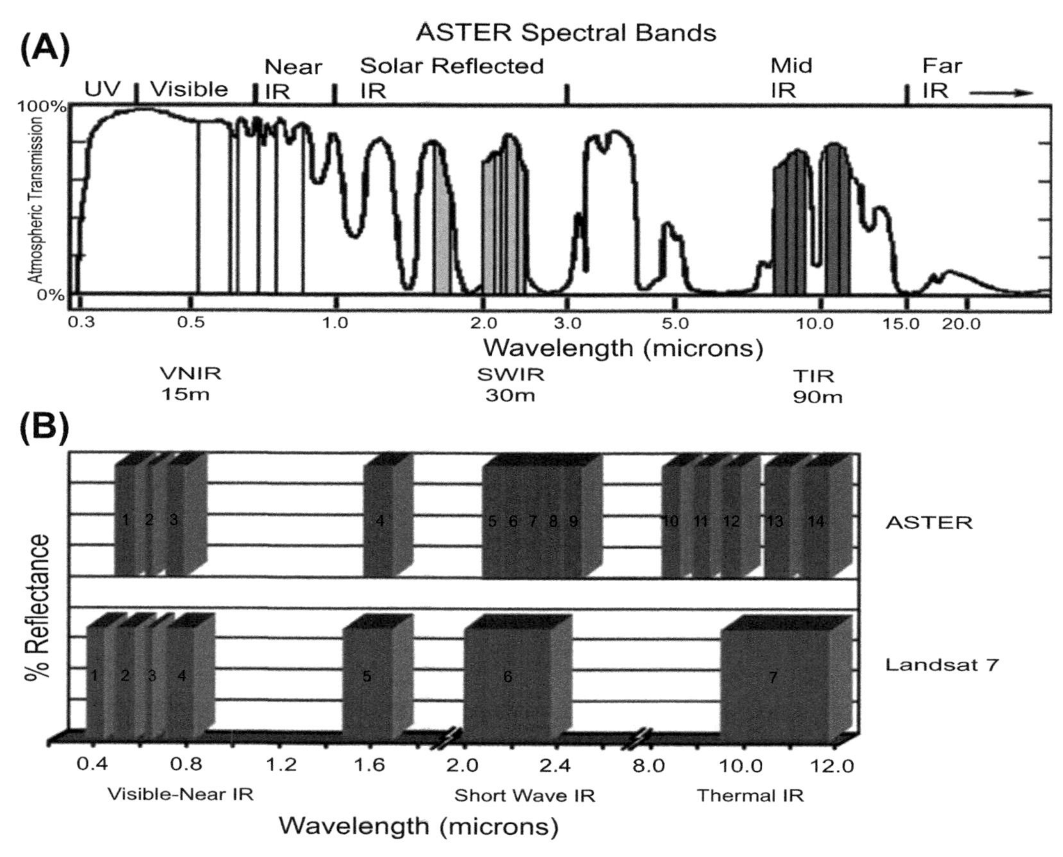

FIGURE 2.7 (A) ASTER spectral bands in the wavelength of the electromagnetic spectrum. (B) The comparison of ASTER spectral bands with Landsat-7 TM/ETM$^+$. *ASTER*, Advanced Spaceborne Thermal Emission and Reflection Radiometer.

with ore mineralizations (Pour et al., 2019a,b,c). The ASTER SWIR bands with sufficient spectral resolution have been widely and successfully used in numerous mineral exploration and lithological mapping projects (Abrams and Yamaguchi, 2019). SWIR spectral bands are unprecedented multispectral data for the discrimination of hydrothermal alteration mineral zones such as phyllic, argillic, and propylitic associated with porphyry copper, hydrothermal gold, and massive sulfide deposits (Mars and Rowan, 2006; Pour and Hashim, 2012a,b). ASTER TIR bands are capable of discriminating silicate lithological groups and carbonates (Ninomiya et al., 2005; Rockwell and Hofstra, 2008; Aboelkhair et al., 2010; Yajima and Yamaguchi, 2013; Guha and K., 2016; Guha and Kumar, 2016; Pour et al., 2018a; Ninomiya and Fu, 2019; Pour et al., 2021a,b). Consequently, ASTER data have widely and successfully been used to identify and map hydrothermal alteration mineral zones and lithological units associated with a variety of ore mineralizations in many metallogenic provinces (Pour and Hashim, 2012a,b; Abrams and Yamaguchi, 2019; Yousefi et al., 2021, 2022).

2.3.4 Advanced Land Imager data

The Earth Observing One (EO-1) Satellite was launched onboard on November 21, 2000. The EO-1 is the first earth-observing satellite of NASA's New Millennium Program (NMP) that developed new technologies for optimizing the feature and reducing the cost of remote sensing data for NASA's future planetary and earth missions (Folkman et al., 2001; Perko et al., 2001; Beck, 2003; Ungar, 2002; Ungar et al., 2003). The EO1 platform carries three sensors onboard, namely, (1) the ALI; (2) Hyperion; (3) Linear Etalon Imaging Spectral Array (LEISA). The ALI is a prototype for a new generation of Landsat 7 TM. The ALI provides multispectral data like that of the Enhanced Thematic Mapper Plus (ETM +) sensor on Landsat 7. The sensor maintains analogous attributes to Landsat 7 with a spatial resolution of 30 m; however, the swath width is 37 km as opposed to 185 km (Hearn et al., 2001; Wulder et al., 2008). ALI is a push-broom sensor and has some additional bands in comparison with the whisk-broom design of the ETM^+ sensor (Thome et al., 2003). The performance attributes of the ALI sensor are shown in Table 2.4. Additional bands in the ALI improved the signal-to-noise ratio (SNR) that is one of the most significant performance aspects of the ALI to increase the quality of data (Lencioni et al., 1999; Thome et al., 2003; Lobell and Asner, 2003). Fig. 2.8 illustrates the comparison of the ALI SNR bands with those of ETM + (Mendenhall et al., 2000; Hearn et al., 2001).

The ALI sensor has 10 channels spanning 0.4–2.35 μm (the VNIR to SWIR), 6 VNIR bands (0.4330.890 μm), and 3 SWIR bands (1.200–2.350 μm), and 1 panchromatic band (0.480–0.690 μm) (see Table 2.4). The VNIR bands can especially be useful for detecting iron oxide/hydroxide minerals from the perspective of geologic mapping and mineral exploration applications. Because of SWIR band center positions, ALI is only capable of discriminating altered zones (OH-bearing minerals and carbonates) from unaltered zones for mineral exploration purposes (Hubbard et al., 2003; Hubbard and Crowley, 2005; Pour and Hashim, 2011b, 2014b, 2015b). Fig. 2.9 shows the comparison of ALI, ETM^+, and ASTER spectral bandpasses for hydrothermal alteration mineral mapping.

Table 2.4 The performance characteristics of the Advanced Land Imager (ALI) sensor.

Sensors	Subsystem	Band number	Spectral range (μm)	Ground resolution (m)	Swath width (km)
ALI	VNIR	Pan	0.480–0.690	10	37
		1	0.433–0.453	30	
		2	0.450–0.515		
		3	0.525–0.605		
		4	0.633–0.690		
		5	0.775–0.805		
		6	0.845–0.890		
	SWIR	7	1.200–1.300		
		8	1.550–1.750		
		9	2.080–2.350		

SWIR, Shortwave infrared; *VNIR*, visible and near-infrared.

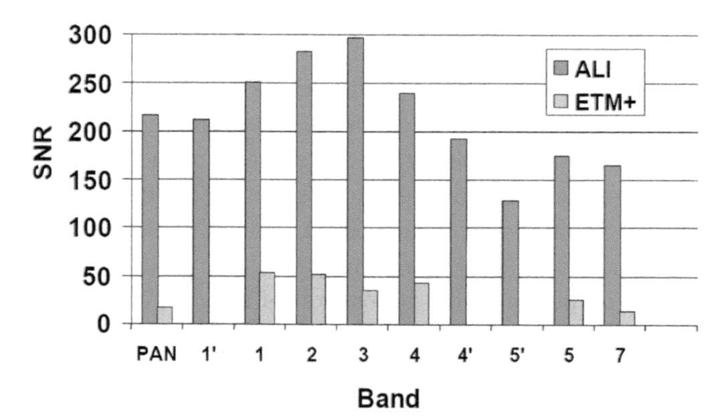

FIGURE 2.8 Comparison of signal-to-noise ratios of the ALI bands and ETM^+ (Hearn et al., 2001). *ALI*, Advanced Land Imager.

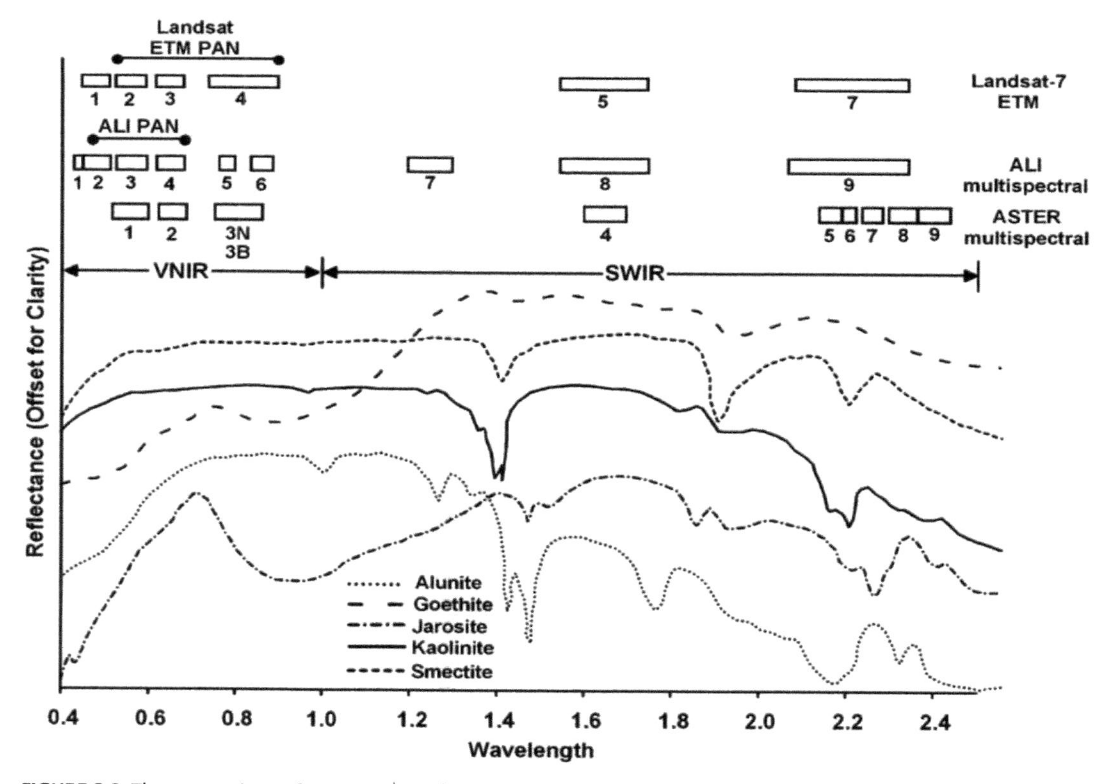

FIGURE 2.9 The comparison of ALI, ETM^+, and ASTER spectral bandpasses for hydrothermal alteration mineral mapping (Hubbard and Crowley, 2005). *ALI*, Advanced Land Imager; *ASTER*, Advanced Spaceborne Thermal Emission and Reflection Radiometer.

2.3.5 Sentinel-2 MSI data

The Sentinel-2 sensor is a multispectral sensor that was launched in 2015. Sentinel-2 has 13 bands covering the VNIR region with 8 bands and the SWIR region with 2 bands having a spatial resolution of 10, 20, and 60 m. The swath width is 290 km. It consists of two multispectral satellites, Sentinel-2A and Sentinel-2B. The satellites orbit at an altitude of 786 km and 180 degrees from each other. In general, the time of complete survey of the earth is repeated every 5 days. The Sentinel-2 images are provided at the L1C level, which is radiometrically and geometrically corrected. Table 2.5 summarizes the spectral and spatial attributes and applications of Sentinel-2 bands.

Sentinel-2 is mainly used for agricultural applications for instance crop monitoring and management, vegetation and forest monitoring, land cover change monitoring for environmental monitoring, observation of coastal areas (marine environmental monitoring, mapping sampling from coastal areas), monitoring inland waters, monitoring glaciers, and others. Nevertheless, valuable information concerning lithological units and alteration minerals can be obtained from bands 2, 4, 5, 6, 7, 11, and 12 of Sentinel-2 (Van der Meer et al., 2014). Considering the relatively appropriate spatial resolution of these bands (10–20 m), color combinations, and advanced image processing techniques can easily be implemented to identify lithological units, iron oxide/hydroxides, OH-minerals alterations, and geological features and structures. Iron oxide/hydroxides and hydroxyl-bearing alteration minerals can be successfully mapped using band ratios of 4/2 and 11/12, respectively (Van der Meer et al., 2014; Van der Werff and Van der Meer, 2016; Zoheir et al., 2019a; Sekandari et al., 2020).

Table 2.5 The spectral and spatial characteristics and applications of Sentinel-2 bands.

Band	Band range (nm)/band center (nm)	Spatial resolution (m)	Purpose in L2 processing context
B1	433–453/443	60	Atmospheric correction
B2	458–523/490	10	Sensitive to vegetation aerosol scattering, iron oxides/ hydroxides detection
B3	543–578/560	10	Green peak, sensitive to total chlorophyll in vegetation
B4	650–680/665	10	Max chlorophyll absorption, iron oxides/hydroxides detection
B5	698–713/705	20	Not used in L2A context
B6	734–748/740	20	Not used in L2A context
B7	765–785/783	20	Not used in L2A context
B8	785–900/842	10	LAI
B8a	855–875/865	20	Used for water vapor absorption reference
B9	930–950/945	60	Water vapor absorption atmospheric correction
B10	1365–1385/1375	60	Detection of thin cirrus for atmospheric correction
B11	1565–1655/1610	20	Soils detection, OH-mineral alteration mapping
B12	2100–2280/2190	20	AOT determination, OH-mineral alteration mapping

LAI, Leaf area index.

Selected bands of Sentinel-2 were used for the exploration of orogenic gold in transpression and transtension zones of the Barramiya–Mueilha area, Egypt, and carbonate-hosted Pb–Zn deposits in the Central Iranian Terrane (CIT), Iran as well as mapping rare metal-bearing granites in the Umm Naggat Area, Central Eastern Desert, Egypt (Zoheir et al., 2019a; Sekandari et al., 2020; Abdelkader et al., 2022).

2.3.6 WorldView data

WorldView is a commercial earth observation satellite retained by DigitalGlobe. The WorldView-1 satellite sensor was launched on September 18, 2007, followed later by the WorldView-2 on October 8, 2009. WorldView-1 provides a high-capacity, panchromatic imaging system that incorporates 0.5 m resolution imagery. The WorldView-2 sensor provides a high-resolution panchromatic band and eight multispectral bands. These bands include four standard colors such as red, green, blue, and near-infrared-1, and other bands, namely, coastal, yellow, red edge, and near-infrared 2. WorldView-2 data can be used for mineral mapping in particular iron oxide/hydroxide groups, spectral analysis, land-use planning, climate change, defense, wildlife monitoring, and disaster relief (Kuester et al., 2015). The WorldView-3 satellite was successfully launched on August 13, 2014. This sensor contains the highest spatial, spectral, and radiation among the multispectral satellite sensors, presently. It has eight bands in the VNIR region with a spatial resolution of 1.24 m and eight bands in the SWIR region with a spatial resolution of 3.70 m. WorldView-3 also has a panchromatic band with 0.31-m spatial resolution (Kuester, 2016). Technical performance and attributes of the WV-3 sensor are shown in Table 2.6. Iron oxides/hydroxide minerals can be comprehensively mapped and discriminated by VNIR bands of WV-3 (Mars, 2018; Pour et al., 2019a; Sekandari et al., 2022). Additionally, SWIR bands of WV-3 contain high

Table 2.6 Technical performance and attributes of WV-3 sensor.

Sensor	Subsystem	Band number	Spectral range (μm)	Ground resolution (m)	Swath width (km)	Year of launch
WV3	VNIR	Coastal blue (1)	0.400–0.450	1.24	13.1	2014
		Blue (2)	0.450–0.510			
		Green (3)	0.510–0.580			
		Yellow (4)	0.585–0.625			
		Red (5)	0.630–0.690			
		Red edge (6)	0.705–0.745			
		NIR1 (7)	0.770–0.895			
		NIR2 (8)	0.860–1.040			
	SWIR	SWIR-1 (9)	1.195–1.225	3.70		
		SWIR-1 (10)	1.550–1.590			
		SWIR-1 (11)	1.640–1.680			
		SWIR-1 (12)	1.710–1.750			
		SWIR-1 (13)	2.145–2.185			
		SWIR-1 (14)	2.185–2.225			
		SWIR-1 (15)	2.235–2.285			
		SWIR-1 (16)	2.295–2.365			

SWIR, Shortwave infrared; *VNIR*, visible and near-infrared.

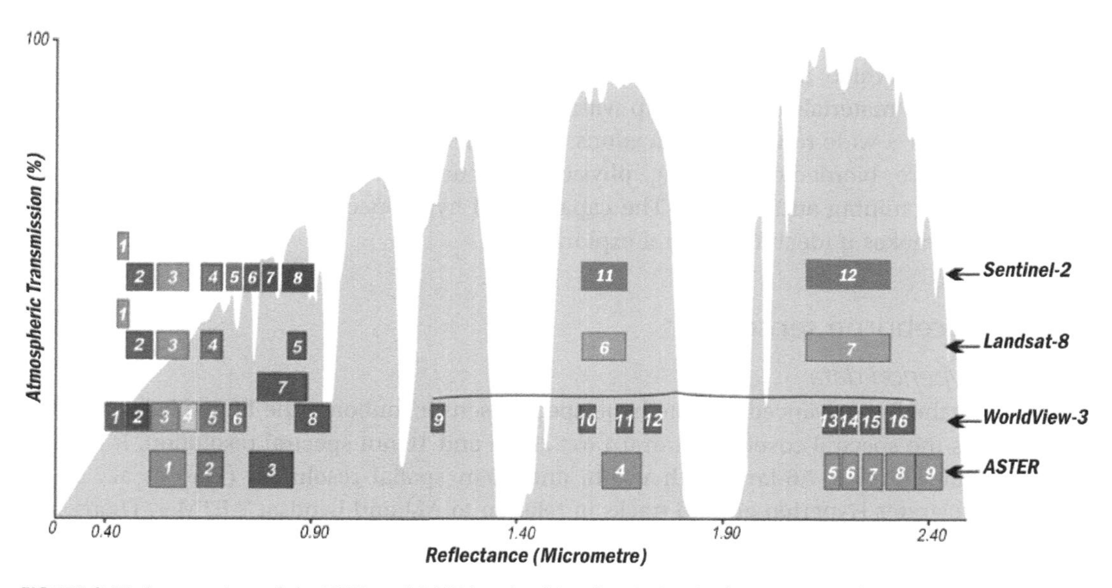

FIGURE 2.10 An overview of the VNIR and SWIR bands of Landsat-8, Sentinel-2, ASTER, and WV-3. *ASTER*, Advanced Spaceborne Thermal Emission and Reflection Radiometer; *SWIR*, shortwave infrared; *VNIR*, visible and near-infrared.

capability for a detailed mapping of Al—OH, Mg—Fe—OH, and Si—OH key hydrothermal alteration minerals (Kruse and Perry, 2013; Kruse et al., 2015; Sun et al., 2017; Mars, 2018). Comparison between the spectral bands of WV-3 with Landsat 8, Sentinel-2, and ASTER emphasizes their priority and a high potential for detailed mapping of alteration minerals in the VNIR and SWIR regions (Fig. 2.10).

WorldView-4 satellite was successfully launched on November 11, 2016. It can discern objects on the Earth's surface as small as 31 cm in the panchromatic and collects four multispectral bands (red, green, blue, and near-infrared) at 1.24-m spatial resolution. WorldView-4 contains a similar resolution as the WorldView-3 satellite sensor. This advanced resolution offers users exceptional and accurate views for 2D or 3D mapping, change detection, and image analysis. WorldView-4 satellite presents exceptional geolocation accuracy, which means that users will be capable of mapping natural and man-made features to better than <4-m CE90 of their actual location on the Earth's surface without ground control points.

2.4 Hyperspectral sensors

Hyperspectral sensors measure energy in narrower and more numerous bands than multispectral sensors. Hyperspectral images can contain as many as 200 (or more) contiguous spectral bands. In hyperspectral imaging, the recorded spectra have fine wavelength resolution and cover a wide range of wavelengths. Hyperspectral imaging measures continuous spectral bands, as opposed to multiband imaging that measures spaced spectral bands (Hagen and Kudenov, 2013). Hyperspectral sensors sense objects using a substantial fraction of the electromagnetic spectrum. Objects have distinctive "fingerprints" in the electromagnetic spectrum, which are well known as spectral signatures. The "fingerprints" allow the detection of the materials that come up with a visualized object. Hyperspectral remote sensing is useful for a wide range of applications such as geosciences, molecular biology, astronomy, agriculture, biomedical imaging, physics, and surveillance. However, it was originally established for mining and geology. The capability of hyperspectral imaging to detect different minerals makes it ideal for mineral exploration.

2.4.1 Spaceborne sensors

2.4.1.1 Hyperion data

Hyperion is the first advanced satellite hyperspectral sensor (onboard the EO-1 platform) operating across the spectral coverage from 0.4 to 2.5 μm and 10 nm spectral resolution. It has 242 spectral bands over a 7.6-km swath width, and 30 m spatial resolution (Liao et al., 2000). Fig. 2.11 illustrates Hyperion ground tracks in relation to ALI and Landsat 7 ETM + (Barry and Pearlman, 2001). The 242 total bands comprise the first 70 bands in the VNIR region and the second 172 bands in the SWIR region, and 21 bands are in a region of bands' overlap between 0.9 and 1.0 μm (Folkman et al., 2001; Pearlman et al., 2003; Ungar et al., 2003).

The system has two spectrometers and a single telescope. Spectrometers operate at the VNIR wavelength (approximately 0.4–1.0 μm) and at the SWIR wavelength (approximately 0.9–2.5 μm), respectively (Ungar et al., 2003; Beck, 2003; Green et al., 2003; Goodenough et al., 2003). Hyperion data have an SNR of about 161/1 in the VNIR region and 40/1 in the SWIR region that somewhat limited the scientific applications of Hyperion data (Folkman et al., 2001; Thome et al., 2003). Table 2.7 shows the performance characteristics of the Hyperion sensor (Folkman et al., 2001). Hyperion data can be used for geology, agriculture

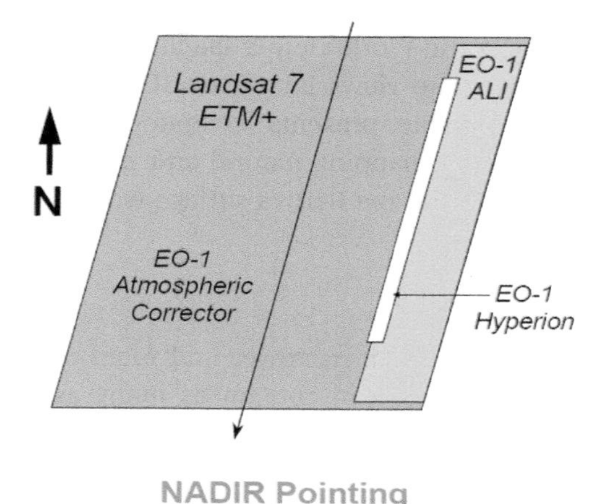

FIGURE 2.11 Hyperion ground track in relation to ALI and Landsat 7 ETM$^+$. *ALI*, Advanced Land Imager.

Table 2.7 The performance characteristics of the Hyperion sensor.

Sensor	Subsystem	Band number	Spectral range (μm)	Ground resolution (m)	Swath width (km)	Radiometric precision (S/N)
Hyperion	VNIR	Continuous	0.400–1.000	30	7.6	161/1
	SWIR	Continuous	0.900–2.500	30	7.6	40/1

SWIR, Shortwave infrared; *VNIR*, visible and near-infrared.

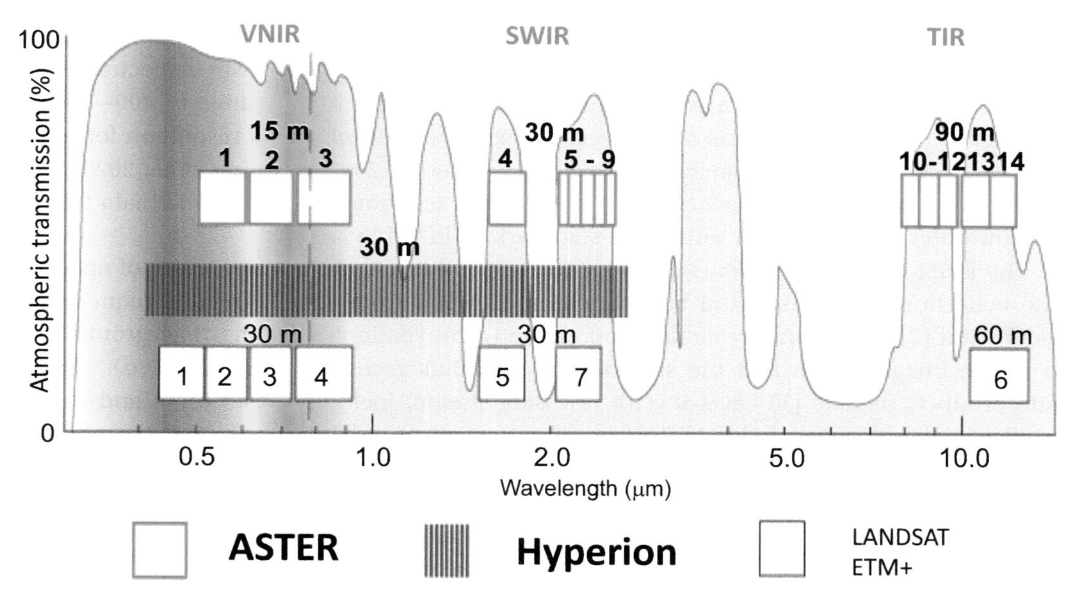

FIGURE 2.12 Hyperion spectral bands compared to ASTER and Landsat 7 ETM$^+$. *ASTER*, Advanced Spaceborne Thermal Emission and Reflection Radiometer.

monitoring, volcanic temperature measurement, the study of reef and coral bay health, and glaciological applications (Barry et al., 2002). Fig. 2.12 illustrates the Hyperion spectral bands compared to ASTER and Landsat 7 ETM + (Waldhoff et al., 2008). Hyperion VNIR bands between 0.4 and 1.3 μm can be used to highlight iron oxide/hydroxide minerals (Bishop et al., 2011; Pour and Hashim, 2011b, 2015b). Hyperion SWIR bands (2.0–2.5 μm) can uniquely identify and map hydroxyl-bearing minerals, sulfates, and carbonates in the hydrothermal alteration assemblages (Kruse et al., 2003; Gersman et al., 2008; Bishop et al., 2011; Pour and Hashim, 2011b, 2015b).

2.4.1.2 Prcursore IperSpettrale della Missione Applicativa data

The new hyperspectral Precursore IperSpettrale della Missione Applicativa (PRISMA) sensor, developed by the Italian Space Agency, was launched in 2019 on a sun-synchronous Low Earth Orbit (620 km altitude) (Loizzo et al., 2018). PRISMA is an Earth observation satellite with advanced electro-optical instrumentation that combines a medium-resolution panchromatic camera with hyperspectral sensors. The benefits of this combination are that plus the classical capability of observation based on the identification of the geometrical features of the scene, there is the one presented by hyperspectral sensors that can reveal the chemical–physical composition of objects that appear on the scene. This presents the scientific community with various applications in the field of environmental monitoring, resource management, crop classification, pollution control, forest types discrimination, mineral detection, and soil fertility mapping (Bedini and Chen, 2020, 2022; Vangi et al., 2021; Heller Pearlshtien et al., 2021; Gasmi et al., 2022). PRISMA captures images in a continuum of 240 spectral bands ranging between 400 and 2500 nm, including (1) hyperspectral sensors in the VNIR: 400–1010 nm (66 bands) and SWIR: 920–2500 nm (173 bands) with a spectral resolution around 10 nm, and (2) a panchromatic camera in the spectral range of 400–700 nm. PRISMA acquires data on areas of 30-km swath width and 30-m spatial resolution for hyperspectral bands and 5 m for panchromatic band (Pepe et al., 2020; Giardino et al., 2020; Vangi et al., 2021). Hyperspectral sensors use a push-broom scanning technique. The main features and attributes of the PRISMA sensor are summarized in Table 2.8.

The PRISMA mission operates in two modes, including (1) the primary mode of operation (the collection of hyperspectral and panchromatic data from specific targets requested by users), and (2) the secondary mode of operation (a conventional ongoing "background" task to obtain imagery such that the satellite and downlink resources are fully used). PRISMA data products include (1) Level 0 (L0): raw data stream, including instrument and satellite ancillary data; (2) Level 1 (L1): TOA (top of atmosphere) radiometrically and geometrically calibrated hyperspectral and panchromatic radiance images; and (3) Level 2 (L2): geolocated and geocoded atmospherically corrected hyperspectral and panchromatic images, atmospheric constituents maps (aerosols, water vapor, thin cloud optical thickness). Typically, PRISMA delivers hyperspectral imagery to the mineral exploration community for the identification of hydrothermal alteration minerals to prospect ore mineralizations. Iron oxide/hydroxide minerals can be mapped using VNIR bands and gypsum and clay mineral groups are detectable by SWIR bands of PRISMA (Bedini and Chen, 2020, 2022; Heller Pearlshtien et al., 2021; Benhalouche et al., 2022).

Table 2.8 The main features and attributes of the Precursore IperSpettrale della Missione Applicativa (PRISMA) sensor.

PRISMA main features and attributes	Spatial resolution		
Hyperspectral range 30 m	VNIR	400–1.10 nm	66 bands
	SWIR	920–2.500 nm	173 bands
Panchromatic spectral range 5 m		400–700 nm	
Average spectral resolution		10 nm	
Swath width		30 km	
Data quantization		12 bit	
Frame rate		230 Hz	

SWIR, Shortwave infrared; *VNIR*, visible and near-infrared.

2.4.2 Airborne sensors

2.4.2.1 Airborne Visible/IR Image Spectrometer data

The AVIRIS is a hyperspectral airborne instrument flown in a NASA high-altitude aircraft. It was the first optical sensor (OPS) (1987) to measure the solar reflected spectrum from the VNIR and SWIR (400–2500 nm) using 224 contiguous bands at 10 nm intervals (Clark et al., 1991; Green et al., 1998). It has nadir-viewing whisk broom detectors with a 12-Hz scanning rate. AVIRIS captured data on four aircraft platforms: NASA's ER-2 jet, Twin Otter International's turboprop, Scaled Composites' Proteus, and NASA's WB-57. The image has 11 km width and up to 800 km length with 20 m spatial resolution. The NASA Earth Research-2 Q-bay aircraft at 20 km altitude covers a width of 11 km and a length of 800 km above of Earth with a spatial resolution of 20 m. Images with 1–4 m resolution were acquired in low attitude.

AVIRIS-NG (Next Generation) has been established to deliver continued access to high-SNR imaging spectroscopy measurements in the VNIR and SWIR ranges. AVIRIS-NG has replaced the AVIRIS-Classic instrument that has been flying since 1986. AVIRIS-NG measures the wavelength range from 380 to 2510 nm with 5 nm sampling in 425 narrow continuous spectral bands. Spectra are measured as images with 600 cross-track elements and spatial sampling from 0.3 to 4.0 m. AVIRIS-NG has better than 95% cross-track spectral uniformity and $\geq$ 95% spectral IFOV uniformity.

AVIRIS radiance data are used to monitor Earth's surface and can be used in a wide range of such as geology, hydrology, ecology, and others. All AVIRIS data from 1989 were archived in JPL. Each AVIRIS scene consists of 677 pixels $\times$ 512 lines $\times$ 224 bands. The data archived from 1993 to 2012 are at the L1B preprocessing level, and the data from 2012 to now are preprocessed at the L2 level. L1B data are radiometrically and ortho-corrected. L2 data are atmospherically corrected by the Atmosphere Removal Algorithm (ATREM) program. A variety of alteration minerals such as iron oxide/hydroxide minerals, clay minerals, carbonates, and ammonium minerals were mapped using AVIRIS data (Baugh et al., 1998; Clark et al., 1991; Rowan et al., 2000; Kruse et al., 2003). AVIRIS-NG data were used for mapping alteration mineral signatures associated with the base metal mineralization in the Pur-Banera area in the Bhilwara district, Rajasthan, India (Guha et al., 2021).

2.4.2.2 HyMap data

"HyMap" (hyperspectral mapper) is an airborne hyperspectral imaging sensor (Manufactured by Integrated Spectronics Pty Ltd, Sydney, Australia) operated by HyVista Corporation. It consists of sensors located on a fixed-wing aircraft typically flown at an altitude of 2.5 km. HyMap data has 126 spectral bands (at approximately 15 nm intervals) spanning the wavelength interval 0.45–2.5 μm (VNIR: 64 bands) and (SWIR: 62 bands), and 5-m spatial resolution (Cocks et al., 1998). This sensor is highly adjusted for mineral exploration tasks and has a high level of performance in spectral and radiometric calibration accuracy, SNR (500–1000 or better), geometric characteristics, and operational stability. HyMap data were used for detailed mapping of hydrothermal alteration minerals associated with Cu–Au–Mo porphyry deposits, carbonatite complex, and metamorphic–hydrothermal U-REE deposit (Bedini, 2009, 2011, 2012; Huo et al., 2014; Salles et al., 2017).

2.5 Unmanned aerial vehicle

An UAV, or drone, is an aircraft deprived of any human pilot on board. UAVs are a component of an unmanned aircraft system (UAS), which also embrace a ground-based controller and a system of communications with the UAV (Hu and Lanzon, 2018). Satellite and airborne platforms typically run on stationary orbits or continuous predetermined paths. However, there is a need for fine-scale remote sensing applications (e.g., geological surveying and mineral exploration) for gathering data along asymmetrical design pathways, closer proximity for specific feature observations, or momentarily altering pathways due to unfavorable environmental conditions. The use of traditional remote sensing platforms is challenging due to a lack of flexibility and high cost and logistic requirements. The data acquisition strategy for remote, high risks, inaccessible sites with traditional platforms is tedious and challenging, such as monitoring polar regions (Leary, 2017), nuclear radiation, volcanoes, and toxic spills (Gomez and Purdie, 2016). Subsequently, remote sensing scientists have gradually recommended and embraced low-altitude unmanned platforms (e.g., kites, balloons, and UAVs) to counteract traditional platforms' inefficiencies (Graham, 1988; Verhoeven et al., 2009; Colomina and Molina, 2014).

UAVs are well known as conventional unmanned vehicles due to their high maneuverability, flexibility, cost-efficiency, data availability, and data quality. To date, the rapid evolution of UAV technology and advancement of UAV sensors (lightweight, inexpensive, miniature in size, and high-detection precision) integrating UAV platforms facilitate the use of UAV remote sensing techniques. The racing to enhance the satellite constellations for more satisfactory spatial and temporal data resolution by satellite data providers will eventually reduce satellite sensors' data acquisition costs. Thus UAVs will be expected to substitute manned airborne platforms and transform into the mainstream for remote sensing applications, together with satellite platforms (Liao et al., 2016). The classification of UAVs based on their range, endurance and weather, wind dependency, and maneuverability is well documented (Eisenbeiß, 2009). With the increasing popularity of UAV remote sensing, several comprehensive review articles are available (Table 2.9) to provide the current state of the technology's art and science of UAV-led applications and discuss future research prospects.

Based on the list of review papers (Table 2.9), it is evident that most of the reviews are based on UAV's hardware development (Colomina and Molina, 2014; Pajares, 2015; Adão et al., 2017; Pádua et al., 2017). UAVs' applications in various fields from agriculture to glaciology (Shahbazi et al., 2014; Bhardwaj et al., 2016; Adão et al., 2017; Pádua et al., 2017; Torresan et al., 2017) are well recorded. For remote sensing applications, Colomina and Molina (2014) and Pajares (2015) only provided comprehensive reviews and related rules and regulations. The regulations in UAVs deployment and operational requirements have been well assessed by the previous literature (Watts et al., 2012; Anderson and Gaston, 2013; Colomina and Molina, 2014; Pádua et al., 2017; Stöcker et al., 2017). While UAV remote sensing technology has proliferated recently, less attention has been given to the innovation of UAV image-processing techniques for geological and mineral exploration studies. Although few articles highlighted specific UAV remote sensing image-processing techniques, image

Table 2.9 Recent review articles related to unmanned aerial vehicle (UAV) remote sensing.

Year	Article title	References
2011	Small-scale UAVs in environmental remote sensing: challenges and opportunities	Hardin and Jensen (2011)
2012	Unmanned aircraft systems in remote sensing and scientific research: classification and considerations of use	Watts et al. (2012)
2012	Development and status of image matching in photogrammetry	Gruen (2012)
2012	Multi- and hyperspectral geologic remote sensing: a review	van der Meer et al. (2012)
2014	State of the art in high-density image matching	Remondino et al. (2014)
2014	Recent applications of unmanned aerial imagery in natural resource management	Shahbazi et al. (2014)
2014	Unmanned aerial systems for photogrammetry and remote sensing: a review	Colomina and Molina (2014)
2015	Overview and current status of remote sensing applications based on UAVs	Pajares (2015)
2016	UAVs as remote sensing platform in glaciology: present applications and future prospects	Bhardwaj et al. (2016)
2016	Recent developments in large-scale tie-point matching	Hartmann et al. (2016)
2016	A review of hyperspectral imaging in close-range applications	Kurz and Buckley (2016)
2017	Hyperspectral imaging: a review on UAV-based sensors, data processing, and applications for agriculture and forestry	Adão et al. (2017)
2017	UAS, sensors, and data processing in agroforestry: a review towards practical applications	Pádua et al. (2017)
2017	Forestry applications of UAVs in Europe: a review	Torresan et al. (2017)
2017	Review of the current state of UAV regulations	Stöcker et al. (2017)
2017	UAVs: regulations and law enforcement	Cracknell (2017)
2017	UAVs and spatial thinking: boarding education with geotechnology and drones	Fombuena (2017)
2019	Mini-UAV-based remote sensing: techniques, applications, and prospects	Xiang et al. (2019)
2019	UAV for remote sensing applications: a review	Yao et al. (2019)
2020	Applications of UAVs in mining from exploration to reclamation: a review	Park and Choi (2020)
2022	A review of machine learning in processing remote sensing data for mineral exploration	Shirmard et al. (2022)

alignment (Gruen, 2012; Hartmann et al., 2016), and dense cloud alignment (Remondino et al., 2014), however, those techniques are not explicitly for UAV image processing. Furthermore, several UAV image-processing technologies (e.g., 3D generation and geometric calibration) are well addressed in Colomina and Molina (2014) and Pádua et al. (2017), but a comprehensive review of UAV image processing technology development is still limited. The potential for deep learning in UAV image processing for mining is also a less studied area.

It is essential to provide a comprehensive literature review on UAV remote sensing, including image-processing techniques, applications, and potential future research. A thorough summary of existing work is essential for further improvement in UAV remote sensing, especially for geological surveyors and mineral exploration scientists that need to use the technology. UAV applications have become an advancing area in remote sensing in recent years. UAV technology bridges the gap among spaceborne, airborne, and ground-based remote sensing data. UAVs are generally characterized based on some allied main attributes such as weight, flying altitude, payload, endurance, and range (Yao et al., 2019). They have better spatial, temporal, and radiometric resolution than any airborne or satellite platform (Colomina and Molina, 2014; Pajares, 2015; Gonzalez-Aguilera and Rodriguez-Gonzalvez, 2017). With multispectral and hyperspectral sensors mounted on UAV platforms, high-resolution, georeferenced data can be acquired for studying spatial and temporal changes in geological applications.

The use of UAVs is becoming progressively significant for geological applications (Colomina and Molina, 2014). Recently, UAVs have been used in the mining industry for various applications from mineral exploration to mine reclamation. One of the most significant benefits of using UAVs is the low operational cost and ease of use (Park and Choi, 2020). UAVs offer incredible efficiency and cost advantages in every part of the mining life cycle such as exploration, planning/permitting, mining operations, and reclamation. Hence, UAVs are becoming important tools for mineral exploration by contributing to the safe, efficient, and sustainable provision of the high-tech metals that are required by modern society (Jackisch, 2020). UAVs can be equipped with optical devices, cameras covering different ranges of the electromagnetic spectrum [multispectral, hyperspectral, short/mid-wave range cameras (e.g., thermal)], and geophysical instrumentation (magnetometer) for mineral exploration purposes such as aerial mapping, resource modeling, and geophysical surveys (Kirsch et al., 2018; Jackisch et al., 2019; Heincke et al., 2019; Walter et al., 2020). Drone-based hyperspectral imaging has been used to map rare-earth-element-rich minerals in Namibia. Furthermore, drone-based magnetic surveys were deployed to identify subsurface ore potential at a fraction of the cost of traditional surveys in Greenland (Jackisch, 2020). Integration of drone-based hyperspectral and magnetic data was used for mapping the Otanmäki Fe—Ti—V deposit in central Finland (Jackisch et al., 2019).

2.6 Synthetic aperture radar

An SAR is an active microwave remote sensing system that can acquire data with high resolution regardless of day or nighttime, cloud, haze, or smoke over a region. SAR sensors provide information different from that of OPSs operating in the VNIR through the SWIR and the TIR regions of the electromagnetic spectrum. For generating an SAR image, successive pulses of radio waves are transmitted to illuminate a target scene, and the echo of each pulse is received and recorded (Fig. 2.13). SAR sensors transmit electromagnetic radiation at specified wavelengths and measure the reflected energy. Radar transmits and detects radiation between 2 and 100 cm, typically at 2.5—3.8 cm (X-band), 4.0—7.5 cm (C-band), and 15.0—30.0 cm (L-band) (Woodhouse, 2006). Longer wavelengths optimize the depth of

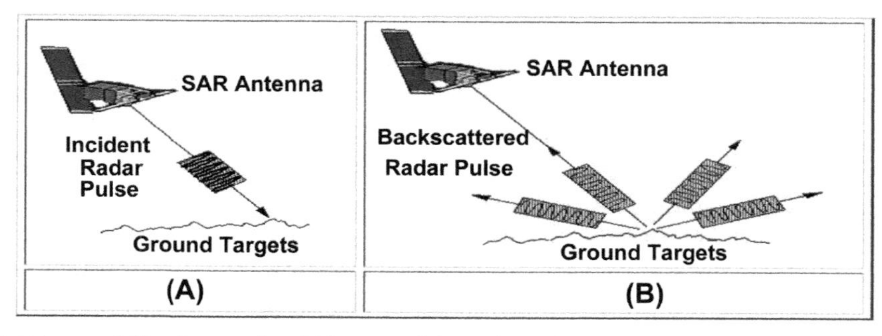

FIGURE 2.13 Data acquisition image of an SAR sensor: (A) A radar pulse is transmitted from the antenna to the ground; (B) the radar pulse is scattered by ground targets back to the antenna. *SAR*, Synthetic aperture radar.

investigation of the radar signal. The long wavelengths allow the radar to have complete atmospheric transmission. Generally, the approximate depth of penetration is equal to the radar's nominal wavelength (Woodhouse, 2006). Single polarization SAR sensors can obtain only the horizontal (HH) or vertical (VV) components of the microwave. Polarimetric SAR sensors can obtain a quad type of data (HH, HV, VH, and VV).

SAR imagery such as RADARSAT, JERS-1 SAR, ERS SAR, PALSAR, and Sentinel-1 data have been extensively used for geological mapping, especially structural analysis related to ore mineralizations (Singhroy, 2001; Paganelli et al., 2003; Pour and Hashim, 2014a,b, 2015a, b,c; Pour et al., 2013, 2014, 2016, 2018a,b; Zoheir et al., 2019a,b,c). RADARSAT data, JERS-1 SAR data, ERS SAR data, and Sentinel-1 SAR data are C-band SAR sensors, on the other hand, PALSAR is an L-band sensor. These sensors have diverse multimode observation functions, multipolarization configuration, variable off-nadir angle, switching spatial resolution, and swath width observation.

The main distortions suffered by an SAR image due to the side-looking architecture are foreshortening, layover, and shadowing (Gelautz et al., 1998; Franceschetti and Lanari, 1999). Foreshortening is inherent to a high-resolution imaging radar, which arises from surface height differences making sections of a rough surface fall into different range bins than they would if the surface were flat. The effect depends on the wave amplitude, wavelength, radar look angle, and resolution (Ouchi, 1988). Foreshortening is a dominant effect in SAR images of mountainous areas. Especially in the case of steep-looking spaceborne sensors, the across-track slant-range differences between two points located on foreslopes of mountains are smaller than they would be in flat areas. Foreshortening is obvious in mountainous areas (top left corner), where the mountains seem to "lean" toward the sensor. This effect results in an across-track compression of the radiometric information backscattered from foreslope areas that may be compensated during the geocoding process if a terrain model is available (Franceschetti and Lanari, 1999).

Layover occurs in those cases where the top of the mountain is closer to the sensor than the bottom, that is, the terrain is sufficiently steep. Layover areas appear in the image as bright regions with the original geometric order being disturbed. Like optical images, those areas that are not illuminated by the radar beam are called shadows. Shadow areas appear in the image as dark regions corrupted by thermal noise (Gelautz et al., 1998). Shadowing areas (i.e., slopes facing away from the sensor) are darker since the energy is spread over a larger area or they are not visible at all. Another cause of distortion in an SAR image is associated with the dissimilar resolutions in range and azimuth. In image formation, the range resolution depends on the bandwidth, while the azimuth one depends on the length of the antenna. As a result, the pixel will not generally be square but rather rectangular on the ground. A rectangular pixel stretches the image in the direction where the resolution is higher (for satellite applications, often this is the azimuth). Due to the severe distortions affecting an SAR image, the latter cannot be overlapped straightforwardly with a map. As a first step, the image must be geocoded to correct the geometric distortions. Subsequently, it must be projected on a coordinate system with a geo-location (Campbel, 2007; Wise, 2002).

2.6.1 RADARSAT data

The RADARSAT program consists of a pair of remote sensing satellites from the Canadian Space Agency (CSA): RADARSAT-1 was an Earth-imaging radar satellite that CSA launched on November 4, 1995 and was decommissioned on March 29, 2013. RADARSAT-2 is an Earth-imaging radar satellite that CSA and MDA (MacDonald Dettwiler Associates Ltd. of Richmond, BC, USA) launched on December 14, 2007 and remains operational. RADARSAT-1 and 2 can image Earth at a single microwave frequency of 5.3 GHz, in the C band (wavelength of 5.6 cm). RADARSAT SAR instrument specifications include the polarization of HH, HV, VH, VV, HH + HV, VV + VH, HH + VV + HV + VH, the spatial resolution of data 1–100 m, and 18–500 km swath widths. RADARSAT-1 and RADARSAT-2 data for research and development consist of ScanSAR Wide and Narrow (single- or dual-pol), Wide (single- or dual-pol), Fine (single- or dual-pol), Standard (single- or dual-pol), and extended high and low (single pol) modes.

Ice and environmental monitoring, marine surveillance, disaster management, resource management, and mapping are among the most important applications of RADARSAT-1 and 2 data. In the geology sector, Canadian radar data are used for both onshore and offshore exploration and mapping and to monitor and detect oil seeps, which reduce the risk and cost of drilling (Singhroy, 2001). It is also used to derive geophysical terrain information such as surface roughness, which is useful for understanding processes such as bedrock weathering and the sorting of unconsolidated solid materials (Paganelli et al., 2003). The RADARSAT-2 advantage over other radar and optical systems for geological applications is its ultrafine resolution and fully polarimetric capabilities. These provide more detailed mapping of terrain features or fine geological structures, better identification of structural features, and an improved discrimination of geologic units (Paganelli et al., 2003).

2.6.2 JERS-1 synthetic aperture radar data

JERS-1 was an environmental Japanese satellite, also known as "Fuyo-1," which launched on February 11, 1992 and ceased operations in October 1998. JERS-1 was a radar/optical mission and was developed by NASDA, the National Space Development Agency of Japan. JERS-1 was launched into a solar-synchronous orbit at an altitude of 568 km and a repeat cycle of 44 days. It stayed in operation for 6.5 years, until contact was lost in October 1998. The JERS-1 carried an L-band (Center Frequency of 1.275 MHz/23.5 cm) SAR, which operated with HH polarization and a fixed 35-degree off-nadir angle, with an image swath width of 175 km (Rosenqvist et al., 2004, 2010). The OPS was a nadir-pointing multispectral imager. Four bands (bands 1–4) in the VNIR (0.52–0.86 μm), as well as four bands in the SWIR (1.60–2.40 μm), were operated in the OPS. An OPS very near-infrared radiometer, JERS-1 SAR Level-1 single-look complex (SLC) image, and JERS-1 SAR Level-1 precision image are the data products of JERS-1. The JERS-1 in commission to observation around the world aimed at resource exploitation, geological structural mapping, national land survey, agriculture, forestry, fishery, environmental protection, disaster protection, coastal monitoring, and others (Rosenqvist, 1996; Metternicht and Zinck, 1998; Rosenqvist et al., 2010; Pour et al., 2014).

2.6.3 European Remote Sensing Satellite synthetic aperture radar data

The ERS (European Remote Sensing Satellite) SAR included ERS-1 and ERS-2. The two spacecraft were designed as identical twins with one important difference—ERS-2 included an extra instrument (GOME) designed to monitor ozone levels in the atmosphere. The ERS-1 satellite launched on July 17, 1991 and ERS-2 followed on April 21, 1995 into the same orbit, providing continuity for ERS-1. The ERS SAR operated in parallel with another instrument—the Wind Scatterometer (WS). Together the SAR/WS configuration was called the Active Microwave Instrument (AMI). AMI had three modes of operation, which allowed the instruments to image land during the day or night or through clouds. The ERS-1 mission ended on March 10, 2000, far exceeding its expected lifespan. The last ERS-2 SAR data was acquired on July 4, 2011 and the ERS-2 mission itself ended on September 5, 2011.

The ERS SAR was operating in C-band with linear vertical polarization (VV Mode), covering a swath width up to 100-km wide. The satellite travels around 4 km. The transmitted pulse with a length of 37.12 μs was intrapulse-modulated using linear frequency modulation (LFM) and a bandwidth of 15.55 MHz. The "ERS tandem mission" doubled the number of measurement points for the radar altimeter and more importantly laid the basis for an accurate, three-dimensional digital map of Earth's land surfaces, which for some of Earth's regions would have been impossible. The ERS-1 and 2 satellites are intended for global measurements of sea wind and waves, ocean and ice monitoring, coastal studies, and land sensing. All ERS SAR products, in accordance with ESA Earth Observation Data Policy, are freely available to users. ERS SAR products include ERS-1/2—SAR Level-1 medium-resolution image product, ERS-1/2—SAR Level-1 Precision Image Product, ERS-1/2—SAR Level-1 Single-Look Complex Image Product, and ERS-1/2—SAR Raw Image Product. All products are provided in Envisat product format generated by the ESA ERS PF-SAR processor. ERS-2 SAR data have been used for the demarcation of different lithological units in the Singhbhum Shear Zone (SSZ) and its surroundings in Jharkhand, India (Pal et al., 2007).

2.6.4 Advanced Land Observing Satellite phased array–type L-band synthetic aperture radar data

Phased array–type L-band SAR (PALSAR) is an active microwave sensor onboard the Advanced Land Observing Satellite (ALOS), which is not affected by weather conditions and operable both daytime and nighttime (Igarashi, 2001; Rosenqvist et al., 2004; ERSDAC, 2006). The PALSAR was developed by the Japanese Ministry of Economy, Trade, and Industry (METI) and Japan Aerospace Exploration Agency (JAXA) for the acquisition of data beneficial to resource exploration and environmental protection. It was launched on January 24, 2006, onboard ALOS (ERSDAC, 2006). PALSAR data can be used in specific fields, including (1) land area basin mapping (geological structural analysis of target areas; collect a database of potential natural resource deposit areas); (2) coastal area basin mapping (extraction of oil exudating areas; monitoring of contamination accompanied by development activities); (3) monitoring of environments and natural disasters (monitoring of disaster such as landslide, volcanic activities, floods and other; environmental monitoring such as forests; international cooperation), and (4) research and development for the processing and application of multipolarimetric SAR data (geological structural analysis on the first stage of resource exploration) (ERSDAC, 2006).

PALSAR sensor is an L-band SAR, with a multimode observation function (Fine mode, direct downlink, ScanSar mode, and Polarimetric mode) of multipolarization configuration (HH, HV, VH, and VV), variable off-nadir angle (9.9–50.8 degrees), and switching spatial resolution (10, 30, 100 m for Fine, Polarimetric, and ScanSar modes, respectively) and swath width observation (30, 70, and 250–350 km for Polarimetric, Fine, and ScanSar modes, respectively) (Igarashi, 2001; ERSDAC, 2006). Full polarimetry (multipolarization), off-nadir pointing function, and other functions of PALSAR improved the accuracy of analyzing geological structure, and distribution of rocks, and are expected to be used for the first stage of ore deposits and hydrocarbon exploration, and environmental protection (ERSDAC, 2006). Consequently, PALSAR data contain outstanding contributions to geological structural analysis, especially in tropical and remote regions (Pour and Hashim, 2014a, 2015c; Pour et al., 2016, 2018b), where OPSs often failed due to bad weather conditions.

Level-1.5 products are such data that are performed the following processing to Level-1.0 (raw) data of high-resolution mode. The processing is (1) range compression using fast Fourier transform (FFT); (2) secondary range compression using range migration compensation; (3) range migration curvature corrections; (4) azimuth compression; (5) multilook processing; and (6) conversion from slant range to ground range (ERSDAC, 2006). The data executed with the previous processing are geo-reference products. The difference between the Level-1.1 product and Level-1.5 product is whether the multilook processing and conversion from slant range to ground range are performed or not. Besides geo-reference products, there exist geo-coded data that are such data that are processed so that the north direction of the observed image corresponds to the upper direction of the image (ERSDAC, 2006). The Level-1.5 product used in this study has a high-resolution mode with 6.25 m pixel spacing and single polarization (HH or VV), which is geo-reference and geo-coded. The nominal incident angle is 7.9–60.0 degrees.

ALOS-2 is the successor of ALOS. ALOS-2 contributes to cartography, regional observation, disaster monitoring, and resource surveys. The ALOS-2 was launched on May 24, 2014, as the successor of ALOS-1 (launched on January 24, 2006, and decommissioned in May 2011) (http://global.jaxa.jp/projects/sat/alos2/). PALSAR-2 of the ALOS-2 has been significantly improved from the ALOS-1's PALSAR in all aspects, including resolution, observation band, and time lag for data provision (http://global.jaxa.jp/projects/sat/alos2/pdf/daichi2_e.pdf). The resolution of the PALSAR was a maximum of 10 m, but that of the PALSAR-2 has been improved to 1–3 m. This improvement became possible by adding a "Spotlight mode" that enables the satellite to change its radio wave radiation direction to its moving direction (vertical direction) while flying to keep observing one specific target location for a long time. A "Dual receiving antenna system" is adopted to secure enough observation bands with high resolution. As a result, the observation band is 50 km with a 3-m resolution in the "Strip Map mode," and 25 km with 1–3 m. In particular, an HV (horizontally transmitted and vertically received) cross-polarization channel is also used in the ScanSAR observation mode, thus increasing the amount of information for geological structural mapping at a regional scale (Pour et al., 2016; Pour and Hashim, 2017). The observable area greatly broadens from 879 to 2320 km. The ALOS was fixed to face the lower right direction toward the moving direction of the satellite; thus it observed only one orientation. To broaden the observable area, the satellite, thus, by changing the satellite's attitude right or left during observations, both sides of the satellite can be observed. The revisit time is reduced from 46 to 14 days (so that the satellite can go back to the target location sooner).

2.6.5 Sentinel-1 synthetic aperture radar data

Sentinel-1 is a C-band SAR sensor, which is the first of the Copernicus Programme satellite constellation conducted by the European Space Agency. Its constellation provides high reliability, improved revisit time, geographical coverage, and rapid data dissemination to support operational applications in the priority areas of marine monitoring, land monitoring, and emergency services. Sentinel-1 potentially images all global landmasses, coastal zones, and shipping routes in European waters in high resolution and covers the global oceans at regular intervals. Having a primary operational mode over land and another over open ocean allows for a preprogrammed conflict-free operation. The main operational mode features a wide swath (250 km) with high geometric (typically 20 m Level 1 product resolution) and radiometric resolutions, suitable for most applications (Attema et al., 2007).

The Sentinel-1 acquires data in four exclusive modes: (1) Stripmap (SM)—A standard SAR strip map imaging mode where the ground swath is illuminated with a continuous sequence of pulses, while the antenna beam is pointing to a fixed azimuth and elevation angle. It features a 5-by-5-m^2 (16 ft by 16 ft) spatial resolution and an 80-km (50 mi) swath. For SM offer, data products are in single (HH or VV) or double (HH + HV or VV + VH) polarization. (2) Interferometric wide swath (IW)—Data are acquired in three swaths using Terrain Observation with Progressive Scanning SAR (TOPSAR) imaging technique. It features a 5-by-20-m^2 (16 ft by 66 ft) spatial resolution and a 250-km (160 mi) swath. In the IW mode, bursts are synchronized from pass to pass to ensure the alignment of interferometric pairs. IW is Sentinel-1 primary operational mode over land. IW offers data products that include single (HH or VV) or double (HH + HV or VV + VH) polarization. (3) Extra wide swath (EW)—Data are acquired in five swaths using the TOPSAR imaging technique. EW mode provides extensive swath coverage at the expense of spatial resolution. It features 20-by-40-m^2 (66 ft by 131 ft) spatial resolution and a 400-km (250 mi) swath. EW offers data products that include single (HH or VV) or doubles (HH + HV or VV + VH) polarization. (4) Wave (WV)—Data are acquired in small strip map scenes called "vignettes," situated at regular intervals of 100 km along the track. It features a 5-by-20-m^2 (16 ft by 66 ft) resolution and a low data rate. It produces 20 km by 20 km (12 mi by 12 mi) sample images along the orbit at intervals of 100 km (62 mi). The vignettes are acquired by alternating, acquiring one vignette at a near-range incidence angle, while the next vignette is acquired at a far-range incidence angle. WV is Sentinel-1 operational mode over the open ocean. WV offers data products only in single (HH or VV) polarization (Attema et al., 2007).

The ESA and European Commission's policies make Sentinel-1 data easily accessible. Various users can acquire the data and use it for public, scientific, or commercial purposes for free. Sentinel-1data products distributed by ESA include (1) raw Level-0 data (for specific usage); (2) processed Level-1 SLC data comprising complex imagery with amplitude and phase (systematic distribution limited to specific relevant areas); (3) ground range detected (GRD) Level-1 data with a multilooked intensity only (systematically distributed); (iv) Level-2 Ocean (OCN) data for retrieved geophysical parameters of the ocean (systematically distributed). Sentinel-1Wave Mode data were used for structural mapping to explore orogenic gold in transpression and transtension zones in the Barramiya–Mueilha Sector, Egypt (Zoheir et al., 2019a).

2.7 Data acquisition

Data acquisition for different types of satellite images is available on several websites (free of charge or purchasable). Some of the recognized websites are presented in this section as follows:

- USGS EarthExplorer (https://earthexplorer.usgs.gov/): The USGS EarthExplorer (EE) tool provides users the ability to query, search, and order satellite images, aerial photographs, and cartographic products from several sources. On this site, a complete set of satellite images, such as Landsat, ASTER satellite images, SRTM digital height images, radar images, and Hyperion hyperspectral images, can easily be downloaded.
- Earthdata Search—NASA (https://search.earthdata.nasa.gov/search): Earthdata Search provides the only means for data discovery, filtering, visualization, and access across all of NASA's Earth science data holdings. It is a Gateway to NASA Earth Observation Data. The Earth Science Data Systems (ESDS) Program provides full and open access to NASA's collection of Earth science data for understanding and protecting our home planet. This website contains an extensive collection of different satellite images. For instance, to download the ASTER datasets, due to the possibility of a better display of image information, it is better to specify the desired images on the USGS Earth Explorer website and download the images by searching their ID in NASA Earthdata Search. Moreover, this site achieved a high-resolution (12.5 m) DEM.
- Copernicus Open Access Hub (https://scihub.copernicus.eu/): The Copernicus Open Access Hub (previously known as Sentinels Scientific Data Hub) provides complete, free, and open access to Sentinel-1, Sentinel-2, Sentinel-3, and Sentinel-5P user products, starting from the In-Orbit Commissioning Review (IOCR). Copernicus is the European Union's Earth Observation Program, looking at our planet and its environment for the benefit of Europe's citizens.
- DigitalGlobe Open Data Program (https://www.maxar.com/): DigitalGlobe is a leading provider of commercial high-resolution Earth imagery products and services. DigitalGlobe delivers 30 cm resolution imagery, delivering clearer, richer images that empower informed decision-making in global coverage. It provides access to a variety of resources, including satellite (i.e., IKONOS, QuickBird, GeoEye-1, and WorldView) and aerial imagery data, metadata, documentation, software, and map data. This website is especially suitable for downloading sample images for the local size of case studies with high resolution.

- NASA Worldview (https://worldview.earthdata.nasa.gov/): The Worldview tool from NASA's Earth Observing System Data and Information System (EOSDIS) provides the capability to interactively browse over 1000 global, full-resolution satellite imagery layers and then download the underlying data. It is a free online visualization tool that is a great launchpad for learners who are new (or veteran) users of satellite data—interactive interface for browsing full-resolution, global, and daily satellite images. Users interact with NASA's daily, global satellite imagery of the Earth within hours of it being acquired. On this website, images with supports time-critical applications such as meteorology and wildfire management are available with a temporal resolution of less than a few minutes.
- NOAA CLASS (https://www.avl.class.noaa.gov/saa/products/catSearch): NOAA has a "classy" way of handling environmental data called the Comprehensive Large Array-data Stewardship System (CLASS). CLASS is an online data management system that gives users faster, easier access to America's environmental data. CLASS is a key component of NOAA's integrated global observation and data management system. CLASS is NOAA's premier online facility for the distribution of data from the following satellite and observing systems: (1) NOAA Polar-orbiting Operational Environmental Satellite (POES); (2) NOAA's Geostationary Operational Environmental Satellite (GOES); (3) NASA's Earth Observing System (EOS); and (4) US Department of Defense Meteorological Satellite Program (DMSP). This website has the option of choosing a resolution for available datasets.
- ALOS—JAXA/EORC (https://www.eorc.jaxa.jp/ALOS/en/index_e.htm): ALOS—JAXA/EORC includes tools for using satellite products, and documents related to product format and other information are available. These tools provide access to Global PALSAR-2/PALSAR/JERS-1 Mosaics and Forest/Non-Forest Maps, Precise Global Digital 3D Map "ALOS World 3D," High-Resolution Land Use and Land Cover Map Products, Global Mangrove Watch, JICA-JAXA Forest Early Warning System in the Tropics, ALOS Ortho Rectified Image Product, Glacial Lake Inventory of Bhutan, PALSAR-2 L2.2 Product, and others.
- NOAA Digital Coast (https://coast.noaa.gov/digitalcoast/): NOAA's Digital Coast provides the data, tools, and training that communities use to manage their coastal resources. Data sets alone are not enough. For data to be truly useful, people often need assistance by way of relevant training and associated tools and information.
- EOS LandViewer (https://eos.com/products/landviewer/): LandViewer is a satellite observation imagery tool that allows for on-the-fly searching, processing, and getting valuable insights from satellite data to tackle real business issues such as On-the-fly processing. It is a central GIS database for a free download of satellite images and access to the most used and up-to-date satellite images. Also, this website allows previewing, ordering, and buying high-resolution satellite images. In addition to downloading images, it also provides users with many tools for data analysis. Some of the satellite images that can be downloaded include Landsats 4–7, MODIS MCD43A4, Sentinels 1–2, Pléiades 1, SPOT 6 and 7, KOMPSAT 2, KOMPSAT 3/3A, SuperView 1, Terrain Tiles images.

- Sentinel Hub (https://www.sentinel-hub.com/): Sentinel Hub services provide long-term analysis in two efficient ways, including EO Browser and Sentinel Playground. EO Browser allows users to visualize satellite data from numerous satellites and data collections instantly. It has several datasets of medium and low-resolution satellite imagery, including a complete archive of all Sentinels, Landsats 5–7, 8, MODIS, Envisat Meris, and Proba-V. Sentinel Playground uses Sentinel Hub technology to enable easy-to-use discovery and exploration of full-resolution Sentinel-1, Sentinel-2, Landsat 8, DEM, and MODIS imagery, along with access to the EO data products. It is a graphical interface to a complete and daily updated Sentinel-2 archive, a massive resource for anyone interested in Earth's changing surface, natural or manmade.
- Google Earth Engine (https://earthengine.google.com/): Google Earth Engine is a planetary-scale platform for Earth science data and analysis. This platform combines a multipetabyte catalog of satellite imagery and geospatial datasets with planetary-scale analysis capabilities. Scientists, researchers, and developers use the Google Earth Engine to detect changes, map trends, and quantify differences on the Earth's surface. Google Earth Engine is a cloud-based geospatial analysis platform that enables users to visualize and analyze satellite images of the Earth. Scientists and nonprofits can also use the Google Earth Engine for remote sensing research, predicting disease outbreaks, natural resource management, mineral exploration, and more. It is now available for commercial use and remains free for academic and research use.

2.8 Preprocessing techniques

Raw satellite images often need correction and restoration in terms of geometry and radiometry. Deterioration of images depends on various reasons such as sensor failure, weather conditions, position and height of the ground surface, the relative motion of the Earth and satellite, changes in the angle of the sun relative to the Earth, and others. As a result, before image enhancement, some of these corrections should be done on the images (Lillesand et al., 2015; Girard et al., 2018). In this section, the atmospheric corrections required in the reflective optical part (0.3–3 μm), including VNIR and SWIR wavelengths in the electromagnetic spectrum, are discussed. When electromagnetic waves pass through the atmosphere, their intensities change due to the reaction with the atmosphere (selective scattering and absorption). Visible wavelengths are more affected by atmospheric scattering. Near-infrared waves and short wavelengths are practically protected from the effects of atmospheric scattering, but in some wavelengths, they have selective absorption characteristics. For longer wavelengths such as TIR, the Earth's atmosphere is only transparent at selected wavelengths. The digital number on satellite images shows the relative brightness value of the target pixel, which must be converted into reflection in the preprocessing stage.

In addition, to identify objects, it is also important to accurately determine their location. Therefore one of the tasks of remote sensing specialists is to prepare maps that show the spatial distribution of objects correctly. Distorted images provide incorrect spatial information.

In some remote sensing applications, a lot of geometric accuracies are required, so that if the geometric error of the satellite data exceeds a certain value, it will not be possible to reach the desired final goal. Therefore the geometric accuracy of remote sensing data is important for creating maps with geographic coordinates and scale for practical purposes.

2.8.1 Noise reduction/correction

Most of the satellite images contain noise. These noises may be of a random type, which in many cases happens due to malfunctioning of mechanical–optical components in the sensor. Such noises cannot easily be fixed without reducing the quality of the image, but by improving the sensitivity of the sensor, the SNR can be kept as high as possible in the image processing chain. Sometimes satellite images will have systematic noises that will be mostly removed during the preprocessing steps. The simplest form of systematic noise is called line or pixel drop-out (Fig. 2.14). In this phenomenon, the values of an image line or a part of it are missing, and all the numerical values in this line are zero. The occurrence of this phenomenon is due to momentary disturbances in the sensor or the receiving station, which leads to the loss of the signal. Most image-processing systems replace the mentioned pixels by taking the average of the digital number (DN) values of the upper and lower lines of the lost data (Legg, 1994).

In some sensors such as ASTER, the detectors that do not work correctly can cause an error that can be corrected in Section 2.8 by knowing the mechanism of this noise generation. For example, in the ASTER sensor, the light leakage from the fourth-band detector affects the light recording by other short-wave infrared band detectors. This problem can be

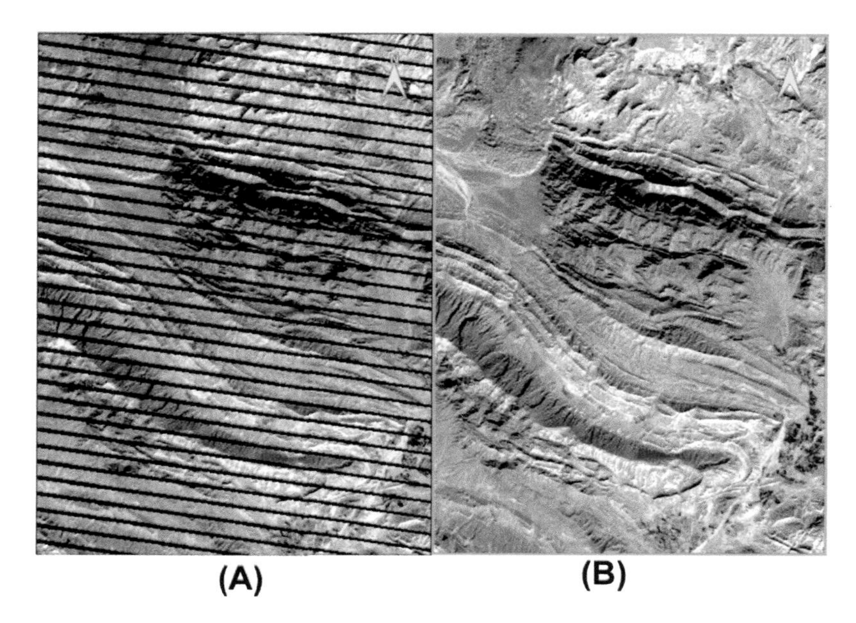

FIGURE 2.14 Removing the lines with zero digital number (DN) values: (A) before correction, (B) after correction.

solved by using Crosstalk correction software (Tonooka and Iwasaki, 2004). Another method for reducing noise from the images is by using the MNF method. MNF is explained in Section 2.9 and is a measure of noise estimation in the spectral bands. The higher MNF band usually shows noisy images. It is possible to invert the MNF images back to the original spaces but leave the noisy MNF images.

2.8.2 Atmospheric corrections

The main goal of atmospheric corrections is to recover the actual values of reflection or emission from the surface of objects in satellite images, which is done by recognizing and eliminating the effects of the atmosphere on the radiation of electromagnetic waves. Before being recorded and used in remote sensing, electromagnetic radiation travels a distance in the atmosphere and reaches the Earth's surface. Therefore along the path of these radiations, they are affected by particles and gases in the atmosphere. When the particles or molecules in the atmosphere cause scattering of the electromagnetic radiation reaching the sensor, they can increase the brightness in the atmosphere. In addition, this phenomenon will increase the ability to scatter in short ultraviolet and blue wavelengths and increases the brightness over terrestrial objects more than expected. This disturbs the process of measuring and identifying phenomena. This disorder and scattering effects can be removed from the images by using image-processing software. If the atmospheric corrections are not applied to the images, we will witness the differences between the image information and the library or field information, which will reduce or completely eliminate the possibility of recognizing the minerals and rocks. Therefore it is necessary to make atmospheric corrections on satellite data before processing them. In general, atmospheric corrections are performed by experimental–statistical methods by using complex radiative transfer models or by integrated methods.

The electromagnetic waves are scattered due to the Mie and Rayleigh scattering phenomena while passing through the Earth's atmosphere. In addition to atmospheric scattering, dark and bright objects that are about each other also affect the amount of reflection of each other (Jensen, 2009). The light produced by these atmospheric scattering affects the spectral characteristics of terrestrial bodies. What reaches the sensor is a combination of scattered waves and spectral characteristics of objects. In the atmospheric correction the main goal is to remove the atmospheric scattering part. ATCOR is a module of Geomatica software, which is used for atmospheric and topographic corrections of satellite images, in the Vis–SWIR (0.4–2.5 μm) and thermal ranges (8–14 μm). ATCOR-3 is an extension of the ERDAS IMAGINE 2014 software, which aims to correct the atmosphere, remove shadows, and modify the atmospheric effects of spectral reflection from surfaces, remove thin clouds, and the effect of fog. The FLAASH model (Fast line of sight Atmospheric Analysis of Spectral Hypercube) is an atmospheric correction module added to ENVI software (Anderson et al., 2002). This module can be used to correct the atmospheric effects of multispectral and hyperspectral satellite images. It is worth mentioning that nowadays, according to the progress of the preprocessing level of raw satellite images, the preprocessing technique is becoming easier for users day by day. The level of preprocessing of each image is clearly defined, and it is easy to determine the same level of preprocessing from the name of the satellite images. Another important point is that performing atmospheric corrections that have a more complex algorithm does not mean a better correction method. In this section a few commonly used atmospheric correction methods are discussed.

2.8.2.1 Atmospheric correction methods

Different methods of atmospheric corrections can be done on satellite images. Indubitably, it is not necessary to do all these corrections at once. Common atmospheric corrections include the dark subtract, the dark dense vegetation, the flat field (FF), the internal average relative reflection (IARR), the empirical line, and the log residual methods.

2.8.2.1.1 Dark subtract

In this method, one should first find an area that has zero reflection in the studied area. Clear and deep waters have this feature in infrared wavelengths. If there is a reflection on the clear water in the infrared wavelength, it is attributed to the path radiance (L_P), which should be subtracted from the pixel values of all the bands. In this method, the shadow can also be used for determining L_P (Chavez, 1988).

2.8.2.1.2 Dark dense vegetation method

When there is a dense vegetation cover in the area, it is expected that the amount of reflection of the red band on the area of vegetation is not more than 2% of the range of pixel values in the red band. If you see more values than this (2% range of pixel values in the red band), it is necessary to subtract the obtained value from the pixel values of all bands (Gillingham et al., 2012).

2.8.2.1.3 Flat field method

In the FF method (Seibert et al., 1998), it is necessary that an area with the following characteristics exists, for using this method:

1. The dimensions of the area should be at least 3 pixel × 3 pixel.
2. The area should have a flat topography.
3. It should have a Lambertian reflection. The sandy desert and beach areas are suitable.
4. Have a stable spectrum over time. Natural vegetation cannot have this feature.

The rocks without vegetation cover, asphalt, artificial grass, desert, and seashore are suitable places to apply this method. Finally, the raw pixel values of the image are divided by the average pixel values in the selected area, which is usually a polygon.

2.8.2.1.4 Internal average relative reflectance

IARR also allows the extraction of spectral information from raw data without the need for ground data (Ben-Dor and Kruse, 1994). In this method, the average reflectance of image pixels is calculated, and the value of each pixel is divided by the average value of image pixels. The number obtained for each pixel will be the relative and normalized reflectance spectrum of that pixel (Ben-Dor and Kruse, 1994). This method especially works well in arid and semiarid terrain. Fig. 2.15 shows the effects of IARR on the raw spectrum that is shown as a line without any prominent reflection/absorption bands (Fig. 2.15A). This is changed to a spectrum that is showing clear absorption/reflection bands (Fig. 2.15B).

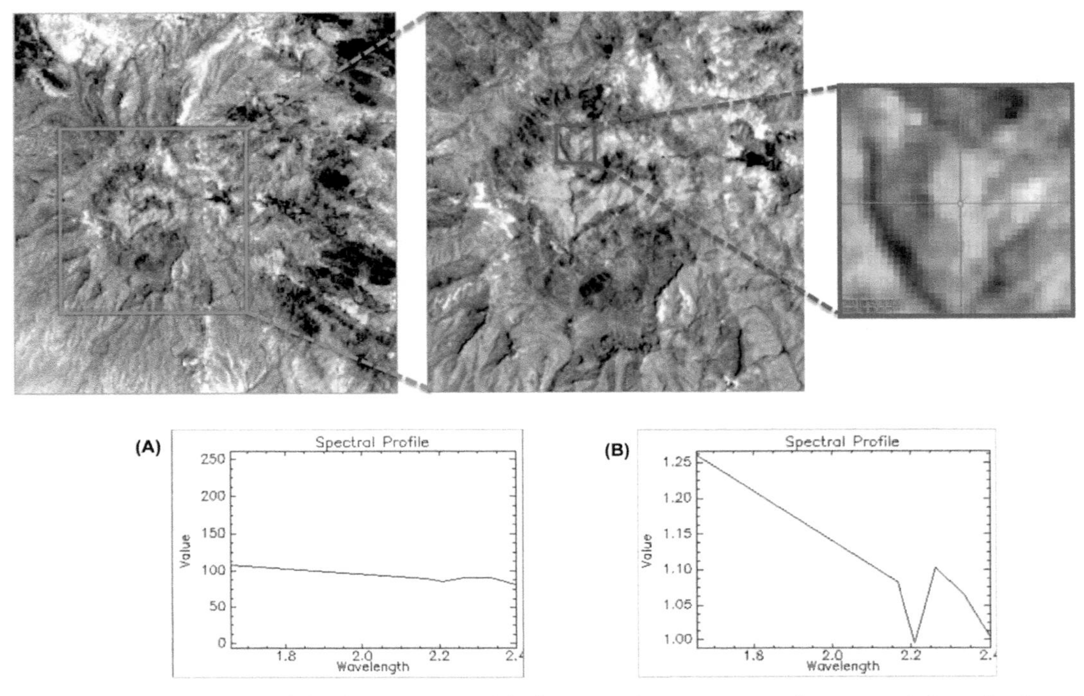

FIGURE 2.15 IARR correction: (A) before correction, (B) after correction. *IARR*, Internal average relative reflection.

2.8.2.1.5 Log residual method

In this method, without the need to use secondary data and only by using the statistical information of the image itself, all the DN values in the image are divided by the geometric mean of the image to achieve the desired result. The geometric mean of a set of N numbers is different from its arithmetic mean. In the geometric mean, instead of addition, multiplication is used, and instead of dividing by N, the Nth root is calculated (Hook et al., 1992). This method is designed to remove solar radiation, topographic, and albedo effects from radiation data. This transformation creates a quasireflection image that is useful for analyzing the absorption properties of rocks and minerals.

2.8.2.1.6 Empirical line method

This method performs atmospheric correction by using ground measurements and relating them to image spectra through a linear relationship. Therefore this method requires measuring the spectra of field samples, awareness, and basic knowledge of the region. In this method, two or more terrestrial phenomena are identified, and the spectrum of their field sample is measured. It is suggested to use dark and white phenomena that have different color tones. To better understand the concept of this method, consider Fig. 2.16. For example, in this figure, two dark and light phenomena are identified in the image and on the ground. The amount of radiation measured for the dark phenomenon by the sensor is A and

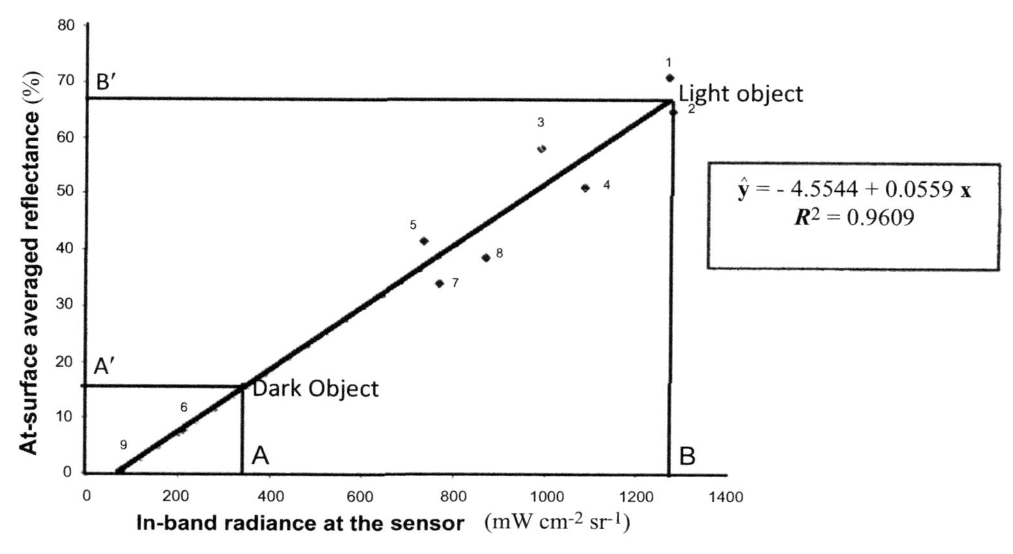

FIGURE 2.16 Empirical linear atmospheric correction method for Ikonus (Karpouzli and Malthus, 2003).

the amount of reflection obtained from the field spectroscopy of this phenomenon is *A′*; in the same way, the amount of radiation measured for the light phenomenon by the sensor is *B* and the amount of reflection obtained from field spectroscopy of this phenomenon is *B′*. Finally, the slope of the line that connects the two points and passes through other points can be calculated by the regression method. The conversion equation of this method is presented in Fig. 2.16 (Karpouzli and Malthus, 2003).

2.8.3 Geometric distortions of satellite images

Based on repeatability or randomness, various geometric distortions are divided into two groups, which are systematic and random. Regular distortions result from the regular relative motions or programmed mechanisms during the data acquisition process. Their effects are predictable and can be corrected. Many systematic distortions are removed in the raw data preprocessing stage. Random distortions are caused by uncontrollable changes and deviations. They are unpredictable and require a more complex process to remove and are generally ignored in the initial preprocessing stage.

2.8.3.1 Geometric correction

The last stage of remote sensing image preprocessing is image geometric correction. Usually, atmospheric preprocessing precedes geometric corrections. Two methods for geometric correction are applied to satellite images, which include systematic and random geometric corrections. The goal of geometric correction is to compensate for the distortions caused by several factors, and the resulting image should have the geometric integrity of a map. The implementation of the first method depends on knowing the orbit of the satellite and its position at the time of image collection. If the exact location, height, and direction of the satellite's movement are known, three-dimensional trigonometry can be used to calculate the geographic longitude and latitude of each pixel. Modern satellites are like stable platforms that send their position and height along with images to the ground so that with the help of this information, accurate geometric correction of the images can be done using professional software. The second method, which was invented in the early years of the birth of remote sensing, is based on the definition of points with specific geographical coordinates, known as control points in the images. The intersection of waterways, the intersection of faults and streams, and road intersections are suitable places for selecting control points in the satellite images (Lillesand et al., 2015). By considering sufficient control points, a more precise correction is made. Both mentioned methods are effective in correcting the images of flat areas, but they do not give perfect results in areas with rough topography (Legg, 1994). In such cases, precise geometrical correction requires the use of DEMs.

Satellite data are often distorted. These data cannot be used as executive maps. The source of these distortions is variable, and the following features can be mentioned:

- change in the height of the imaging platform;
- change in the speed of the imaging platform;
- changing in the direction of the imaging platform;
- the curvature and the rotation of the Earth; and
- feature displacement.

2.8.3.2 Systematic image rectification

Currently, according to the preprocessing level of satellite data, this correction is done on most of the images. Satellites that orbit in polar orbits are usually in the path of elliptical orbits. Since the Earth is also rotating during the rotation of the satellite, the received images are not located in the north–south direction. The rotational movement of the globe from west to east causes that every time a line is scanned by the sensor, which covers an area approximately west of the first scanned line. Here, each of the scanned lines should be moved to the west (Fig. 2.17). With this operation, a square or rectangular image turns into a parallelogram image. Converting a basically square image to a parallelogram image that more accurately shows the reflection of the satellite's path on the Earth's surface is done by a process called resampling of pixels. This function does not change the initial values of the pixels and will have a little visual effect on the image features (Legg, 1994).

In the process of geometric correction, it is usually necessary to specify the datum and the map projection. The datum can be selected locally (such as the sea level of open water) or based on a mathematical model suggested by geodesists based on accurate satellite information. In recent years, the World Geodetic System (WGS84) model has been suggested as the standard level that can be used globally. For the map projection, we can refer to conical, cylindrical, and planar systems.

2.8.3.3 Random geometric correction

Auxiliary control points should be used for random geometric correction. In the geometric correction operation, which is also called georeferencing, several processes are performed in one image, which includes translation, scale change, and rotation (Fig. 2.18).

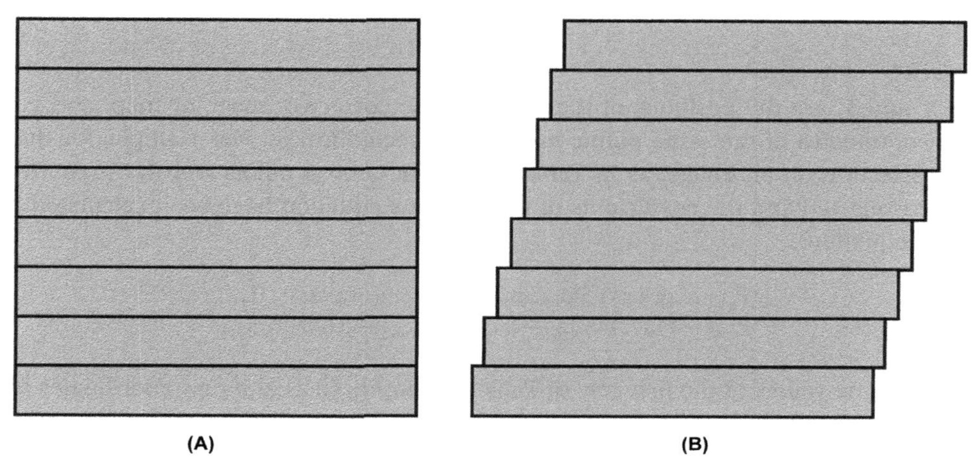

FIGURE 2.17 Systematic geometrical corrections on the images for the Earth and satellite rotations: (A) before correction, (B) after correction. The boxes are the scan lines for the detectors.

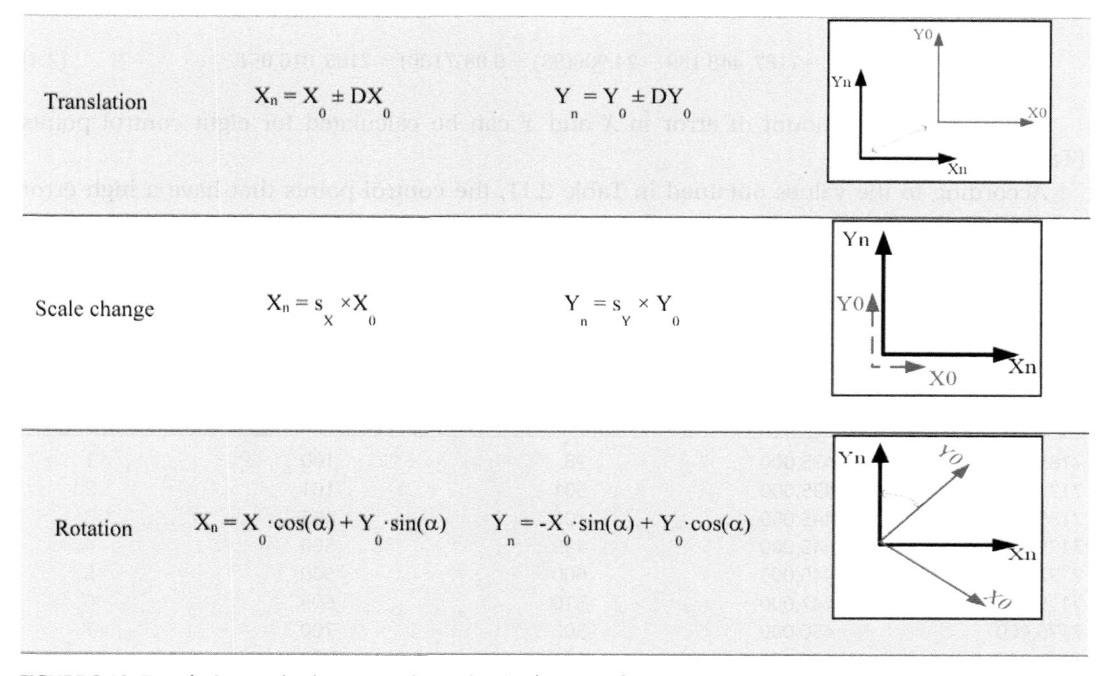

FIGURE 2.18 Translation, scale change, and rotation in the georeferencing process.

The coordinates of the control points should be used in a regression analysis to obtain coefficients for two coordinate transfer equations that can connect the coordinates on the map with the coordinates on the satellite image (Lillesand et al., 2015), as shown in the following equation:

$$\begin{aligned} X' &= a_0 + a_1 x + a_2 y \\ Y' &= b_0 + b_1 x + b_2 y, \end{aligned} \tag{2.1}$$

where X' and Y' are the positions of the points in the corrected image or map, and x and y are the coordinates of the same points in the uncorrected image. For example, for the geometric correction on an image, as in Table 2.10, eight control points with UTM coordinates have been selected and the coefficients of the previous equation have been calculated in the following equation:

$$\begin{aligned} X_{map} &= 432,511.259 + 25.010\left(X_{pixel}\right) \pm 0.074\left(Y_{pixel}\right) \\ Y_{map} &= 7187,448.139 - 24.906\left(Y_{pixel}\right) + 0.087\left(X_{pixel}\right). \end{aligned} \tag{2.2}$$

If we put the values of the first row of Table 2.10 in Eq. (2.2), the new coordinates for the map will be obtained. These coordinates are slightly different from the captured coordinates that were obtained. These errors are attributed to the measurement and wrong location of the point in the image:

$$X_{map} = 432,511.259 + 25.010(100) \pm 0.074(98) = 435,005.007 \tag{2.3}$$

$$Y_{map} = 7187,448.139 - 24.906(98) + 0.087(100) = 7185,016.051. \tag{2.4}$$

In this way, the amount of error in X and Y can be calculated for eight control points (Table 2.11).

According to the values obtained in Table 2.11, the control points that have a high error can be removed. For example, control points, number 4 and 6, have a high error. Of course, the amount of these errors is determined according to the scale of the map. Table 2.12 shows the amount of standard error defined by the US Mapping Organization.

Table 2.10 Selected control points for image geometric correction.

Y Map (GPS)	*X* Map (GPS)	Pixel *Y* (row)	Pixel *X* (col)	GCP no.
7185,000	435,000	98	100	1
7175,000	435,000	501	101	2
7185,000	445,000	100	499	3
7175,000	445,000	499	500	4
7172,500	445,000	600	500	5
7175,000	447,000	510	605	6
7175,000	450,000	500	700	7
7172,500	450,000	601	700	8

Table 2.11 The amount of error in the control points.

*Y*Error	*Y* Map (predicted)	*Y* Map (GPS)	*X*Error	*X* Map (predicted)	*X* Map (GPS)	Pixel Y (row)	Pixel *X* (col)	GCP no.
−16.051	7185,016.051	718,5000	−5.007	435,006.007	435,000	98	100	1
21.209	7174,978.791	7175,000	−0.069	435,000.059	435,000	501	101	2
−0.695	7185,000.695	7185,000	16.018	444,983.982	445,000	100	499	3
−63.140	7175,063.140	7175,000	20.660	444,979.340	445,000	499	500	4
−47.597	7172,547.597	7172,500	28.166	444,971.834	445,000	600	500	5
201.741	7174,798.259	7175,000	−104.607	447,504.607	447,000	510	605	6
−55.545	7175,055.545	7175,000	18.668	449,961.332	450,000	500	700	7
40.002	7172,540.002	7172,500	26.174	449,973.826	450,000	601	700	8

Table 2.12 The amount of standard error according to the scale of the map.

Scale	Standard error (ft)	Scale	Standard error (ft)
1:1200	± 3.33	1:12,000	± 33.33
1:2400	± 6.67	1:24,000	± 40.00
1:4800	± 13.33	1:63,360	± 105.60
1:9600	± 26.67	1:100,000	± 166.67
1:10,000	± 27.78		

The more control points are selected, the better the coordinate accuracy of the resulting image. The following relationship can be used to estimate the error:

$$\text{RMS}_{\text{error}} = \sqrt{\left(X' - X_{\text{orig}}\right)^2 + \left(Y' - Y_{\text{orig}}\right)^2}, \tag{2.5}$$

where X_{orig} and Y_{orig} are the coordinates of the control points in the raw image, respectively, and X' and Y' are the calculated coordinates of the image, respectively.

In the process of geometric correction, all pixels must be moved to the new locations. The new values of the pixels are calculated by the three methods of the nearest pixel in the neighborhood, intensity interpolation, and cubic convolution. Each pixel is moved to a new coordinate and based on the values of adjacent pixels, which assumes a new pixel value that is calculated based on basis of the intensity interpolation relationship:

$$B_v = \frac{\sum \left(Z_k / D_k^2\right)}{\sum \left(1/D^2\right)}, \tag{2.6}$$

where B_v is the brightness values of the corrected image pixels, Z_k is the values of the surrounding and neighboring pixels, and D is the distance among the pixels to the desired position. This relation is for four pixels that are placed around the transition pixel. The same relationship is used in the cubic convolution method, but the number of pixels used in calculating the transition pixel value is 16. In the nearest neighbor method, each point takes the value of the nearest pixel in its neighborhood. The raw image is corrected during this process and transferred to new coordinates.

2.9 Image-processing algorithms

Digital image processing is an important stage of the remote sensing process in which the surface features of the Earth are detected, and information is extracted in the form of new maps by using a computer. The preprocessing of satellite images can be considered a part of the image-processing process. An image-processing system is divided into two parts, namely, software and hardware, which means personal computers and the software section is related to remote sensing software available in the global market. Some of the image-processing software is an open source and freely available. In this section the common methods of image processing for extracting geological information from satellite images are explained.

2.9.1 Color combinations

Since most of the satellite images are available in the form of multiple spectral bands, the analysis of one band alone cannot provide suitable and sufficient information to the user. Understanding the spectral properties is very important in diagnosing phenomena and the type of area coverage. This insight will be useful in using more than one spectral band in the image-processing systems and producing multiband composite color images. In this process, the three-color channels such as blue, green, and red are assigned to different bands so that they can be seen on the color display screen. White light, which is produced from a light source such as the sun or any other artificial source, is a mixture of all the colors in the spectrum, which can be divided into three main colors, including red, green, and blue, or RGB. Colors can be combined. This combination allows the production of a wide range of colors. By combining the three spectrums of red, green, and blue light in equal and maximum amounts, white light is obtained again, which is known as additive colors. Computer monitors, televisions, and industrial displays use the RGB model to display color images. Contrary to additive colors, subtractive colors are the ones that produce black color by combining them (Fig. 2.19).

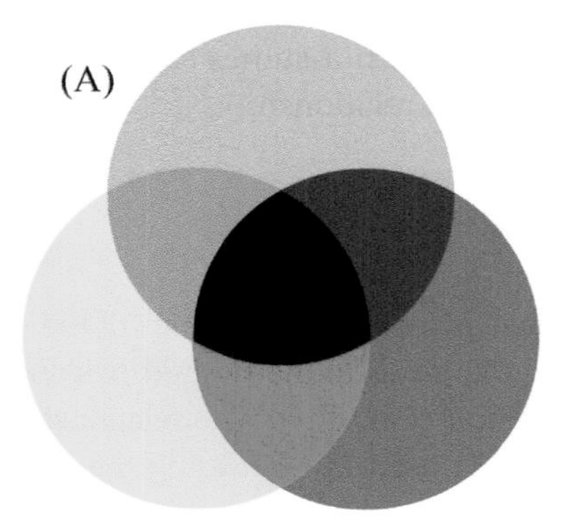

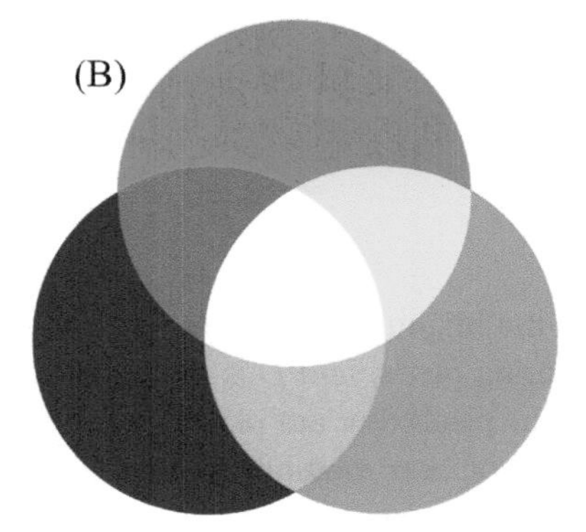

FIGURE 2.19 (A) Subtractive and (B) additive colors.

Before we continue with the color composition of images, let us examine the human visual system in relation to color perception. A part of the electromagnetic waves, which cover a certain wavelength range, passes through the lens of the eye and the clear liquid inside it, and reaches the cylindrical and cone cells of the retina. Retinal cells are sensitive to light, and the interaction of light with these cells, through a complex process, creates an electric signal that reaches the brain cells through the nerves and the image is perceived by the brain. Cylindrical cells are sensitive to low light and do not record information related to color, while cone cells record the intensity and saturation of colors. There are three types of cone cells in the human eye, each of which is sensitive to one of the main colors such as blue, green, and red. The human eye works like an advanced camera and by combining the light that reaches these cone-shaped cells, which produces various colors (Mather and Koch, 2011). Various color models have been used to improve the quality of digital images, of which RGB and Intensity, Hue, and Saturation (IHS) models are the most famous ones. The RGB model is based on the principle that red, green, and blue are the three primary additive colors. This model is depicted with the three-dimensional Cartesian coordinate system, in which the three-color axes of red, green, and blue starting from black or zero exist (Fig. 2.20). The equal combination of all three colors is placed on the gray axis, and the combination of different values of all three colors creates composite colors. For example, the combination of blue and red creates magenta.

Fig. 2.21 shows how to create color images using the RGB model and three arrays stored in the graphics memory. The upper array contains digital numbers in the range of 0–255, which represent 256 levels of red (in an eight-bit dynamic range). Similarly, the middle and bottom arrays show the distribution of green and blue levels, respectively. The values of these arrays are converted from digital to analog form with three digitals to analog converters before being displayed on the screen (Fig. 2.21).

There are many color compositions that can be obtained from multiband images. For example, 6 color compositions can be obtained from a 3-band set, such as SPOT multispectral data,

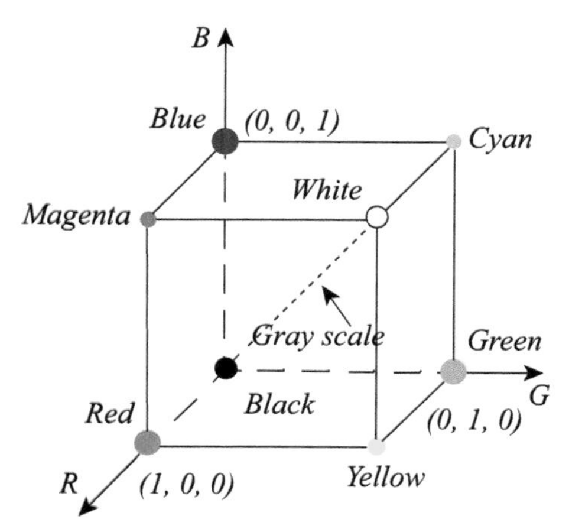

FIGURE 2.20 RGB color model. *RGB*, Red–green–blue.

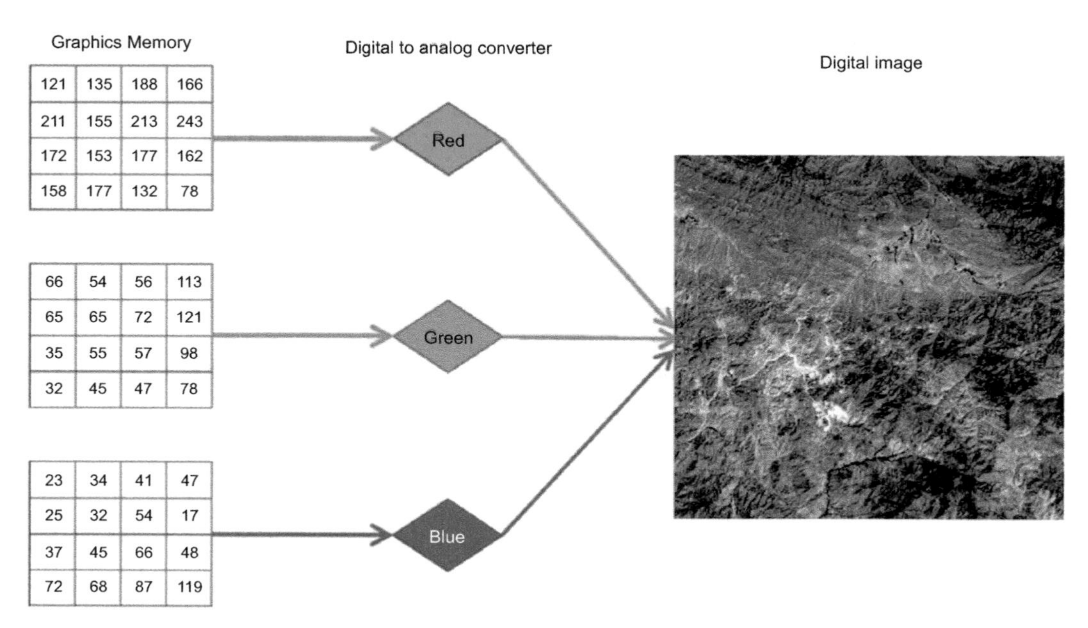

FIGURE 2.21 Creating a color composition using three arrays stored in the graphics memory.

and 84 color compositions can be obtained from a 7-band set, such as Landsat TM data. Of course, to have an interpretable image, known color compositions are used. For example, we can refer to the compositions of the second (blue), third (green), and fourth (red) bands of the ETM + sensor, which shows vegetation in red and water in blue, or the 4–6–8 ASTER band composition, which shows the altered regions in pink. Color images are created in two ways: true color composite and false color composite. In true color compositions, colors and wavelengths are compatible, which means that the band that is imaged in the red wavelength will be placed in the red channel, the band that is imaged in the green wavelength will be placed in the green channel, and the band that is in the blue wavelength will be placed in the blue channel, while in the false color compositions, the images with different wavelengths will be placed in the red, green, and blue channels (colors and wavelengths are not compatible) (Fig. 2.22).

Chavez et al. (1984) used the optimal index factor (OIF) to select the most appropriate color compositions. Based on this, about 20 compositions of three bands from 6 Landsat bands can be made. Of course, this is applicable to any remote sensing data. This equation is based on the value of total standard deviation and correlation coefficients among different bands. The relationship calculated based on three bands is as follows:

$$\mathrm{OIF} = \frac{\sum_{k=1}^{3} S_k}{\sum_{j=1}^{3} \mathrm{Abs}(r_j)}, \tag{2.7}$$

where S_k is the standard deviation for band k, and r_j is the correlation coefficient between any two bands out of the three bands that have been used. The three combined bands with

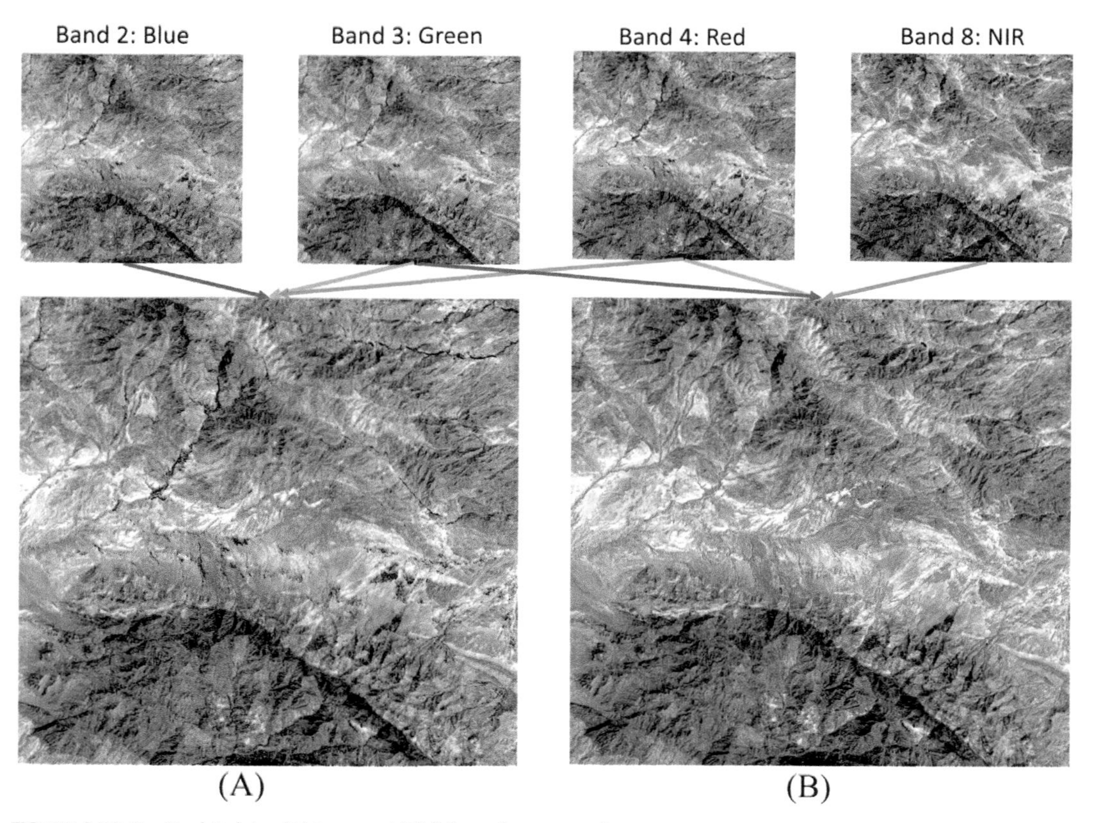

FIGURE 2.22 Sentinel-2 data: (A) true and (B) false color composites.

the largest OIF value usually have the highest information. The OIF helps in choosing the right band combination for creating a color image (Jensen, 2005). It is worth mentioning that combining bands based on the trial-and-error method is more common. In this method, based on the objectives and the known spectral characteristics of the surface phenomena, the spectral bands are used accordingly (Gupta, 2018).

2.9.2 Color space transforms

Color compositions usually consist of three spectral bands corresponding to different wavelengths or three different images for an area, in which the mentioned bands will be placed in three colors of blue, green, and red. The color image is generated based on Fig. 2.23. Another way of generating the color images is using intensity, hue, and saturation (Fig. 2.23). (In different texts, you may see this color space with the names of IHS, ISH, or HSV.) This display method is more compatible with the way of perception of colors by human eyes.

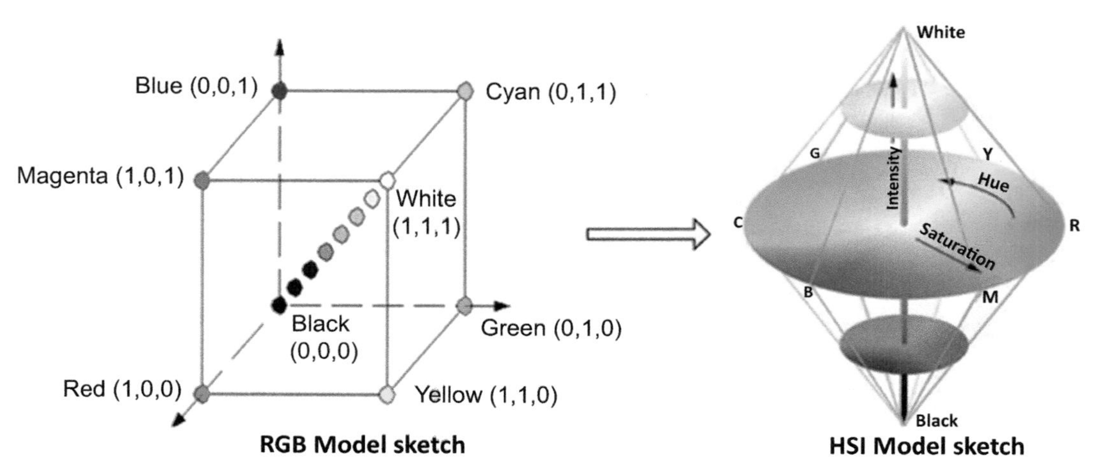

FIGURE 2.23 Diagram of the color space model from RGB to HSI (Cao et al., 2019). The intensity axis overlaps the gray axis in the RGB model. *HSI*, Hue–saturation–intensity; *RGB*, red–green–blue.

Intensity is an estimate of brightness, hue is the expression of color, and saturation estimates the depth of the color of each pixel, which is calculated based on the following relationships, and the RGB color model can be converted to the ISH color space:

$$I(r,g,b) = \frac{1}{\sqrt{3}}(r+g+b)H(r,g,b) = \arccos\frac{2b-g-r}{2V},$$

where

$$V = \sqrt{(r^2+g^2+b^2)-(rg+rb+gb)}S(r,g,b) = \arccos\frac{r+g+b}{\sqrt{3(r^2+g^2+b^2)}}. \tag{2.8}$$

Image-processing systems can convert RGB images to IHS images and vice versa. In ISH images, which are more compatible with color perception by the human eyes, lithological changes can be seen well. Fig. 2.24A and B shows the color conversion from RGB space to ISH space. Fig. 2.24C shows the average intensity image of three 7–4–1 bands, but in the saturated image (Fig. 2.24D), vegetation is more visible. In the grayscale image, complications become more obvious, which are not easily seen in the other two images. For example, roads can be seen better in Fig. 2.24E.

2.9.3 Contrast stretching

Satellite images often have a low range of pixel values, which makes the image appear dark. In the method of stretching the contrast in an 8-bit image, this range is converted into 0 and 255 ranges, which results in a brighter image. Contrast stretching is performed linearly and nonlinearly. In the linear contrast method, pixel values are converted with the following equation:

$$BV_{out} = \left[\frac{(BV_{in} - BV_{min})}{(BV_{max} - BV_{min})}\right] \times 255 BV_{out} = a \times BV_{in} + b. \tag{2.9}$$

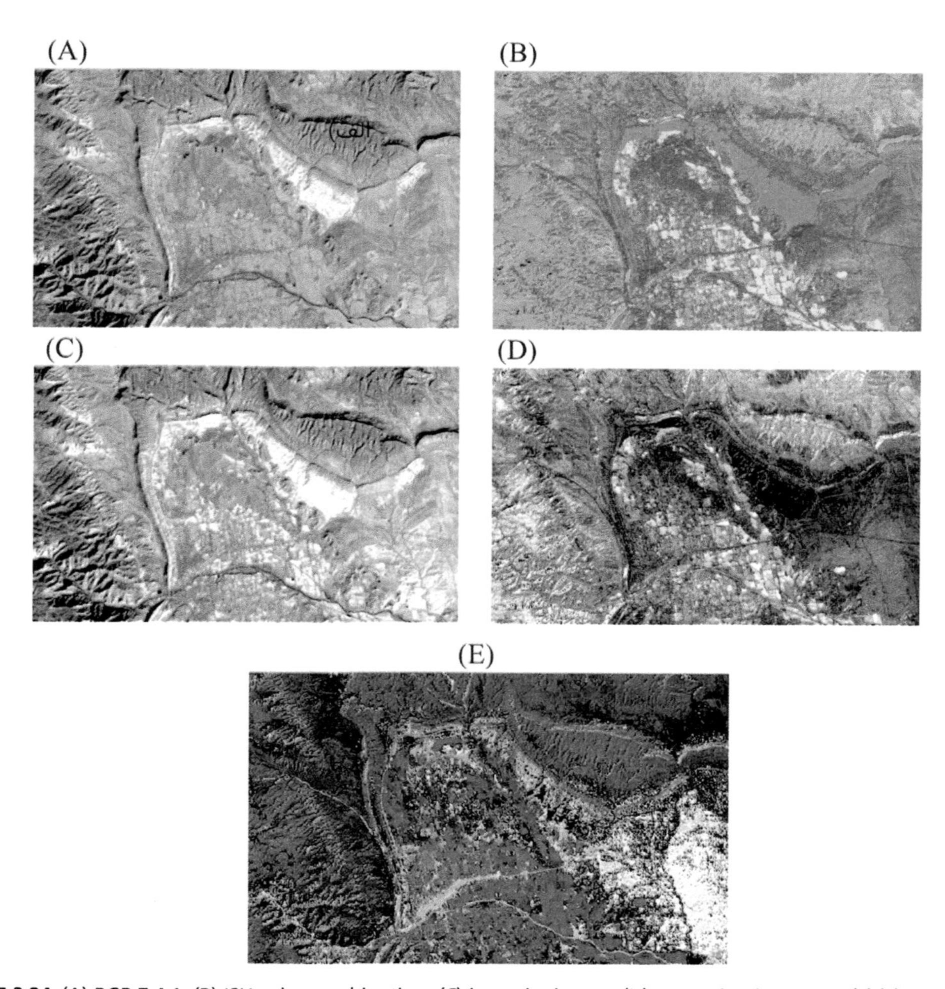

FIGURE 2.24 (A) RGB:7,4,1, (B) ISH color combination, (C) intensity image, (D) saturation image, and (E) hue image. *ISH*, Intensity, saturation and hue; *RGB*, red–green–blue.

In Fig. 2.25A, the Landsat 4 image is shown. The dynamic range of the image is very low, and the image looks dark. Fig. 2.25B shows the histogram related to the Landsat 4. The range of pixel values is in the range of 4–79, although most of these values are in the range of 10–55. The narrow peak at values 14 and 15 represents water, while the wider and main peak represents the land. The number of pixels with a value greater than 60 is too low, so the image is dark. In fact, the range of pixel values is not large (approximately between 8 and 60) and as a result, the contrast of the image is low. Fig. 2.26A shows the result of linear contrast stretching on the image. Fig. 2.26B shows the histogram of this image.

Although the image in Fig. 2.26A is clearer than the original image, it is still relatively dark, and this is because x_{min} and x_{max}, which are the minimum and maximum brightness

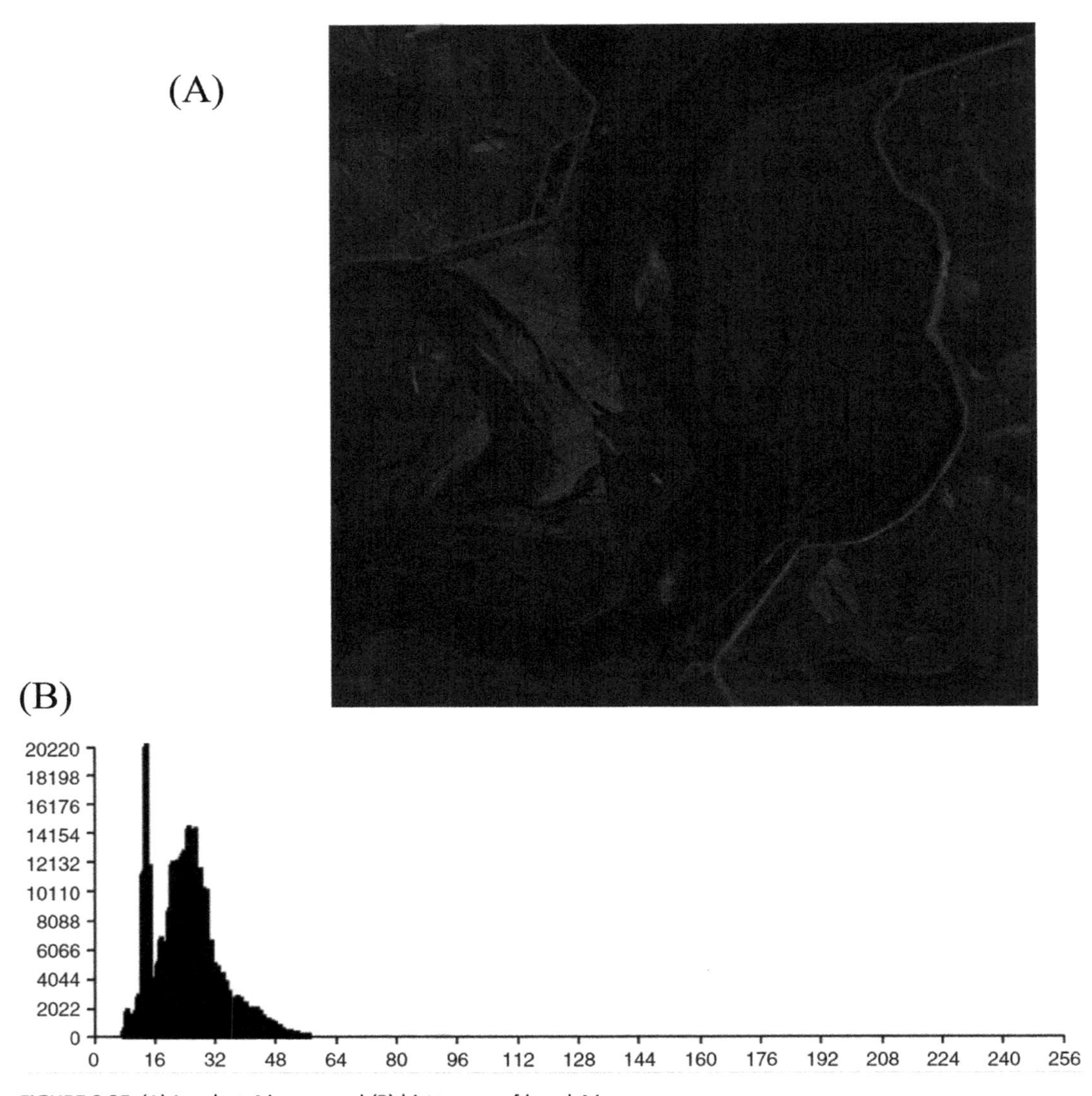

FIGURE 2.25 (A) Landsat-4 image and (B) histogram of band 4 image.

values of the pixels, do not have a suitable distribution in the range of 0–255. To solve this problem, very low and high pixel values can be removed from the original raw image. Visually, the limit of pixel values seems to be 8 and 58, while the lowest value is 4 and the highest value is 79. These relatively low and relatively high values have affected the linear stretch. Therefore if 5% of the low values and 5% of the high values of the pixels are removed, we will see an improvement in the brightness of the image resulting from the contrast stretching technique. As seen in Fig. 2.27A and B, the image is brighter and has more contrast than the image in Fig. 2.25A.

(A)

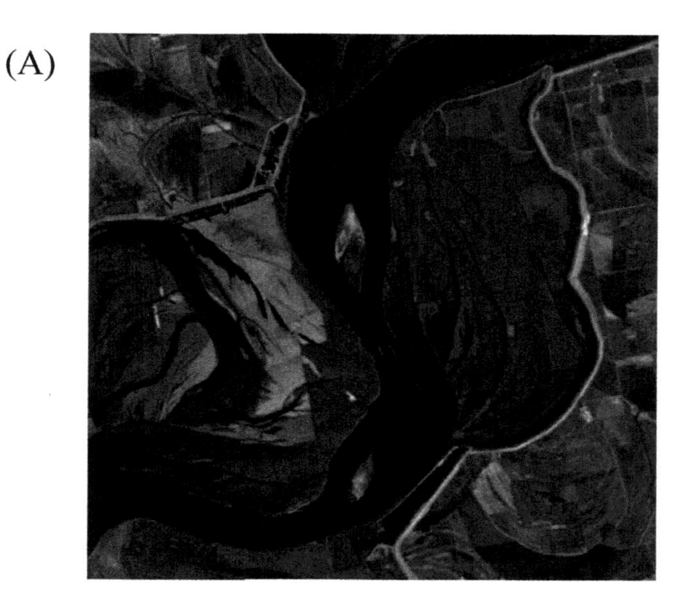

(B)

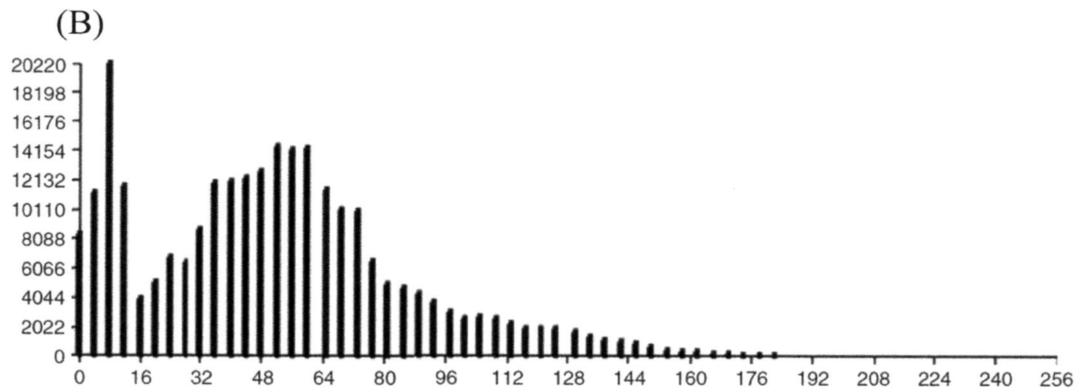

FIGURE 2.26 (A) Landsat-4 image after applying linear contrast stretching and (B) histogram of the image.

2.9.4 Nonlinear contrast stretching methods

Power stretching method: In this method, the values of each pixel are raised to the power of two and then stretched linearly between 0 and 255. As a result, higher DN values become brighter.

Logarithmic stretching method: In this method, the logarithm of the pixel values is calculated and then they are stretched between 0 and 255. As a result, the pixel values in the lower half become brighter.

Exponential stretching method: In this method, based on the following relationship, the values of the pixels are calculated and then stretched between 0 and 255:

$$BV_{\mathrm{out}} = ae^{(BV_{\mathrm{in}})}. \tag{2.10}$$

(A)

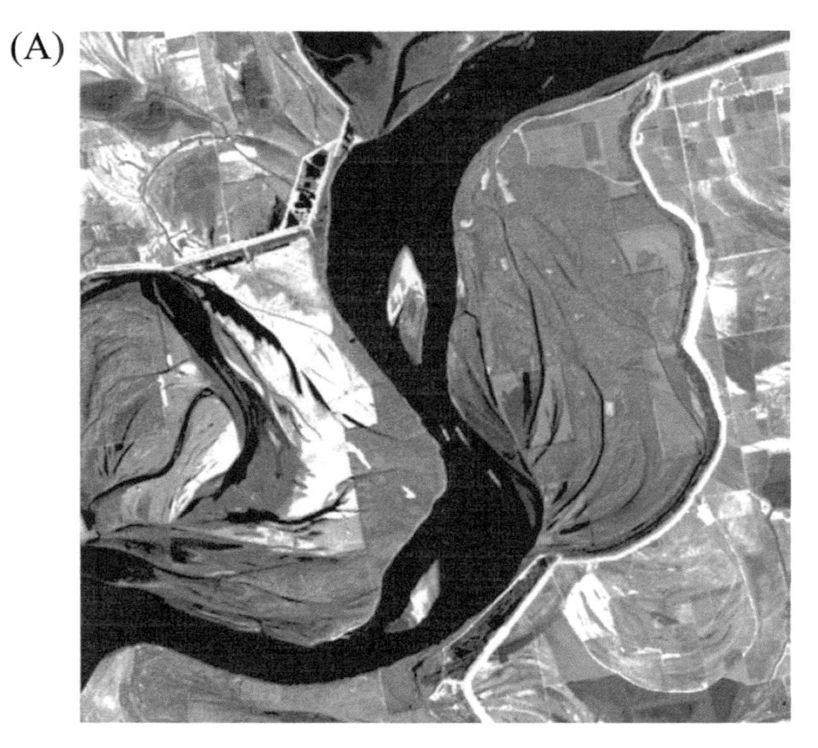

(B)

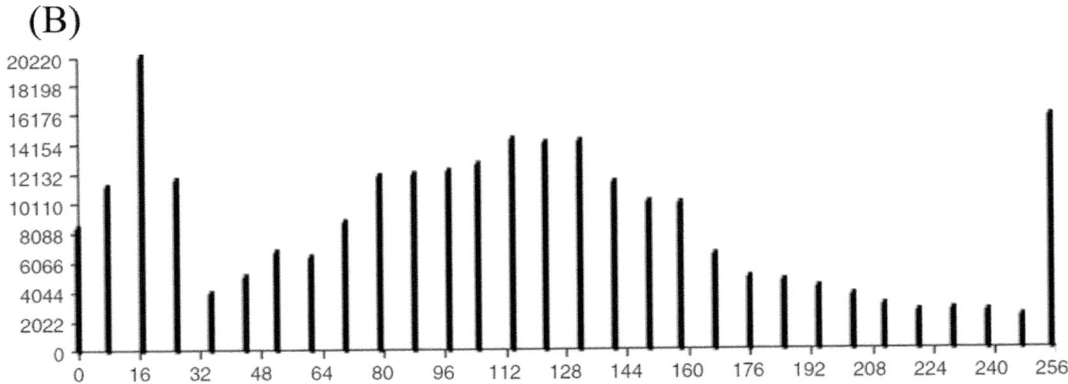

FIGURE 2.27 (A) Applying percentage linear contrast stretching on the image of Fig. 2.25A. (B) Histogram of the image.

Square root stretching method: In this method, the square root of the pixel values is taken, then they are normalized between 0 and 255. As a result, smaller pixel values become brighter.

Gaussian stretching method: In this image-stretching method, parts of the image that are dark or bright become clearer by losing information from the middle parts (in the range of 1, 2, or 3 times the standard deviation).

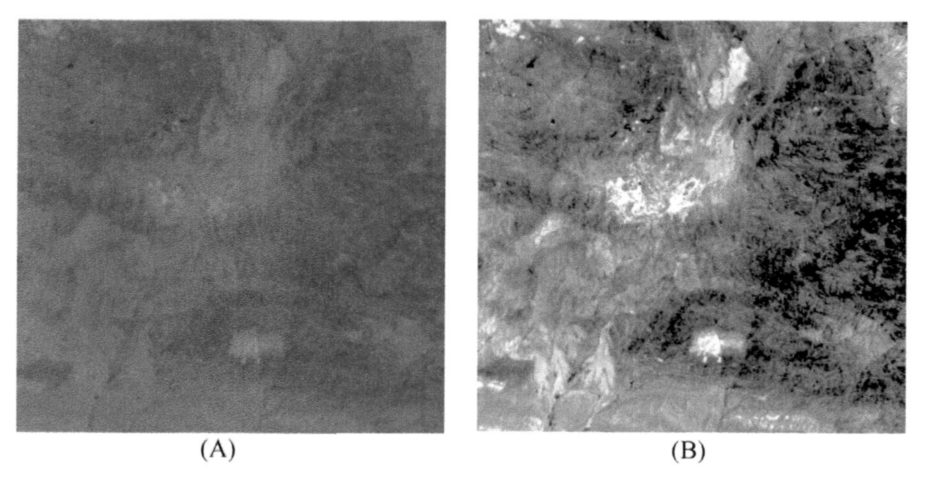

FIGURE 2.28 (A) Sarcheshmeh copper mine raw RGB image and (B) after applying contrast stretching, histogram equalization for three bands. *RGB*, Red–green–blue.

Histogram equalization method: In this method, the histogram distribution of the raw image is changed in such a way that an almost uniform distribution is obtained. Here some categories should be combined. As a result, the number of color shades in the final image is reduced. In this method, the middle parts of the histogram are stretched and become clearer. Contrast stretching can be applied to all three bands of true or false color composite. This is illustrated in Fig. 2.28A and B.

2.9.5 Density slice

The human eye shows a much greater ability to understand the difference between subtle color changes and slight changes in grayscale levels. To highlight differences in a single-band image, it is often useful to assign a color to each gray level or to a group of DN values. This method is called density slicing. Each simple density slice assigns a uniform color to each interval of gray levels defined by the user. By doing this, the difference between the land covers will be shown, but the intra-group changes will still remain hidden (Legg, 1994). In more practical methods of density slicing the user can assign colors to the upper and lower limits of each slice. This method is successful in revealing features with outstanding differences, but it does not show the details of the original image. The reason for removing details is to convert a black-and-white image with 256 levels to an image with a limited number of color layers. Fig. 2.29 shows an ASTER band 4 (SWIR) image of the Sarcheshmeh Copper Mine in southeastern Iran. Alteration zones intensely show reflection at 1.6 μm and appear as light tones (Fig. 2.29A). In Fig. 2.29B, the conversion of the black and white image into a color image using the density slicing method can be seen. Color values are assigned to each DN range. This image conversion is done manually and in each step, a color is assigned to the DN range selected by the user.

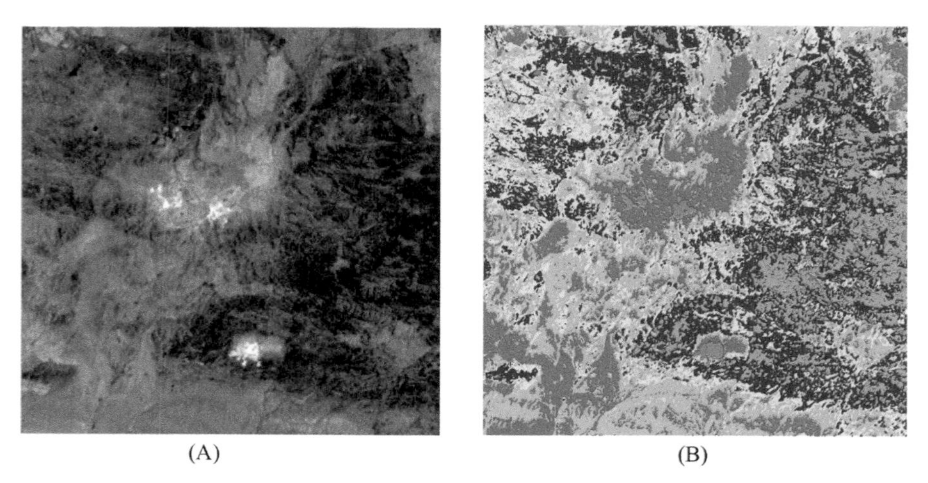

FIGURE 2.29 (A) ASTER band 4 image of the Sarcheshmeh copper mine and (B) converting the black and white image "A" to a color image using the slicing technique. *ASTER*, Advanced Spaceborne Thermal Emission and Reflection Radiometer.

2.9.6 Band ratios

Band ratios are useful methods for detection of the geological features in multiband images. This method is used to reduce the effect of sunlight, and topography, and highlights spectral information in the images. A new digital image is created based on the division of the corresponding pixel in two or more images or the input band based on the following equation:

$$DN_{new} = m\left(\frac{DN_A \pm k_1}{DN_B \pm k_2}\right) + n, \tag{2.11}$$

where DN_A and DN_B are the values of the corresponding pixels in the two input images, k_1 and k_2 are factors depending on the brightness of the path of the two images, and m and n are the scaling factors obtained for the black and white image (Gupta, 2018). The obtained image can also be stretched again or used as a component in the preparation of a color image (Gupta, 2018). The most important advantage of using band ratio is the preparation of an image that is completely independent of lighting conditions (Gupta, 2018). The result of dividing the two bands for a feature on both sides of the mountain is almost the same. As a result, applying the band ratio method removes the shadow effect (Fig. 2.30). Also, this method is used to map the vegetation, sediments, rocks, and soils and to identify the hydrothermal alteration zones.

2.9.6.1 Vegetation detection

The presence of vegetation in the areas for exploratory studies using satellite images is one of the main problems. Indubitably, in some cases, the investigation of vegetation cover helps

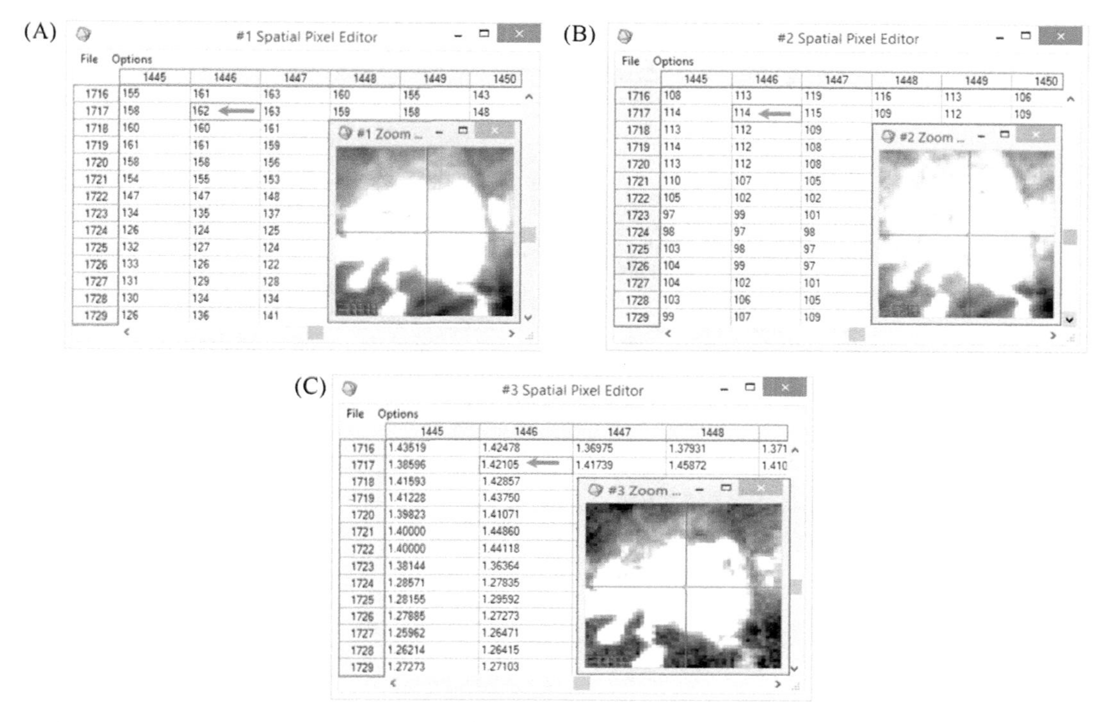

FIGURE 2.30 Applying the band ratio on ASTER data: (A) image of band 4, (B) an image of band 6, and (C) the result of a band ratio of 4/6. *ASTER*, Advanced Spaceborne Thermal Emission and Reflection Radiometer.

exploratory studies for detecting faults. The presence of similar spectra between vegetation and clay minerals requires special methods to eliminate the effect of vegetation, especially in low spectral resolution satellite images. There are different methods for detecting and eliminating the effects of vegetation. The commonly known index used is the normalized difference vegetation index (NDVI), which is defined as follows:

$$\mathrm{NDVI} = \frac{\mathrm{NIR} - R}{\mathrm{NIR} + R}, \tag{2.12}$$

where NIR is the band in the near-infrared range, and R is the band in the red wavelength range. Fig. 2.31 shows the NDVI image of the Sarcheshmeh area.

2.9.6.2 Minerals detection

To detect alteration minerals and lithologies using the band ratio method, it is necessary to know about their spectral behavior. Here are some examples of the ability of band ratio to identify alteration minerals. To use the band ratio, it is suggested that the spectra be resampled based on the wavelength range of the bands. For example, as shown in Fig. 2.32, to detect iron oxides using Landsat 8 images, a band ratio of 4/2 is used, and the reason for

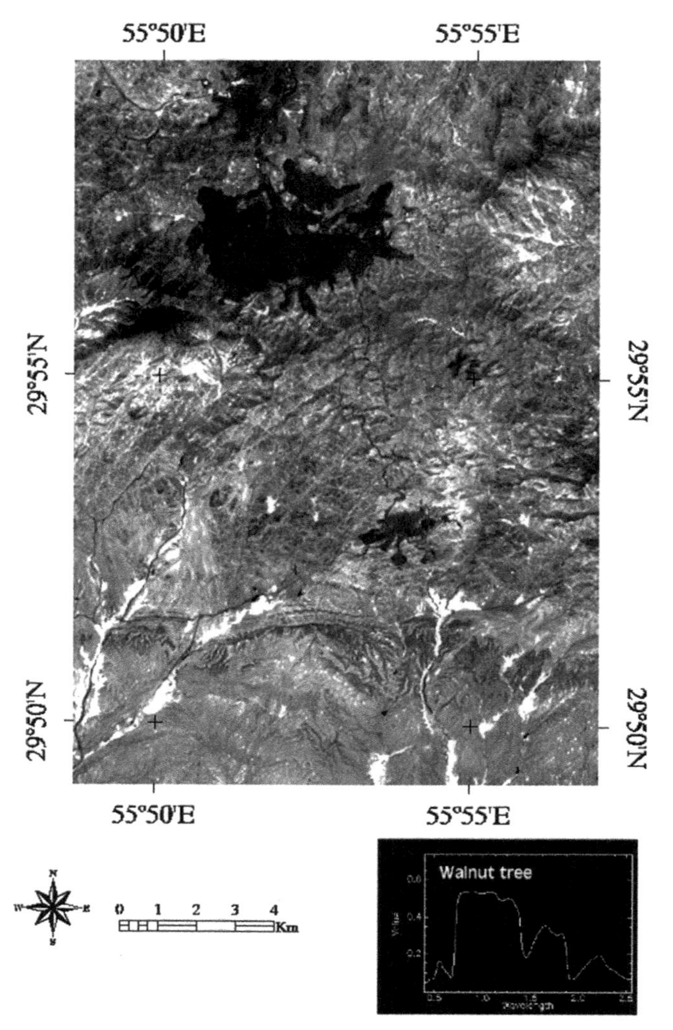

FIGURE 2.31 The spectral curve of vegetation (walnut tree) and NDVI index image. *NDVI*, Normalized difference vegetation index.

using these bands is the high reflection of iron oxides in band 4 and the higher absorption in band 2 of Landsat 8. Bands 4 and 2 of Landsat 8 are equivalent to bands 3 and 1 of Landsat 7, respectively. In Fig. 2.32 the brighter pixels indicate the most likely presence of iron oxides in the scene.

A band ratio of 6/7 (Landsat 8 data) is used to detect minerals related to hydrothermal alteration that contain kaolinite and muscovite. Considering that these minerals have a reflection in band 6 and absorption in band 7 (Fig. 2.33), this band ratio can be used. Bands 6 and 7 of Landsat 8 are equivalent to bands 5 and 7 of Landsat 7, respectively. As can be seen in Fig. 2.33, in addition to altered areas, vegetation has also been enhanced, which is

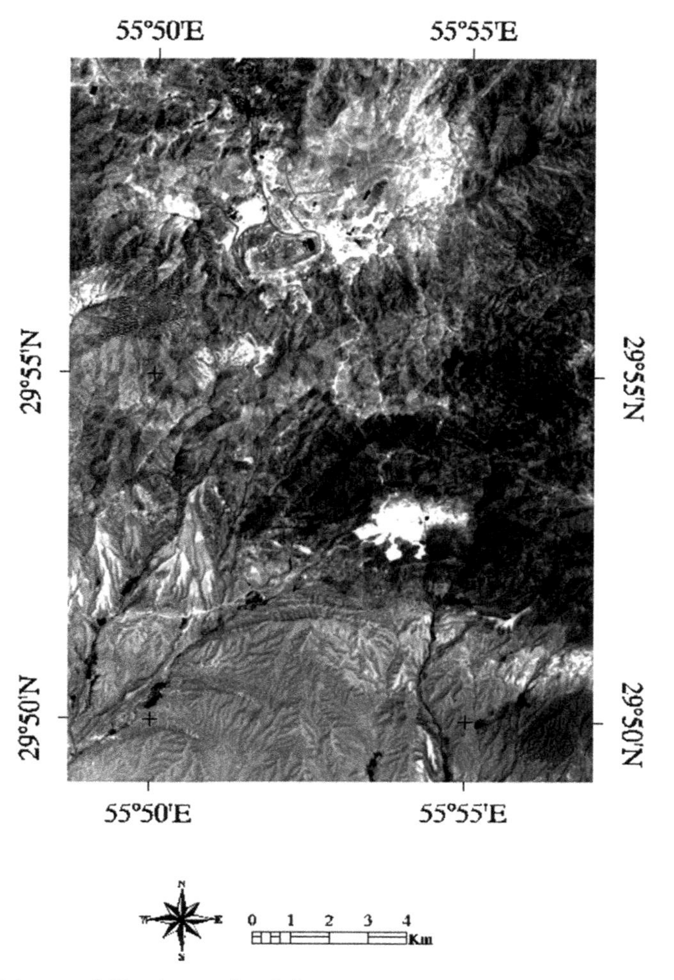

FIGURE 2.32 Band ratioimage of 4/2 using Landsat 8 data.

due to the spectral similarities of vegetation and OH-bearing minerals, related to the hydrothermally altered zones.

2.9.7 Absorption band depth

Based on alteration minerals' spectral properties, absorption band depth (ABD), a common method is used to enhance alteration minerals. In this method, the sum of two reflectance bands is divided by an absorption band. In such a way, reflectance properties could be amplified. ABD is calculatable for relative and true situations (Fig. 2.34A and B). Laboratory reflectance spectra resampled to ASTER VNIR + SWIR bands for three different minerals are presented in Fig. 2.35.

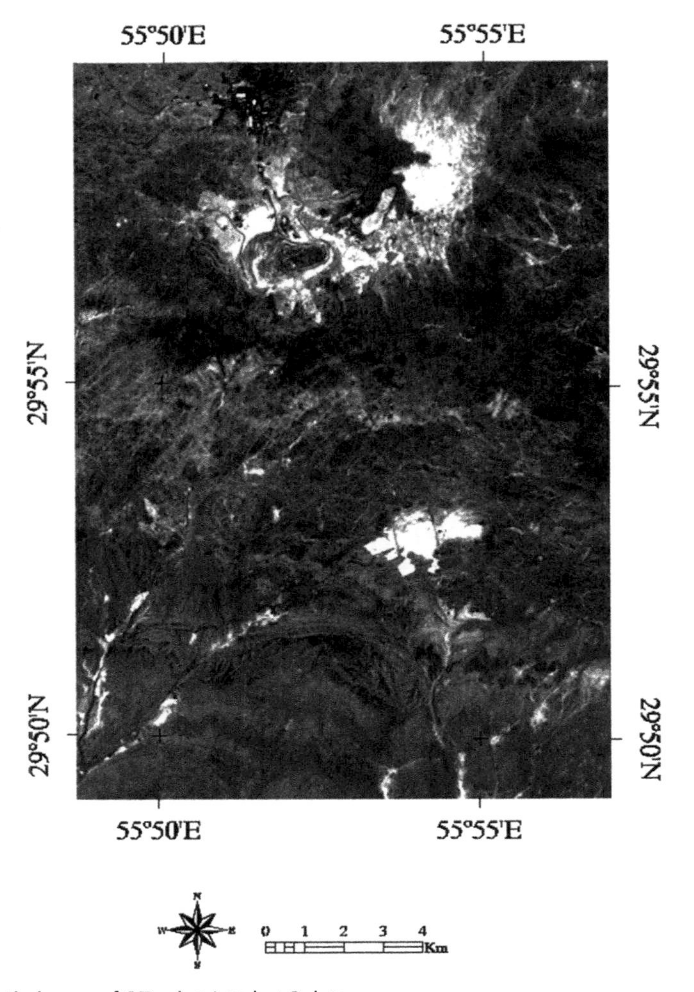

FIGURE 2.33 Band ratio image of 6/7 using Landsat 8 data.

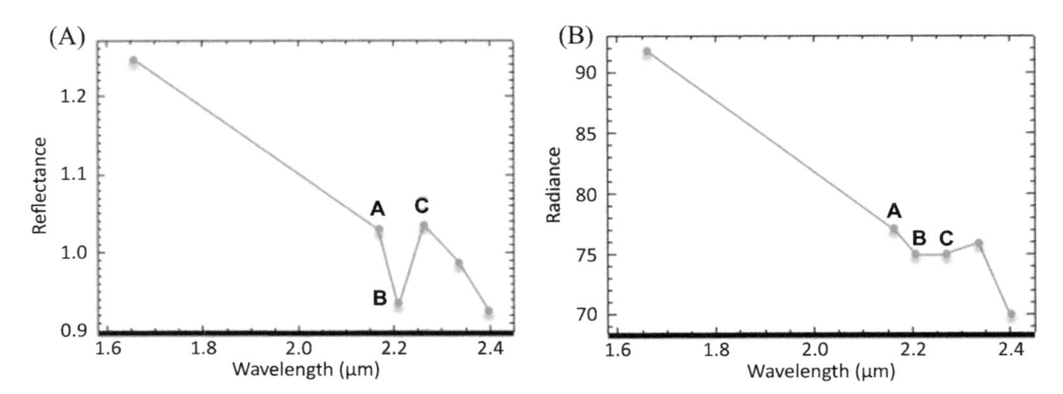

FIGURE 2.34 Absorption band depth in (A) absolute and (B) relative conditions.

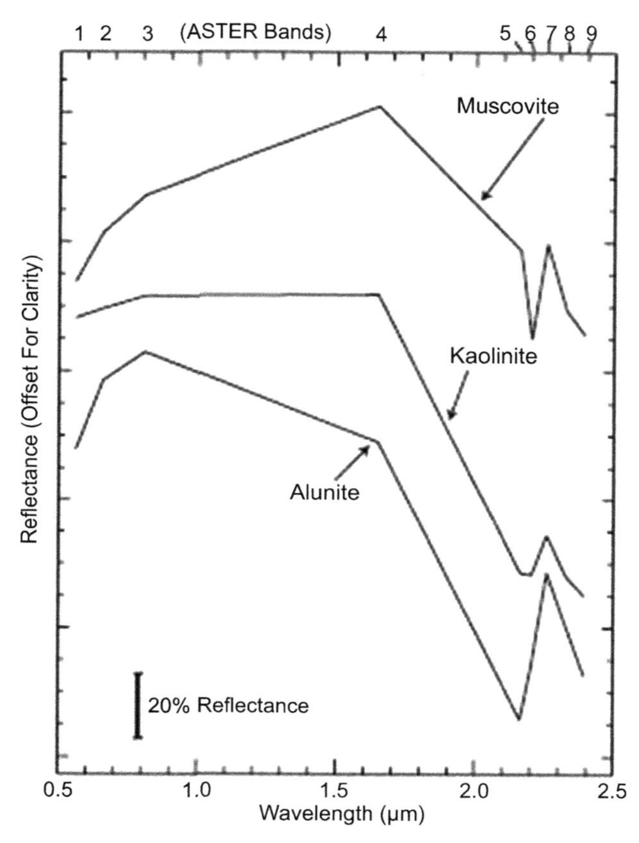

FIGURE 2.35 Laboratory reflectance spectra resampled to ASTER VNIR + SWIR bands for muscovite, kaolinite, and alunite. *ASTER*, Advanced Spaceborne Thermal Emission and Reflection Radiometer; *SWIR*, shortwave infrared; *VNIR*, visible and near-infrared.

Muscovite is recognizable by using the true ABD (TABD) and relative ABD (RABD) as follows:

$$\text{TABD} = \frac{\text{Band5} + \text{Band7}}{\text{Band6}}, \tag{2.13}$$

$$\text{RABD} = \frac{\text{Band4} + \text{Band6}}{\text{Band5}}. \tag{2.14}$$

Fig. 2.36A–D shows ASTER scene of the Abdar porphyry copper deposit. In this figure, argillic alteration (kaolinite as indicator was considered), phyllic alteration (muscovite as indicator was considered), and propylitic alteration (chlorite indicator was considered) are enhanced and presented using RGB color composition in which phyllic alteration appeared in magenta, and propylitic alteration is visible in green color tone (Fig. 2.36D).

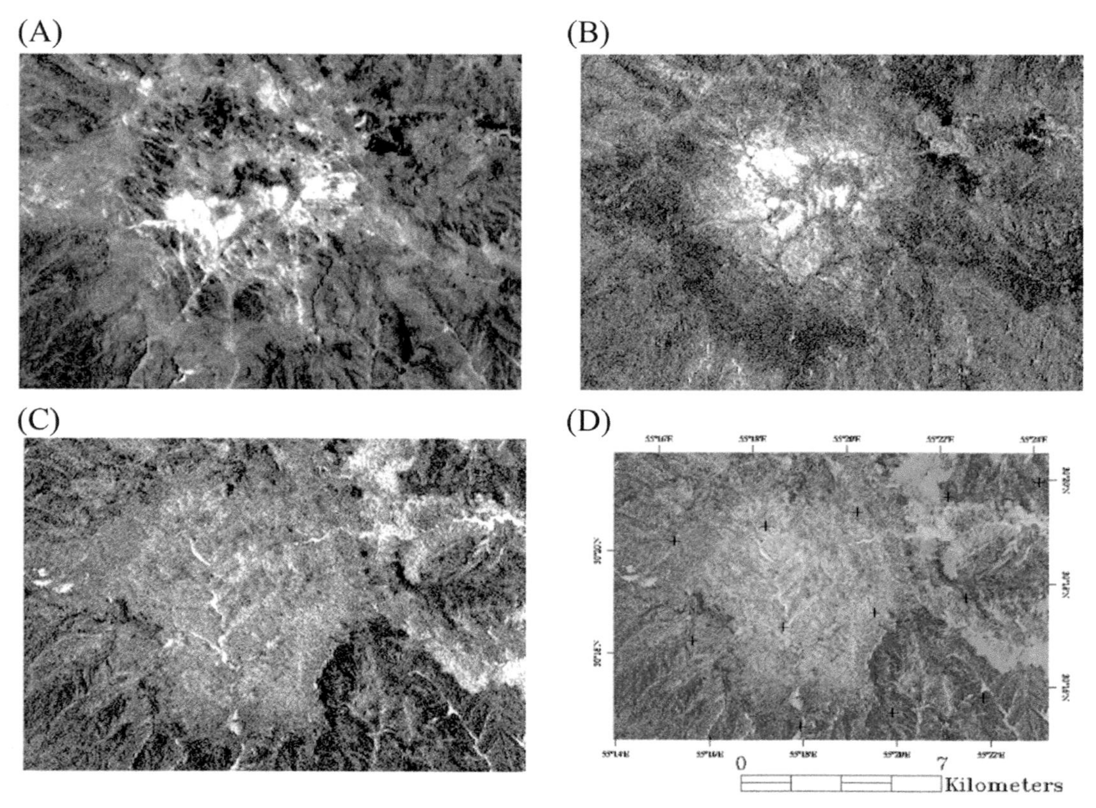

FIGURE 2.36 Results of RABD method for ASTER scene of Abdar porphyry copper deposit: (A) kaolinite, (B) muscovite, and (C) chlorite (D) RGB image of muscovite (R)–chlorite (G)–kaolinite (B). *ASTER*, Advanced Spaceborne Thermal Emission and Reflection Radiometer; *RABD*, relative absorption band depth; *RGB*, red–green–blue.

In the band ratio technique, we can divide band 4 into band 5 (ASTER data) to enhance kaolinite zones. In Fig. 2.37 the band ratio method is used to enhance kaolinite in the ShirKouh granites in Yazd province, Iran. Note that the vegetated areas are also enhanced along with kaolinite, because of their spectral signature similarities (Fig. 2.37).

2.9.8 Logical operators

One of the widely used methods in investigating and detecting alteration zones is the use of logical operators. This method includes specific algorithms, which are defined on the basis of band ratios, logical operators, and thresholds obtained from the images. One of the comprehensive and applied studies using this method was conducted by the US Geological Survey in the Urmieh-Dokhtar magmatic arc of Iran, the results of which were published in 2006 and 2014 (Mars and Rowan, 2006; Mars, 2014). The purpose of this study was to investigate and highlight argillic and phyllic alteration zones to explore porphyry and epithermal deposits. To use the method of logical operators, the following algorithms can be implemented (Table 2.13).

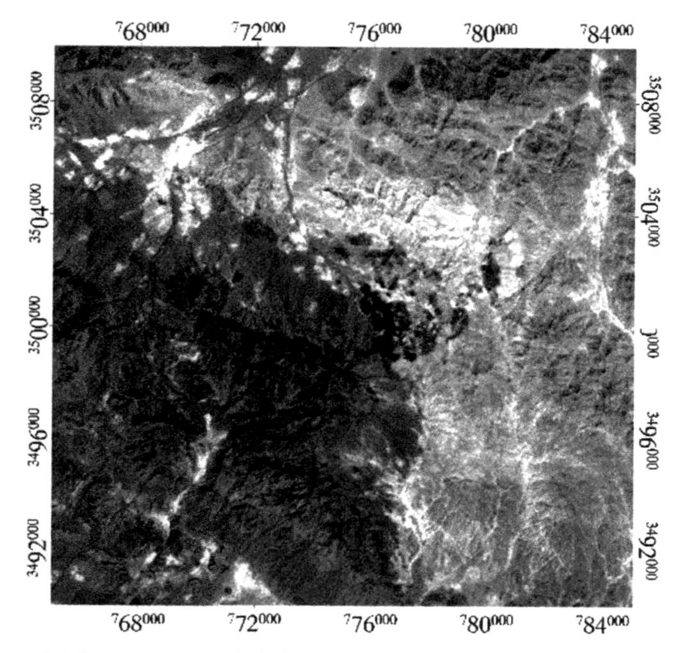

FIGURE 2.37 Kaolinite as bright zones in the Shirkuh granites in Yazd province, Iran (applying simple band ratio to ASTER data). *ASTER*, Advanced Spaceborne Thermal Emission and Reflection Radiometer.

Table 2.13 Logical operators for hydrothermal alteration mapping.

Target	Algorithm	References
Argillic alteration	((float(b3)/b2) le?) and (b4 gt?) and ((float(b4)/b5) gt?) and ((float(b5)/b6) le?) and ((float(b7)/b6) ge?)	Mars and Rowan (2006)
Phyllic alteration	((float(b3)/b2) le?) and (b4 gt?) and ((float(b4)/b6) gt?) and ((float(b5)/b6) gt?) and ((float(b7)/b6) ge?)	Mars and Rowan (2006)
Silica-rich hydrothermal alteration	((float(b3)/b2) le?) and (b4 gt?) and ((float(b4)/b7) ge?) and ((float(b13)/b12) ge?) and ((float(b12)/b11) lt?)	Mars (2014)
Propylitic alteration (carbonate)	((float(b3)/b2) le?) and b4 gt? and (float(b6) / b8 gt?) and (b5 gt b6) and (b7 gt b8) and (b9 gt b8) and ((float(b13) / b14) gt?)	Mars (2014)
Propylitic alteration (chlorite/epidote)	((float(b3)/b2) le?) and b4 gt? and (float(b6) / b8 gt?) and (float(b5) / (float(b4) + b6) gt 0.456) and (b5 gt b6) and (b6 gt?) and (b7 gt b8) and (b9 gt b8) and ((float(b13) / b14) le?)	Mars (2014)
Alunite	((B3/B2)*(B1/B2) lt? and ((B4/B5)*(B7/B5))^0.3 ge? and (B5/B6) lt? and ((B7/B5)*(B6/B5))^2 ge? and DPC 3 (B3/B2; B4/B6; B5/B6; B7/B5) ge?)	Tözün and Özyavaş (2020)
Kaolinite/halloysite	((B3/B2)*(B1/B2) lt? and ((B4/B5)*(B7/B5))^0.3 ge? and ((B4/B5) *(B4/B6)) ^0.3 ge? and (B5/B6) ge? and ((B5/B6)*(B8/B6))^0.5 ge?)	Tözün and Özyavaş (2020)
Fe^{3+}	((B3/B2)*(B1/B2) lt? and (B2/B1)^2 ge? and (B2/B1)/(B3/B2)^2 ge?)	Tözün and Özyavaş (2020)

Note: Instead of the question mark values, the threshold values calculated from the image should be replaced. *ge*, greater or equal to; *lt*, less than; *le*, less than or equal to; *gt*, greater than.

Three important points regarding the implementation of logical operators are as follows:

1. If atmospheric corrections are made on input data, there is no need to use the float function in the above algorithms, such as the algorithms introduced in the study of Tözün and Özyavaş (2020).
2. To implement logical operator algorithms in ENVI software, you can use the Band Math (BM) tool.
3. To determine the thresholds for the formulas in the logic operators algorithm, you can use mean + 2 × standard deviations.

2.9.9 Principal component analysis

In satellite images, there is a lot of correlation between the spectral bands, so most of the areas that are bright or dark in one band have the same characteristic and appearance in other bands. In fact, this similarity and correlation between bands cause a series of additional and redundant information. If this additional information is reduced, the data in the multispectral images will be compressed. The principal component analysis (PCA) is a method to remove or reduce such redundant information in image processing. The PCA method can be performed on multispectral/hyperspectral data containing any number of bands. Therefore in short, the analysis of the main components leads to the reduction of the size of the main data. In other words, the aim of this method is to compress all the information obtained in a main composite data set of n bands into n new components. For this reason, the first principal components have the most information compared to other components.

If a color composite is created by using the first three PCs, the information is more than that created in the simple color combination of three normal spectral bands and shows more details of the spectral difference of different geological phenomena such as rock units. The use of principal components helps geologists in distinguishing the lithological boundaries that cannot be recognized in the color composite images obtained from the raw bands. Also, when the three components with high degrees are displayed as color images, it is possible for the user to identify small areas that are spectrally completely different from the whole scene (for instance, altered zones) (Vincent, 1997). The theory of PCA states that the ultimate goal of this method is to transfer the pixel values of the input bands from the initial coordinate axes to the new coordinate axes. For a better understanding of this point, pay attention to Fig. 2.38.

If the number of input bands is two (Fig. 2.38A), first the pixel values of both bands are plotted. Now, to determine the axes of the new coordinates (new components), the first axis is drawn along the cloud of points in such a way that it has the largest variance around it. The second axis for PC_2 is drawn perpendicular to the PC_1 axis. In multispectral images, the resulting PC axes are orthogonal to each other. Therefore the y_1 axis is drawn in such a way that the maximum variance of the data is placed around this axis. The second axis (y_2) is perpendicular to the y_1 axis and has a lower variance. The length of these axes is proportional to the variance percentage of the same component, which is also called the eigenvalue. The coordinates of these axes are obtained using mathematical operations based on the

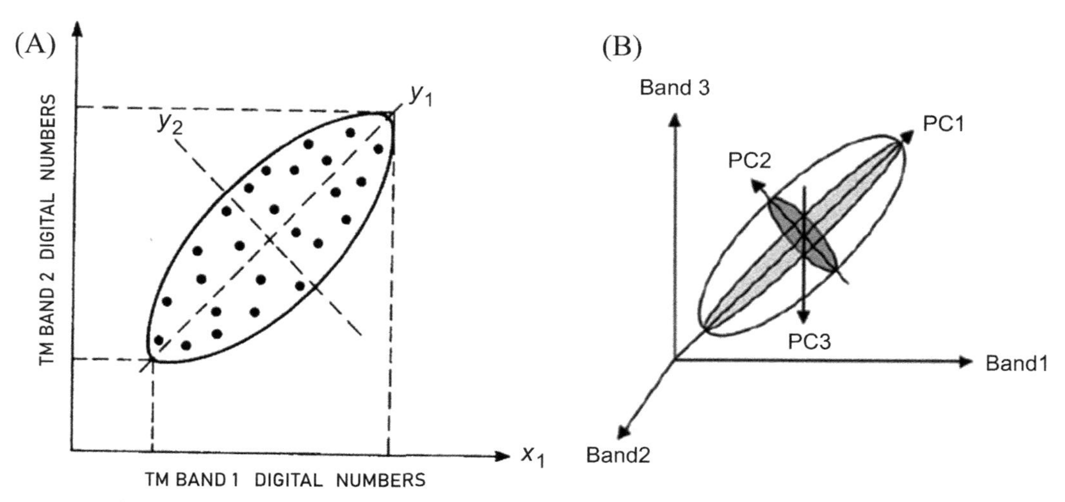

FIGURE 2.38 (A) DN scatterplot of two separate bands (bands 1 and 2 of TM as *X*1 and *X*2 axes). Y_1 and Y_2 are the newly transformed axes after PCA (Sabins, 1987). (B) Three principal components of three different bands. *DN*, Digital number; *PCA*, principal component analysis.

variance–covariance matrix or the correlation coefficient matrix, which are called eigenvector values. The mathematical relationship of this coordinate axis transfer is as follows:

$$PC_1 = a_{11}x_1 + a_{12}x_2, \quad PC_2 = a_{21}x_1 + a_{22}x_2. \tag{2.15}$$

In the previous equations, the value of a represents the coefficients or loading, x_1 and x_2 are the DN values of the pixels in the original bands, and PC_1 and PC_2 are the new values of the pixels. For example, if we want to write the first component for Table 2.1, we have the following:

$$PC_1 = 0.26(b1) + 0.34(b2) + 0.49(b3) + 0.34(b4) + 0.53(b5) + 0.42(b7). \tag{2.16}$$

For more familiarity with the calculation of eigenvalues and eigenvectors of a matrix, it can be referred to Davis and Sampson (1986). For multispectral data (such as Landsat, ASTER, and Sentinel-2), the first principal component (PC_1) contains the highest percentage of the total variance of the image, and the next components (PC_2, PC_3, . . ., PC_n) each contain a smaller percentage of variance or image information (Fig. 2.38B). In addition, due to the fact that each of the subsequent components is chosen to be perpendicular to the previous component, the main components have a minimal correlation between themselves (Sabins, 1987; Lillesand et al., 2011).

Each PCA image has information from all spectral bands and can be displayed as a black-and-white image in the output. Each principal component can show a specific phenomenon. In the first step, the user should have a correct image of the spectral characteristics of a phenomenon in mind, so that he or she can find out the absorption and reflection wavelengths corresponding to the used spectral bands and according to the vector loadings and signs. The vector loadings, and the mathematical signs for different bands in each component,

Table 2.14 Eigenvector matrix of six bands of Landsat-7 derived from principal component analysis for the Sarcheshmeh copper mining region in Kerman, Iran.

Eigenvector	Band 1	Band 2	Band 3	Band 4	Band 5	Band 7	Eigenvalue	% Variance
PC 1	0.26	0.34	0.49	0.34	0.53	0.42	1193	88.04
PC 2	−0.49	−0.40	−0.39	0.01	0.54	0.39	115	8.50
PC 3	0.20	0.18	0.02	−0.91	0.13	0.29	26	1.90
PC 4	0.02	−0.21	0.09	0.11	−0.62	0.74	12	0.88
PC 5	0.63	0.14	−0.73	0.20	0.05	0.13	7	0.52
PC 6	−0.51	0.79	−0.26	0.05	−0.16	0.13	2	0.15

specify that a component represents a particular phenomenon. For example, Table 2.14 shows the results of PCA on six bands of Landsat 7.

As shown in this table, the first three components contain 98.44% of the information of the six bands used. The first principal component (PC_1) has positive loadings for all bands and contains the information of all bands. The PC_1 is also called an albedo image (Fig. 2.39). The PC_2 has negative loadings for visible bands and positive loadings for the near-infrared and short-wave infrared bands, which can be related to the areas with hydrothermal alteration (Fig. 2.39). In the case of the PC_3, due to having a loading of −0.91 for the reflective band of vegetation (band 4), it is expected that the vegetation to be shown as dark in the resulting image of this component (Fig. 2.39). The PC_4 shows the areas with hydrothermal alteration with a dark tone. Since the loading of band 5 is negative (band 5 of Landsat corresponds to the reflective band of clay minerals), the pixels associated with the hydrothermally altered areas are seen as dark (Fig. 2.39). In PC_5, band 3 (reflective band of iron oxide) has a high loading and a negative sign, but band one that is the absorption band of iron oxide, has a high loading but with a positive sign. As a result, the PC_5 shows areas containing iron oxide with a dark tone (Fig. 2.39). In PCA analysis, usually, the last component shows the noise in the images. Accordingly, the PC_6 has the texture of salt and pepper that is attributed to the noise (Fig. 2.39). However, this PC shows the areas containing argillic alteration in the northeast of Sarcheshmeh (Sereidoon) with a dark tone due to the negative loading of band 5.

Cŕosta et al. (2003) introduced a modified PCA in which the appropriate bands are selected based on the spectral properties of the desired minerals. Landsat 7 images have been used to highlight areas containing iron oxide out of bands 1−3. To highlight areas with vegetation cover, bands 2−4, and to highlight areas with hydrothermal alteration bands 3, 5, and 7 are used, respectively. In Table 2.15, bands used for highlighting iron oxides, clay minerals, and vegetation are presented. If you use the ASTER data to highlight the minerals, you can use the directed PCA method, based on the standard spectra of the minerals to choose appropriate bands. For example, bands 5−7 are used to highlight muscovite, and bands 7−9 are used to highlight chlorite. These bands can easily be determined from the spectral properties of minerals (absorption and reflection bands) (Honarmand et al., 2012).

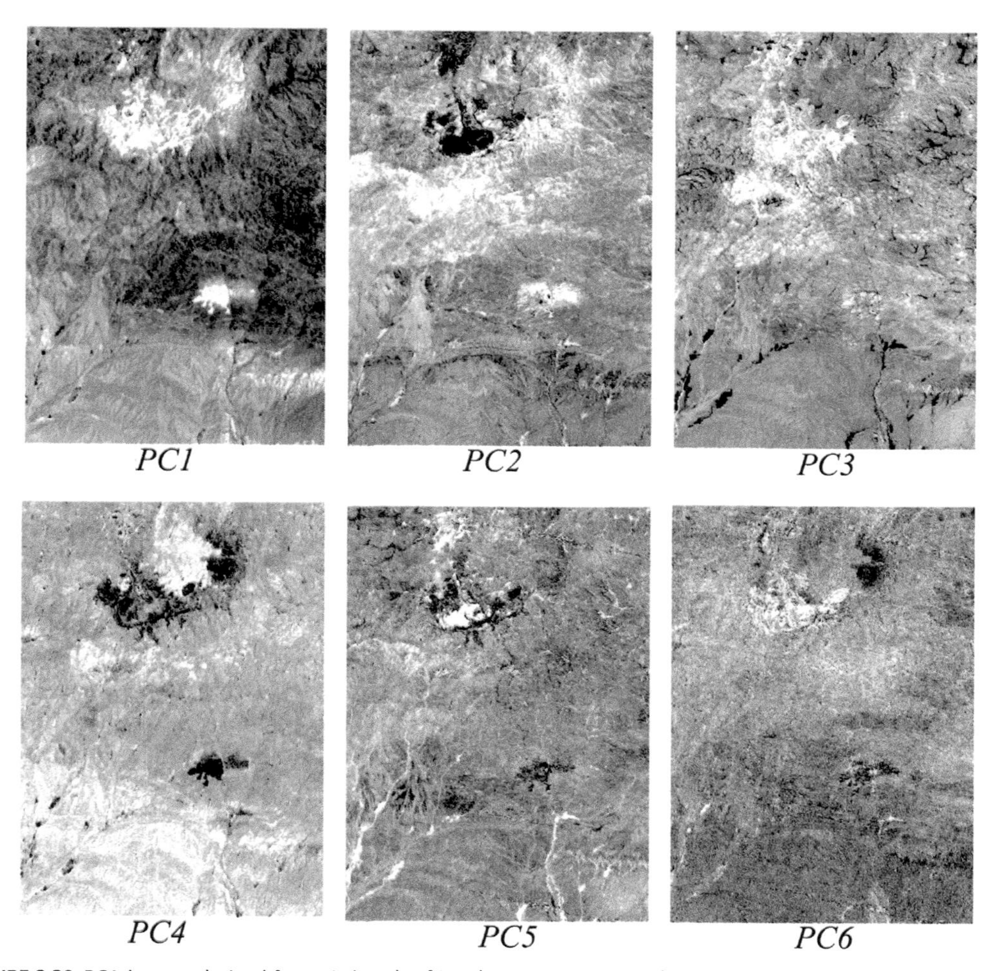

FIGURE 2.39 PCA images derived from six bands of Landsat 7. *PCA*, Principal component analysis.

Table 2.15 Principal component analysis results and eigenvector loadings based on Crosta technique for Landsat-7 bands.

Eigenvector		PC_1	PC_2	PC_3
Band1	Fe oxide	0.507	−0.834	−0218
Band 2		0.458	0.046	0.888
Band 3		0.731	0.549	−0.405
Band 2	Vegetation	0.366	−0.510	−0.778
Band3		0.593	−0.517	0.617
Band4		0.717	0.687	−0.904
Band 3	Hydrothermal alteration	0.331	0.920	0.210
Band 5		0.843	−0.388	0.372
Band 7		0.424	0.054	−0.904

2.9.10 Independent component analysis

Independent component analysis (ICA) is a computational signal-processing method that aims to describe random variables as a linear combination of statistically independent components (e.g., Comon, 1994; Hyvärinen and Oja, 1997; Stone, 2004). This is achieved by the decomposition of a mixed signal using the assumption that each constituent component has a non-Gaussian probability distribution. This assumption is based on the premise that the sum of a sufficient number of non-Gaussian probability distributions tends toward a Gaussian distribution (the central limit theorem), so that a strongly non-Gaussian component is unlikely to be produced by a combination of different sources.

The assumption of statistical independence employed by ICA makes it possible to find a unique solution to the decomposition of a mixed signal, in a similar way that the assumption that sources are uncorrelated is the basis of PCA. As ICA retrieves sources by maximizing statistical independence (rather than signal variance, as in PCA) it is appropriate for the extraction of low-magnitude signals, even where noise is high, without a priori assumptions beyond the independence of the components (Hyvärinen et al., 2004). Statistical independence is assessed by non-Gaussianity. This can be quantified by using different properties of the random variables, of which kurtosis and negentropy are widely used. Kurtosis describes the relative contribution of extreme deviations to a probability distribution ("tailedness") and is normally measured as the absolute value of the fourth standardized moment of the data, with a Gaussian distribution taking a value of 3. The calculation of kurtosis is simple, but in practice, it is more sensitive to outliers than negentropy, which is more widely used for ICA. Negentropy is a concept from information theory that describes the difference in entropy—a measure of the unpredictability of information content—relative to the Gaussian distribution of the same mean and variance. This is based on the result that a Gaussian distribution has the highest value of entropy of all the possible random variables with the same variance. To avoid the challenging estimation of the probability density function, most algorithms use an approximation of negentropy to assess Gaussianity (e.g., Hyvärinen and Oja, 2000).

2.9.11 Minimum noise fraction

The minimum noise fraction (MNF) method proposed by Green et al. (1988) is a well-known data dimension and noise reduction technique. This linear transform is widely used for feature extraction, noise whitening, and spectral data reduction of remote sensing imagery, especially for hyperspectral data (e.g., Kruse et al., 2003; Khan et al., 2007; Pengra et al., 2007; Pournamdari et al., 2014; Assiri, 2016). The transform consists of two separate PCA rotations, in which the noises in the data are readjusted and separated in the first transformation. In the first rotation, the data noises are decorrelated and rescaled, and in the second phase, the PCs are derived from the noise-whitened data. Through the two steps of affine transforms, a set of decorrelated PCs is generated, which contain steadily decreasing useful information with the increasing component numbers (Boardman and Kruse, 1994).

2.9.12 Spectral angle mapper

The spectral angle mapper (SAM) is an efficient method for mapping a specific mineral or a rock by comparing the spectrum of a pixel to an image-derived spectrum or a standard (library) spectrum. The algorithm of this method calculates the similarity between the two spectra by a parameter called spectral angle. The spectra are converted to vectors in a space with the dimensions of the number of bands. The spectral angle between the vectors is calculated. For a better and simpler understanding of this topic, two references (standard) and the studied spectrum can be considered in two-band or two-dimensional space (Fig. 2.40). For calculating the angle, the direction of the vectors is important and not their length, so the illumination of the ground does not affect the final result (Mather and Koch, 2011).

To obtain the angle (α) between the two vectors obtained from the images or laboratory spectrum (t) and the reference spectrum (r) (Fig. 2.40), the following relationship is used (Kruse et al., 1993):

$$\alpha = \cos^{-1}\left(\frac{\vec{t}\cdot\vec{r}}{|\vec{t}|\cdot|\vec{r}|}\right). \tag{2.17}$$

If n bands are used to identify the desired phenomenon, the following relationship is used to obtain the following angle:

$$\alpha = \cos^{-1}\left(\frac{\sum_{i=1}^{nb} t_i r_i}{\left(\sum_{i=1}^{nb} t_i^2\right)^{1/2}\cdot\left(\sum_{i=1}^{nb} ri2\right)^{1/2}}\right), \tag{2.18}$$

where nb is the number of bands.

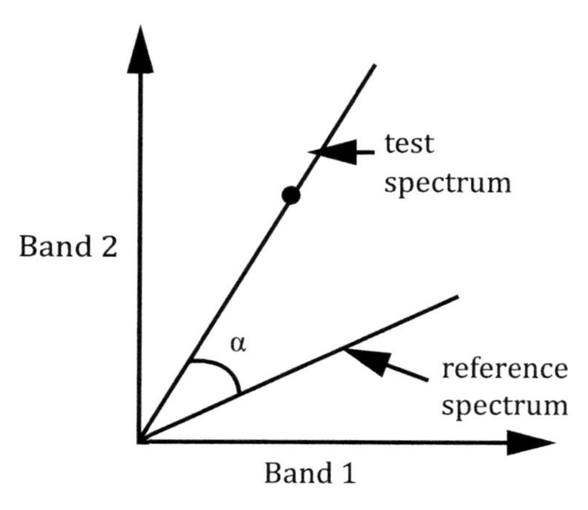

FIGURE 2.40 The vectors were obtained from the laboratory spectrum and the reference spectrum (Kruse et al., 1993).

In fact, the lower the angle (between 0 and 1 radians), the greater the similarity between the two spectra. This method was first presented by Kruse et al. (1993).

2.9.13 Linear spectral unmixing

The linear spectral unmixing (LSU) is a subpixel sampling algorithm (Boardman, 1989, 1992; Adams et al., 1993). The reflectance at each pixel of the image is assumed to be a linear combination of the reflectance of each material (or end-member) present within the pixel. For example, if 25% of a pixel contains material A, 25% of the pixel contains material B, and 50% of the pixel contains material C, the spectrum for that pixel is a weighted average of 0.25 times the spectrum of material A, plus 0.25 times the spectrum of material B, plus 0.5 times the spectrum of material C (Research Systems, 2008). The LSU is used to determine the relative abundance of end-members within a pixel based on the end-members' spectral characteristics. This algorithm assumes that the observed pixel reflectance can be modeled as a linear mixture of individual component reflectance multiplied by their relative proportions. Mathematically, the LSU can be represented as

$$R_i = {}^{N}\sum i = 1 F_e R_i + E_i, \tag{2.19}$$

where R_i is the surface reflectance in band i of the sensor; F_e is the fraction of end-member e; R_e is the reflectance of end-member e in the sensor wave band i; N is the number of spectral end-members; and E_i is the error in the sensor band i for the fit of N-end-members (Adams et al., 1995).

2.9.14 Matched filtering

Matched filtering (MF) is used to find the abundance of user-defined end-members using partial unmixing. This technique maximizes the response of the known end-member and suppresses the response of the composite unknown background, thus "matching" the known signature. The results of the MF appear as a series of grayscale images (fraction maps), one for each selected end-member. These fraction maps have values that range from 0 to 1, where 0 represents a nonmatch to the end-member (training) spectrum and 1 represents a perfect match. Thresholds can then be identified to create binary maps from the fraction maps to show areas with relatively good matches to the end-member spectra, or they can be combined with other fraction maps as ternary images to produce a lithological–compositional map. Unlike linear unmixing, MF does not require knowledge of all the end-members within the scene. Thus in areas of highly mixed rocks, where identification of all the end-members is difficult, MF may be a better choice for classification (Harris et al., 2005).

2.9.15 Mixture-tuned matched filtering

The mixture-tuned matched filtering (MTMF) algorithm is a hybrid method based on the MF method with physical constraints imposed by mixing theory (Boardman, 1998; Chang, 2003;

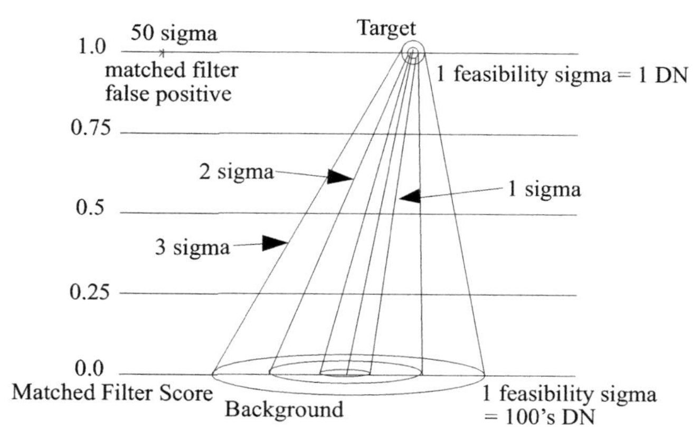

FIGURE 2.41 Diagram showing the performance of the MTMF method. *MTMF*, Mixture-tuned matched filtering.

Boardman and Kruse, 2011). This method uses linear spectral mixing theory to constrain the result to feasible mixtures and reduce false alarm rates (Research Systems, 2008; Boardman and Kruse, 2011). From MF, it inherits the advantage of its ability to map a single known target without knowing the other background end-member signatures, unlike traditional spectra mixture modeling. From spectral mixture modeling, it inherits the leverage arising from the mixed pixel model, the constraints on feasibility, including the unit-sum and positivity requirements, unlike the MF that does not employ these fundamental facts. As a result, MTMF can outperform either method, especially in cases of subtle, subpixel occurrences (Boardman, 1998).

Two sets of images are presented as MTMF results, including (1) MF score images (values from 0 to 1.0) that assess the relative degree of match to the reference spectrum and approximate subpixel abundance (where 1.0 is a perfect match) and (2) infeasibility images, where highly infeasible numbers show that mixing between the composite background and the target is not feasible. The greatest match to a target is acquired when the MF score is high (near 1) and the infeasibility score is low (near 0) (Research Systems, 2008). This method performs similarly to MF and also adds an infeasibility image to the results. The infeasibility image is used to reduce the number of false positives that are sometimes found when using MF. Pixels with a high infeasibility are likely to be MF false positives. The infeasibility values are in noise sigma units that vary in DN scale with an MF score (Fig. 2.41) (Research Systems, 2008).

2.9.16 Constrained energy minimization

The constrained energy minimization (CEM) is a target signature-constrained algorithm that restrains the desired target signature with a particular attain, minimizing effects made by other unknown signatures derived from the background (Harsanyi, 1993; Chang and Heinz, 2000; Chang et al., 2000; Jiao and Chang, 2008). For implementing this algorithm, just preceding information about the preferred target signature is necessary. It accomplishes a partial unmixing of spectra to determine the abundance of end-member (user-defined) for a number of reference spectra from image or laboratory spectra (Chang and Heinz, 2000). The CEM algorithm uses a finite impulse response (FIR) filter to pass through the desired target while minimizing its output energy from a background other than the desired targets (Harsanyi, 1993; Harsanyi et al., 1994; Johnson, 2003).

2.9.17 Adaptive coherence estimator

The adaptive coherence estimator (ACE) is a target detection algorithm that carries out a partial unmixing approach to isolate features of interest from the background, and its input is a single score (abundance of the target) per pixel (Bidon et al., 2008). It is generated from the generalized likelihood ratio (GLR) approach, which is a homogenously most powerful invariant detection statistic (Kraut et al., 2001; Alvey et al., 2016). The ACE is invariant to the relative scaling of input spectra and has a constant false alarm rate for such scaling (Warren, 1982). Geometrically, it determines the squared cosine of the angle between a known target vector and a sample vector in a whitened coordinate space. The space is faded based on assessing the background statistics, which straightforwardly influences the presentation of the statistic as a target detector (Hall et al., 2002). The standard formulation of the ACE detection statistic is defined as follows:

$$\mathrm{ACE}(X) = \frac{\left[(t-\mu)^{TS}\sum^{-1}(x-\mu)\right]^2}{\left[(t-\mu)^{T}\sum^{-1}(t-\mu)\right]\left[(t-\mu)^{TS}\sum^{-1}(x-\mu)\right]}, \tag{2.20}$$

where t is a known target signature (reference spectra from a spectral library signature), and x is a data sample. The background is assumed to be a Gaussian distribution parametrized by μ and Σ that represent the mean and covariance, respectively. The ACE statistic is a number between 0 and 1, which can be interpreted as a measurement of the presence of t in x. The ACE can be estimated as the square of the cosine of the angle between x and t, in a coordinate space transformed by the background estimation. For example, if *ACE* produces 0.85, it indicates a relatively strong presence of t in x. The key to effective ACE performance is accurate background estimation. Furthermore, the ACE does not need information about all the end-members within an image scene.

2.9.18 Spectral feature fitting

The spectral feature fitting (SFF) algorithm eliminates the continuum of absorption features from the spectra of the reference spectral library and also from each spectrum of the image dataset. The SFF compares the continuum-removed spectrum of the image with the continuum-removed spectrum of reference spectra and operated the least square fitting. The selection of the best fitting material based on spectral features or group of features is done by comparing the correlation coefficient of fits (Clark et al., 1993a,b).

The process of feature extraction involved the first step continuum removal (CR) that is used to identify the individual absorption features carrying the spectrum (Kruse, 1993). The convex hull using straight line segments fitted on the top of the spectrum that joins the maxima of local spectra is known as a continuum. For the removal of the continuum, the original spectrum (S) for every pixel in an image is divided by the (C) continuum curve (Harris Geospatial Solutions, 2017):

$$S_{cr} = \frac{S}{C}, \tag{2.21}$$

where S_{cr} is the continuum-removed spectra, S is the original spectra, C is the continuum curve.

Initially, the subtraction of the continuum removed spectra by one is done to generate the Scale image that helps for the predictions; thus inverting spectra and continuum is made to zero. The scale is used for the determination of the depth of the absorption feature in each pixel. The scale factor represents the abundance of spectral features. Band-by-band calculation is performed for the computation of least squares fit. A higher scale value represents the strong absorption in the mineral's spectral band. Mineral abundance is highly related to the scale value (Van der Meer, 2004; Verdel et al., 2001). Brighter pixels in the scale image represent better matching of the pixel spectrum from the reference spectrum of the mineral. RMS (root mean square) error image is also produced for each end-member to compute the total RMS error. Low RMS pixels appear black (darker) in the image, and the shape of the spectra is highly similar to the reference spectrum of the mineral. RMS is a method, which is used for the examination of the existence and absence of minerals (Van der Meer, 2004; Verdel et al., 2001). Those pixels that are having low RMS error and high scale factor are closely matched. An equal number of images are generated by the scale and RMS. The Fit images are generated by making a ratio between the scale image and RMS error image to measure the similarities (match) between the unknown spectrum and reference spectrum based on pixel-by-pixel (Van der Meer and De Jong, 2003).

2.9.19 Image classification

The main goal of image classification is to place pixels in separate groups based on their spectral properties. In this method the pixels related to each object such as soil, water, and rocks are grouped and displayed in different colors. But there are certain limitations in this field. For example, the soil and vegetation cover on the outcrops (rocks) makes it impossible to classify the rocks. The spectral pattern in each pixel is used as a basis for classification. Two objects may have the same reflection in one band, while their reflection in other bands is different. As a result, the spectral classification of the image is based on the spectral properties of the pixels and not on their geometrical position. In another classification method the spatial pattern of the pixels and their placement relative to each other are considered. In this classification method, size, image texture, shape, orientation, proximity to other pixels, and others are the basis of classification.

When the pixels are classified in only one specific category, the term hard classification is used. In this method, it is possible that a pixel is covered by two or more types of coverages, but it is placed in a category that has more coverage within the pixel's surface. The soft classification method is used when a pixel consists of two or more different features. When such pixels are classified, the effect of each phenomenon appears in the output image based on the area of the pixel that each feature is covered. Assume that a pixel contains 60% basalt, 30% granite, and 10% vegetation. In the hard classification, it is placed in the basalt category, but in the soft classification, it shows the percentage of each cover. The fuzzy classification method is included in this category. To classify satellite images, it is not possible to be satisfied with only one solution. There are two general methods for classifying images, which are (1) supervised classification and (2) unsupervised classification.

In the supervised classification, the user defines several different groups such as water, plants, soil, urban areas, or different lithological units in the form of training areas for the software. Then the classification process is done, while in the unsupervised classification method, the software classifies the image according to the number of groups given to it. In this method, after classifying the image, the user specifies the names of the categories, while in the supervised classification method, the names of the categories are specified before the classification.

The supervised classification includes the steps of image preprocessing, training and separation of categories, image classification, data display, and finally evaluating the accuracy of the classification.

a. The training and separation of categories

In this step, the number of categories is determined and the areas that belong to these categories are marked in the images. This information can be obtained from existing data, such as geological maps or actual ground data.

b. Image classification

At this stage, with a series of mathematical and statistical operations, the pixels are classified into their respective groups. There are many image classification methods such as minimum distance to the mean, parallelepiped, maximum likelihood classifier (MLC), SAM, neural networks, fuzzy, and support vector machine (SVM). Here, we explain minimum distance to mean, parallelepiped, MLC, SAM, and SVM.

c. Data display

At this stage, the pixels of the image are classified and displayed in different colors. The colors are determined according to the user's choice.

d. Accuracy assessment

At this stage, the accuracy of the classification is evaluated using ground truth or auxiliary data and is usually expressed as a percentage.

2.9.20 Supervised classification

2.9.20.1 Minimum distance to mean classification

In this method, the mean of each category is calculated. If we want to place an unknown pixel in any of these categories, we can calculate its distance with the means of the categories (Fig. 2.42), then this pixel will be placed in the category that has the least distance with the mean of that category.

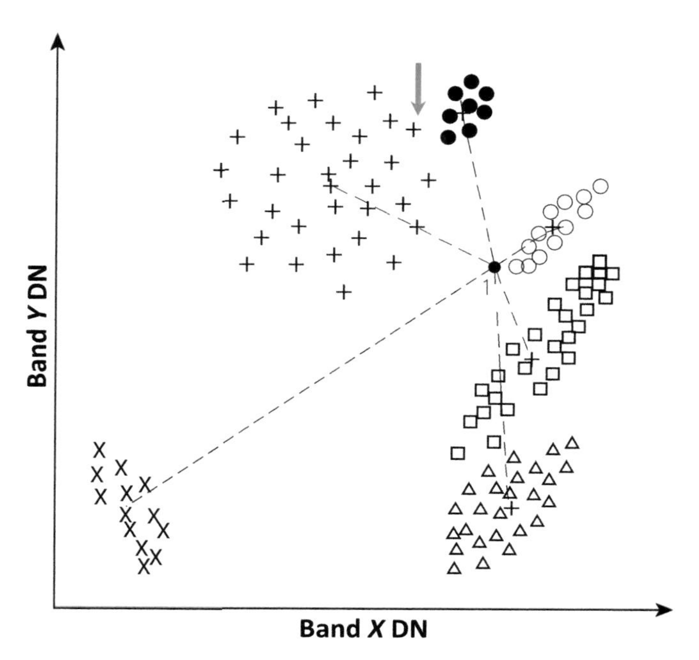

FIGURE 2.42 The concept of the minimum distance to the mean method (here the values of the pixels are displayed only for two bands on the *x* and *y* axes).

The distance can be calculated with the following equation:

$$D_{12} = \sqrt{(x_1 - x_2)^2}, \quad (2.22)$$

where D is the Euclidean distance between the unknown pixel value (x_2) and the average of the training class (x_1).

This method has simple calculations and does not consider the variance of the groups' data. This method is not used when the spectral groups are similar and have a large variance. When the categories have a low and homogeneous variance, it is considered a suitable classification method. If a pixel is in the edge position of the group indicated by " + " (the position of the pixel is indicated by the red arrow), it is placed in the group indicated by the solid circle. The reason is that it has a smaller distance than the mean of this group, though it should be in the " + " category.

2.9.20.2 Parallelepiped

In this method, the sensitivity to the variance of the data of the categories can be increased. In this regard, it is possible to consider the range of specific pixel values for each category. This range can be the maximum and minimum values in each category. As a result, the

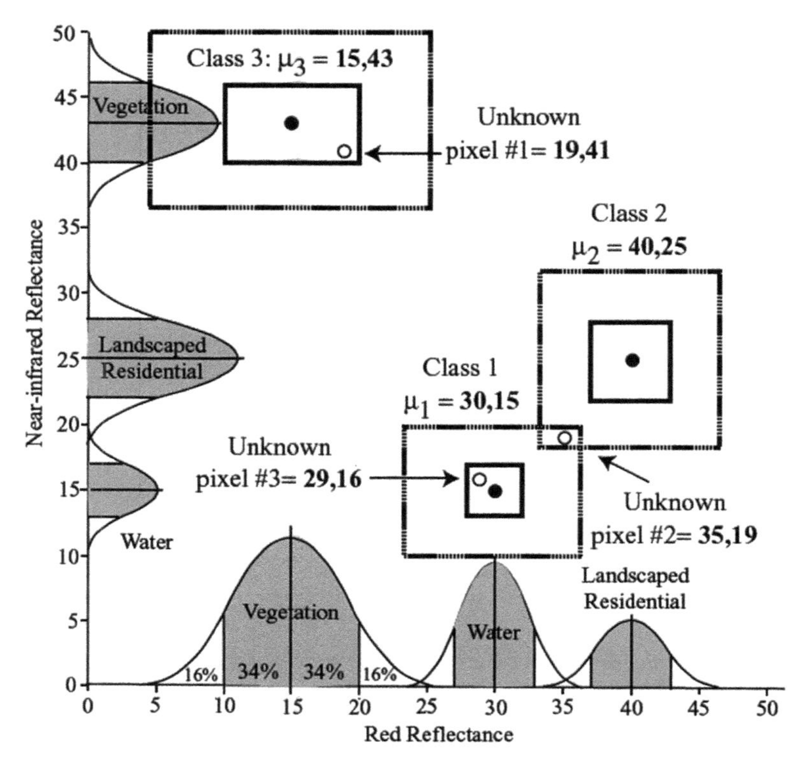

FIGURE 2.43 The general scheme of the parallelepiped classification (the border of the squares is determined based on the variance border of each category).

border of the groups is surrounded by rectangles (Fig. 2.43). Each unknown pixel is classified in the same category if it is located within the border of any of these rectangles. The pixels that are outside these rectangles are not classified. For this reason, in this classification method, a large number of pixels may not be classified. In cases in which the rectangles overlap in this method (Fig. 2.43), the user is allowed to alter the value range of the categories and make decisions and categorize them in stages.

2.9.20.3 Maximum likelihood classification

In this method, the variance and covariance of the categories are considered and the probability of each unknown pixel belonging to a category is calculated and evaluated. Here, we must assume that the pixels of each training area have a normal or Gaussian distribution. The software calculates the probability of each pixel belonging to each category, according to the probability function, and finally, the unknown pixel is transferred to the category that has the most similarity (highest probability value) (Fig. 2.44). As a result, the mean and variance of the data should be known.

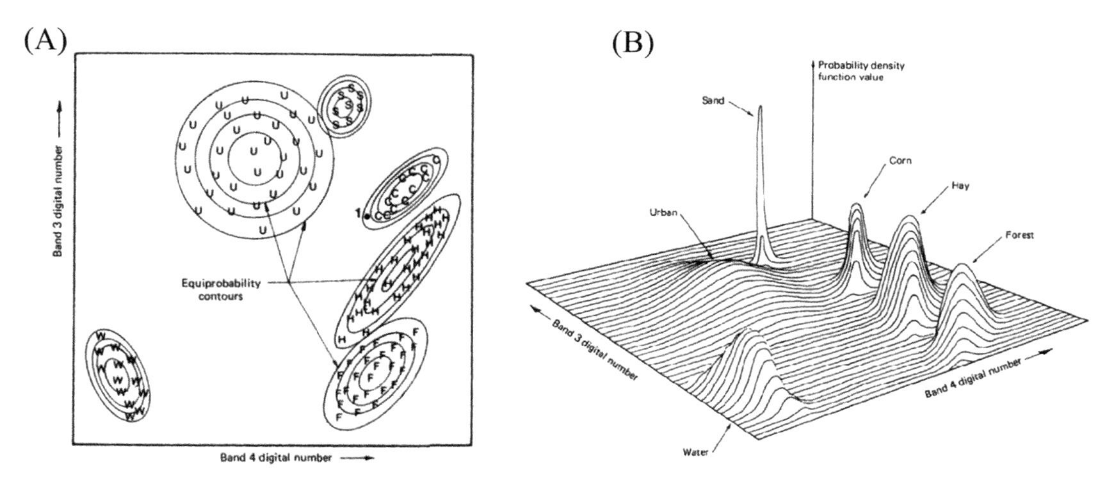

FIGURE 2.44 (A) Equiprobability contours are defined by a maximum likelihood classifier, and (B) probability density functions are defined by a maximum likelihood classifier (Lillesand et al., 2011).

2.9.20.4 Support vector machines

SVMs have received particular attention in the classification of hyperspectral remote sensing data (Melgani and Bruzzone, 2004; Camps-Valls and Bruzzone, 2005; Pu, 2017) because SVMs can robustly classify high dimensional and noisy datasets. Classic classification approaches could not efficiently classify a limited number of training hyperspectral samples with high dimensions. In SVMs, kernel methods are used to map the data from the original nonlinear space to the higher dimension and linear space. The SVM methods are popular to classify two-class separation problems; however, they can be extended to multiclass separation as well (Melgani and Bruzzone, 2004). In the following a two-class problem classification for linear and nonlinear cases has been explained.

2.9.20.4.1 Linear support vector machine

Assume that, there are two classes (x_i, y_i) for $(i = 1, 2, \ldots, N)$, in this case x_i is the training sample and y_i shows the label of the ith sample with values $+1$ or -1, N is the number of training samples. SVMs find a hyperplane that separates the two classes. Each class is located on one side of the hyperplane, and the minimum distance of the closest training samples (these samples are called the support vectors) to either side of the hyperplane should be maximized (Fig. 2.45A).

If the training samples are linearly separable, the general linear hyperplane is defined as (Pu, 2017):

$$\begin{cases} w \cdot x_i + b \geq +1 y_i = +1 \\ w \cdot x_i + b \leq -1 y_i = -1 \end{cases}, \tag{2.23}$$

where and are the hyperplane parameters and show the direction and position of the hyperplane, respectively. The inequalities in Eq. (2.23) can be integrated into one inequality as

$$y_i(w \cdot x_i + b) - 1 \geq 0, i = 1, 2, 3, \ldots, N. \tag{2.24}$$

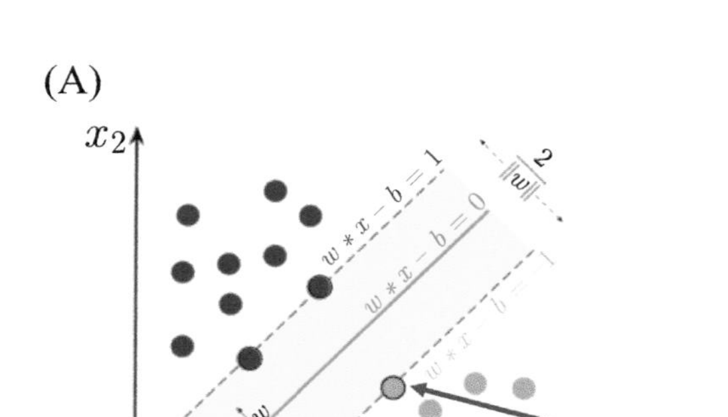

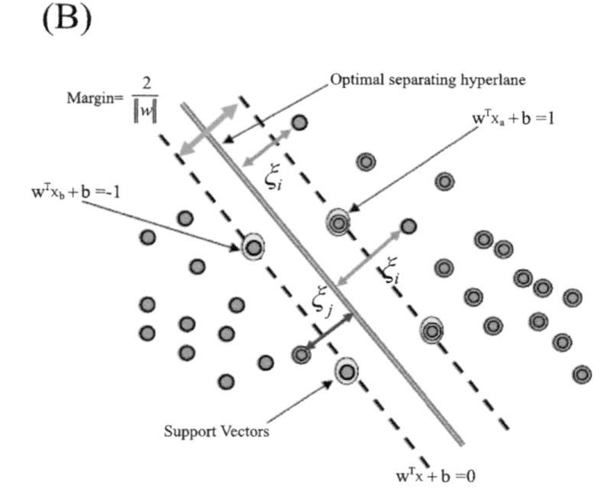

FIGURE 2.45 (A) The optimal separating hyperplane in a linearly separable case. (B) The optimal separating hyperplane in a linearly nonseparable case (Bekkari et al., 2013).

The aim of SVM classifiers is to find the optimal values of w and b, which maximize $\frac{2}{||w||}$ or minimize the norm-2 ($\frac{1}{2}||w||^2$). Therefore the following optimization problem should be solved:

$$\begin{cases} \min\left\{\frac{1}{2}||w||^2\right. \\ y_i(w.x_i + b) \geq -1, i = 1, 2, 3, \ldots, N \end{cases} \tag{2.25}$$

To solve Eq. (2.25), the dual problem based on the Karush–Kuhn–Tucker (KKT) conditions is calculated and Eq. (2.25) is obtained as follows:

$$\begin{cases} \max_\alpha\left\{ \sum_{i=1}^{N} \alpha_i - \frac{1}{2}\sum_{i=1}^{N} \sum_{j=1}^{N} \alpha_i\alpha_j y_i y_j (x_i x_j) \right\} \\ \sum_{i=1}^{N} \alpha_i y_i = 0 \text{ and } \alpha_i \geq 0 i = 1, 2, 3, \ldots, N, \end{cases} \tag{2.26}$$

where some of unknown Lagrange multipliers will be zeros and the training samples with nonzeros are called the support vectors (Fig. 2.45A).

If the classes are not completely linearly separable due to the noise or mixture during training data collection, a regularization parameter and slack variable are added into the constraints in Eq. (2.20) to consider the noise or error in the dataset due to misclassification (Fig. 2.45B). The optimization problem for the linearly nonseparable case becomes

$$\begin{cases} \min\left\{\frac{1}{2}||w||^2 + C\sum_{i=1}^{N} \zeta_i\right\} \\ y_i(w x_i + b) \geq 1 - \zeta_i, \zeta_i \geq 0 \quad i = 1, 2, 3, \ldots, N \end{cases} , \tag{2.27}$$

where C is a penalty factor and controls the misclassification error. The dual optimization problem is obtained as

$$\begin{cases} \max_{\alpha}\left\{ \sum_{i=1}^{N} \alpha_i - \frac{1}{2}\sum_{i=1}^{N}\sum_{j=1}^{N} \alpha_i\alpha_j y_i y_j (x_i x_j) \right\} \\ \sum_{i=1}^{N} \alpha_i y_i = 0 \text{ and } 0 \leq \alpha_i \leq C i = 1, 2, 3, \ldots, N \end{cases} \tag{2.28}$$

The only difference between Eqs. (2.26) and (2.28) is the Lagrange multipliers, α_i are bounded by the penalty value C.

2.9.20.4.2 Nonlinear support vector machine

Most cases in hyperspectral data classifications are nonlinear. The following popular kernel functions can be used to map the nonlinear data into higher linear feature space (Fig. 2.46):

1. Linear kernel: $X \cdot X_i$
2. Polynomial kernel with order p: $(X \cdot X_i + 1)^p$
3. Radial function kernel: $e^{-\frac{\|(x-x_i)\|^2}{2\sigma^2}}$
4. Sigmoid kernel: $\tanh(\alpha(X_i, X_j) + \varphi)$

The original data (x_i, x_j) with the inner product are converted to $[\varphi(x_i), \varphi(x_j)]$ using Mercer's theorem in which there exists a kernel k such that (Vapnik, 1998)

$$k(x_i, y_i) = \varphi(x_i) \cdot \varphi(x_j). \tag{2.29}$$

In this case, the optimization problem Eq. (2.29) can be considered as follows:

$$\begin{cases} \max_{\alpha}\left\{ \sum_{i=1}^{N} \alpha_i - \frac{1}{2}\sum_{i=1}^{N}\sum_{j=1}^{N} \alpha_i\alpha_j y_i y_j K(x_i x_j) \right\} \\ \sum_{i=1}^{N} \alpha_i y_i = 0 \text{ and } 0 \leq \alpha_i \leq C i = 1, 2, 3, \ldots, N \end{cases}. \tag{2.30}$$

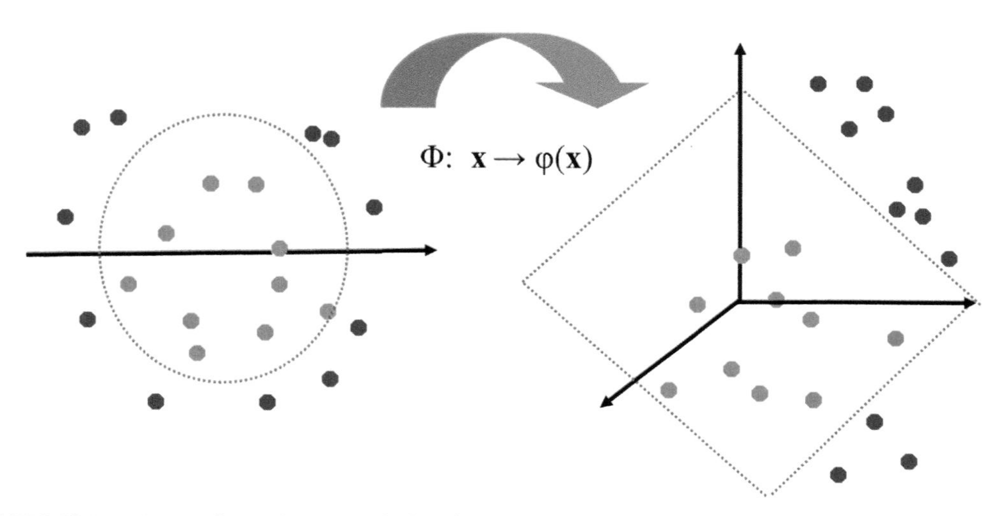

FIGURE 2.46 Mapping nonlinear data into a higher dimension and linear feature space.

The kernel and regularization parameters have a large effect on the classification accuracy they should appropriately select based on the dataset. Some algorithms such as *k*-fold cross-validation can be used (Anguita et al., 2009).

2.9.20.4.3 Nonlinear support vector machine

In general, the hyperspectral dataset includes more than two classes. Recently different approaches have been introduced to classify multiclass problems such as the one against all, one against one, binary hierarchical tree balanced branches, and binary hierarchical tree one against all algorithms (see Melgani and Bruzzone, 2004; Vapnik, 1998 for more details).

2.9.21 Methods of increasing the spatial resolution of satellite images

Usually, multispectral sensors have a panchromatic band with high resolution along with a number of multispectral bands with lower spatial resolution. For example, the QuickBird satellite has a spatial resolution of 2.5 m for multispectral bands, while it has a panchromatic band with a spatial resolution of 60 cm. It is possible that a color image created with a pixel size of 2.5 m can be upgraded to an image with a pixel size of 60 cm. Combining images with a low spatial resolution with images with a high spatial resolution is called improving spatial resolution or image sharpening (Du et al., 2007; Alimuddin et al., 2012; Pushparaj and Hegde, 2017). This operation is done to insert the spatial resolution of some images, such as single-band panchromatic images, into multispectral images. It is possible that if the previous conditions are met, satellite images from different sensors can be combined together. Fig. 2.47 shows panchromatic and multispectral images belonging to the Kahang area, located in Isfahan province. It should be noted that the improvement of spatial resolution should be done according to the conditions and characteristics of the images.

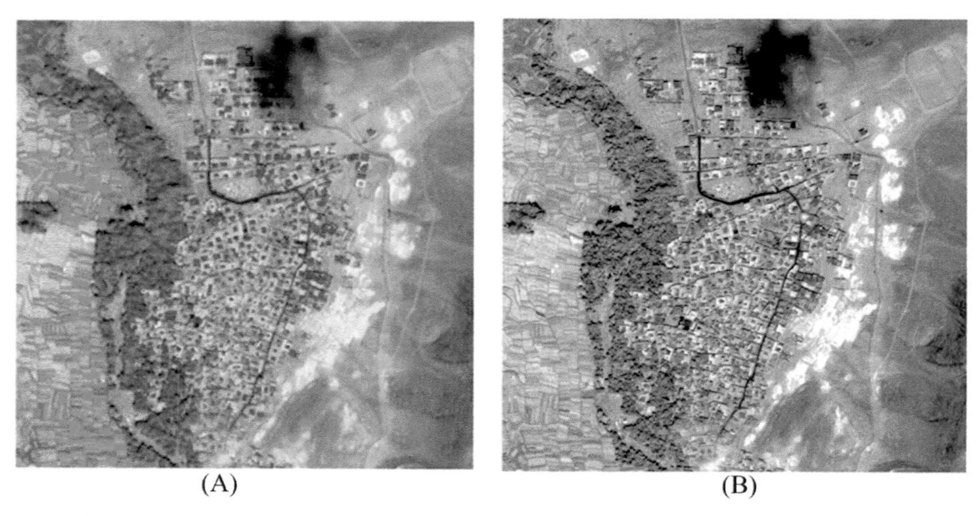

FIGURE 2.47 (A) Multispectral data with 2.4 m resolution and (B) panchromatic data with 0.6 m resolution.

For instance, simultaneity, precise geometric matching, and the sameness of geometrical conditions (Sun angle) are among these conditions. To increase the power of spatial separation of multispectral images, Brovey, IHS, principal component, and color normalization (CN) methods can be used.

2.9.21.1 Brovey method

The Brovey method is perhaps the fastest and easiest method of combining images with different resolutions. In this method, first, the bands of the multispectral image are normalized and then they are multiplied into the image with a higher resolution. Brovey's method can be summarized as follows:

$$DN_{fi} = \left(\frac{DN_{bi}}{DN_{b1} + DN_{b2} + DN_{b3}}\right) \times DN_p, \tag{2.31}$$

where DN_{bi} is the gray level of the pixel in the ith band of the multispectral image, DN_p is the gray level of the pixel in the image with higher resolution and DN_{fi} is the digital number obtained for the pixel in the output image (Fig. 2.48A). In this way, more spatial information is added to the multispectral image, and the result will be an image with better visual quality (Price, 1999). One of the disadvantages of this method compared to other methods is that the improvement of spatial resolution is applied only to the three bands from which the color composition is made, not to all bands.

2.9.21.2 Intensity, Hue, and Saturation method

The application of IHS conversion in image sharpening is that, first, the three input bands that are displayed as RGB are converted to IHS space. Then, the image with a higher spatial resolution is matched with an intensity (I) image in terms of the histogram. After this stage, the mentioned image replaces the intensity image. The reverse conversion from IHS space to RGB is done in the next step. The resulting color composition will contain more spatial details than the original composition of the multispectral image (Vrabel, 1996). One of the main drawbacks of this method is that it can be performed only on three bands (Fig. 2.48B) and is suitable for visual studies. In the case of other methods such as principal components and CN, this operation is applied to all bands.

2.9.21.3 Principal component analysis method

In the PCA method, improving the power of spatial resolution is done based on principal components analysis. Several principal components are obtained based on the variance-covariance matrix or the correlation coefficient matrix. The first principal component has the highest data variance. The panchromatic band is replaced with the first PC and then the images of the principal components are converted back into raw bands (Vrabel, 1996). Fig. 2.48C shows the image improved by the PC method.

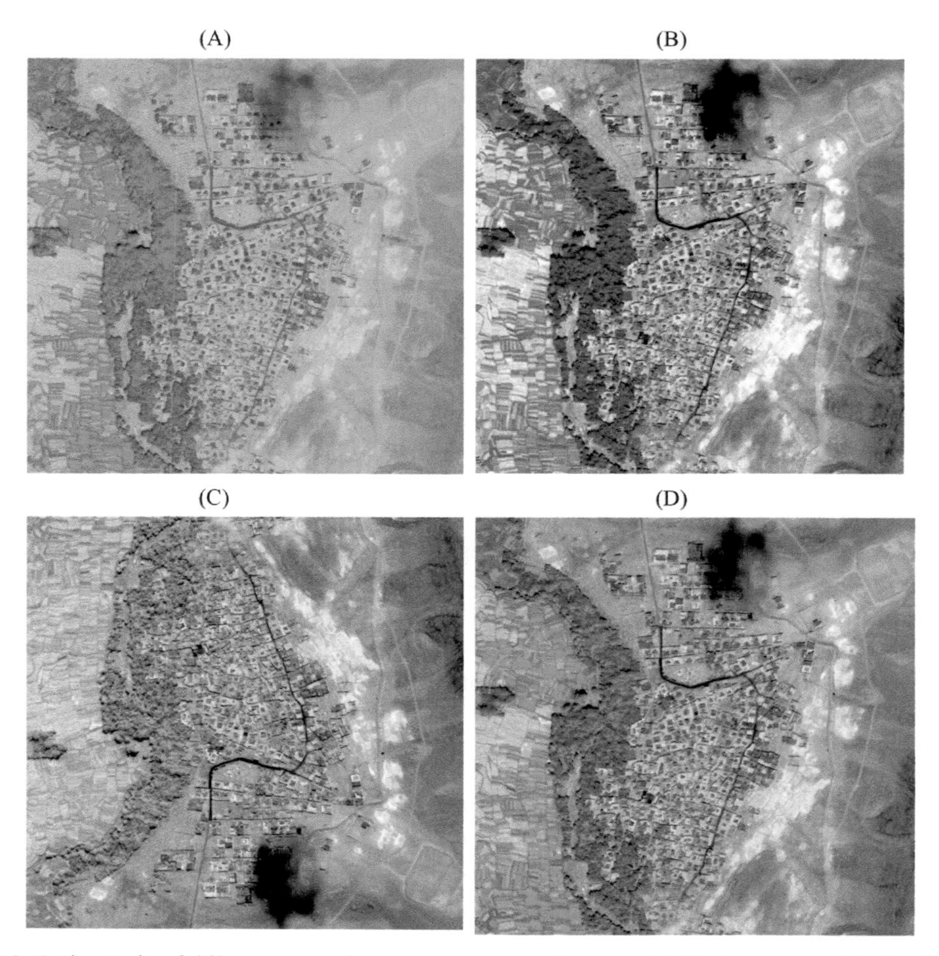

FIGURE 2.48 The results of different image-sharpening methods: (A) Brovey, (B) IHS, (C) PCA, and (D) CN. *CN*, Color normalization; *IHS*, Intensity, Hue, and Saturation; *PCA*, principal component analysis.

2.9.21.4 Color normalization method

The CN algorithm uses bands with a higher spatial resolution (but lower spectral resolution) to increase the spatial resolution of bands with a higher spectral resolution of the input images. The input images are sharpened if the wavelengths of the input bands fall within the range of the spectral width of the high-resolution images, and all other input bands that do not fall within this range remain unchanged in the output bands (Fig. 2.48D). The bands of the input images are grouped into spectral sections, which are defined by the spectral range of the panchromatic bands. This means that if the spectral range of the pan band is between 0.4 and 0.9 μm, all the bands that are in this wavelength range are placed in one group. Each input band is multiplied by the panchromatic band, then it is normalized by dividing by the sum of the input bands in the spectral range of the panchromatic band (Eq. 2.32).

$$\text{CN\_Sharpened\_Band} = \frac{\text{InputBand} \times \text{SharpeningBand} \times (\text{Num\_Bands\_In\_Segment})}{(\sum \text{Input\_Bands\_In\_Segment}) + (\text{Num\_Bands\_In\_Segment})}. \quad (2.32)$$

2.9.22 Image filtering

Image filtering is a very important domain of digital image processing. All filtering algorithms involve so-called neighborhood processing as they are based on the relationship between neighboring pixels rather than a single pixel in point operations. Filtering is widely used to sharpen images. However, care should be taken because filtering is not so "honest" in retaining the information of the original image. It is advisable to use filtered images in close reference to color composite images for interpretation (Liu and Mason, 2016).

In the image filter method the brightness of the pixels is enhanced or reduced based on the contrast they have with the neighboring pixels. Images are presented in grayscale (black and white) or pseudocolor. This is to emphasize or strengthen some desired features in the image for the user. For example, in the discussion of geology and exploration of mineral deposits, this method is used to enhance faults, geological boundaries, land and water boundaries, and roads. Image filters are used in various cases, especially for smoothing an image, highlighting the edge, etc. Digital filtering can be applied either by "box filters" based on the concept of convolution in the spatial domain or using Fourier transform (FT) in the frequency domain. In the practical applications of remote sensing, convolution-based box filters are the most useful for their computing efficiency and reliable results. In this text, box filtering is explained.

Herein, the concept of spatial frequency needs to be introduced to the reader. Changes in brightness values of pixels relative to location in images are called spatial frequency. If these changes are small, the image has a low spatial frequency, and if these changes are large, the image has a high spatial frequency (Fig. 2.49).

The background pattern with slow changes in the brightness of pixels in an image can be considered a two-dimensional waveform with a long wavelength or low frequency. Therefore the filter that separates this component with slow changes from the remaining information in the image is called a low-pass filter. On the other hand, details with faster changes such as two-dimensional waveforms with short wavelength or high frequency are known. A filter that highlights this type of detail is called a high-pass filter. These two types of filters are used separately. Low-frequency information makes it possible to identify the background pattern. The details in the output image of this type of filter are smoothed or removed from the main (input) image. This type of filter is used as an image retouch filter and some unwanted details of the image can be removed with this type of filter. Therefore the low-pass filter is a type of image blurring. High-frequency information causes the separation or highlighting of local details such as faults or lithology boundaries (Mather and Koch, 2011).

2.9.22.1 Low-pass filters

Fig. 2.50 shows the performance of a low-pass filter on a pixel row of a digital image. The original image data (input) is in dark blue. The output image data resulting from the application of the 5 × 5 moving average filter is shown in red. As can be seen, the sharp changes in the values of the pixel number in the red diagram have been removed. Therefore compared to the raw data curve, a smoother graph has been obtained. In this case the trend of the data can be seen more clearly. Sharp changes in the raw curve indicate the high-frequency component of the data and may be the result of local or spurious features. The low-pass filter includes mean, mode, median, and Gaussian.

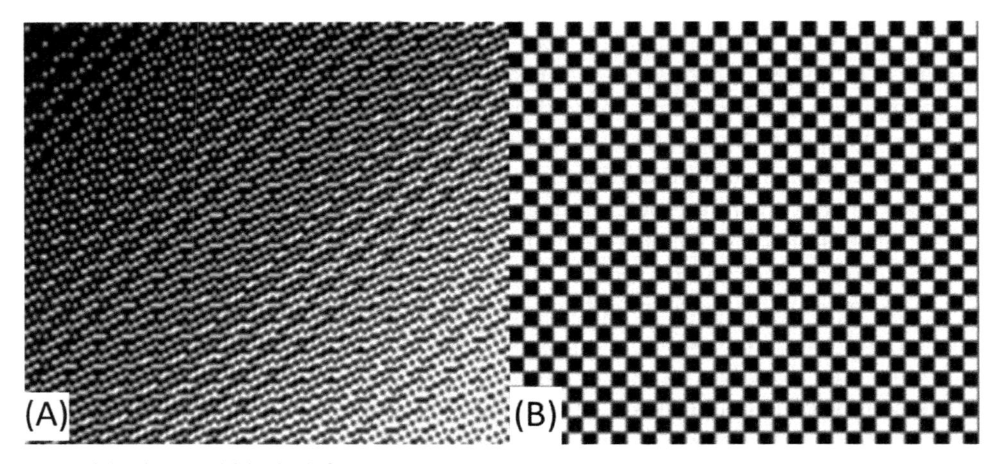

FIGURE 2.49 (A) A low- and (B) a high-frequency image.

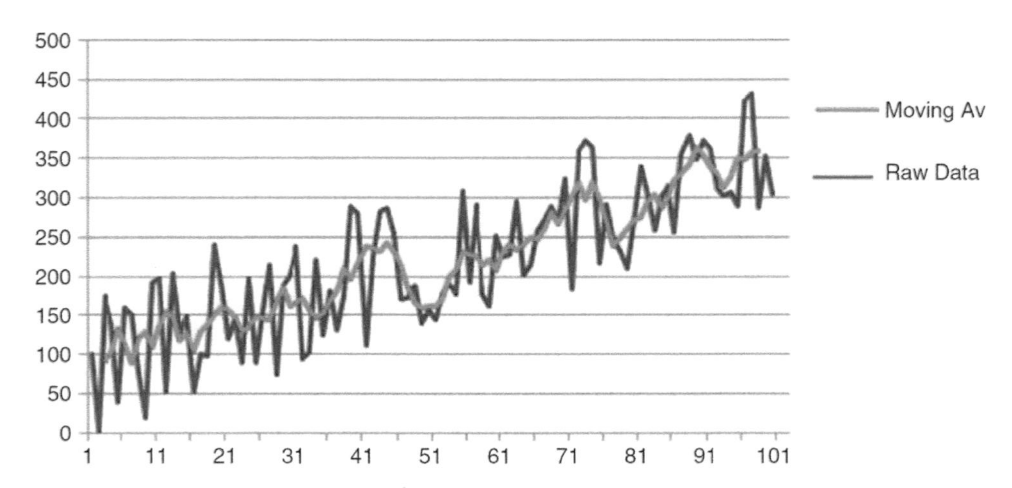

FIGURE 2.50 The digital number of a row of pixels of a digital image before and after applying the filter (Mather and Koch, 2011).

2.9.22.2 Mean filter

A two-dimensional moving average filter or kernel is defined as its horizontal and vertical dimensions. These dimensions are odd, positive, and integer. A two-dimensional moving average filter is defined according to its size (e.g., 3 × 3 pixels). In a matrix (image) the location of a pixel is determined by its row (vertical, y) and column (horizontal, x) coordinates. The origin is defined as the upper left corner of the matrix or image. The central element of the filter is located at the intersection of the middle row and column of the kernel window $n \times m$. Therefore for a 3 × 3 window, the central element is placed at the intersection of the second row and the second column. To start, the kernel is placed in the upper left corner of the filtered image (Fig. 2.51).

2.9.22.3 High-pass filters

High-pass filters are filters that enhance the sudden changes in the image. Image quality is improved by selectively increasing the distribution of its high-frequency components. These changes include faults and streets. High-pass filters include (1) filters that enhance features in all directions and (2) filters that enhance features in a specific direction.

2.9.22.3.1 Filters that enhance features in all directions

In this type of filter the kernel consists of a 3 × 3 or 5 × 5 matrix. The constituent numbers of this kernel include even and odd numbers whose algebraic sum is zero. The Laplace filter is one of these filters (Fig. 2.52).

(A)

35	25	27	31	25	46	45	50	35	40
36	25	30	34	26	35	27	28	29	41
26	27	22	23	25	27	26	27	20	21
35	36	38	41	45	28	55	53	41	32
32	42	47	32	33	45	25	27	42	35
60	35	20	22	25	27	28	26	27	30
36	25	30	34	26	35	27	28	29	41
35	36	38	41	45	28	55	53	41	32

(B)

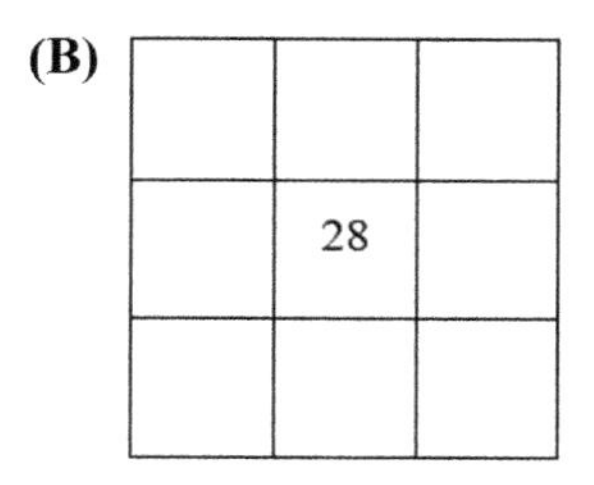

FIGURE 2.51 (A) Performance of a mean filter. (B) The pixels in the gray part are averaged and assigned to the pixel in the center of the kernel.

1	1	1
1	-8	1
1	1	1

1	-2	1
-2	4	-2
1	-2	1

FIGURE 2.52 Two types of directional filters (Laplacian filter).

0	0	0		0	-2	0		-2	0	0		0	0	-2
-2	4	-2		0	4	0		0	4	0		0	4	0
0	0	0		0	-2	0		0	0	-2		-2	0	0
North-South				East-West				Northeast-Southwest				Northwest-Southeast		

FIGURE 2.53 Four types of directional filters.

The function of this filter is that the kernel moves on the image and at each stop, the values of the pixels are multiplied by the numbers in the kernel, and the algebraic sum of the products is added to the value of the pixel in the middle of the kernel. In this method, pixels with high frequency are sharpened in all directions.

2.9.22.3.2 Filters that enhance features in a specific direction

In these types of filters, the numbers are arranged in a certain direction. As its name suggests, this type of filter highlights linear features in certain directions. Fig. 2.53 shows four types of directional filters.

This filter functions like the Laplace kernel. At each stop, the pixels of the image are multiplied by the kernel numbers, and the algebraic sum is added to the central pixel in the kernel. In Fig. 2.54, after applying the NE–SE filter the value of 50 is increased to 104 ($-46 \times 2 - 27 \times 2 + 45 \times 0 + 50 \times 0 + 28 \times 0 + 26 \times 0 + 50 \times 0 + 35 \times 0 + 50 \times 4 = 54$). If the NW–SE kernel is applied to this part of the image, the value of the pixel is remained unchanged ($46 \times 0 - 27 \times 0 + 45 \times 0 - 50 \times -2 + 28 \times 0 + 26 \times 0 + 50 \times -2 + 35 \times 0 + 50 \times 4 = 0$).

2.10 Accuracy assessment techniques

2.10.1 Principle of accuracy assessment

Accuracy assessment is a crucial step in any remote sensing–based classification exercise, given that classification maps always contain misclassified pixels and, thus, classification errors. The impact of these errors depends on many factors, including

- the type of input data to the classification process,
- the quality of the training signatures,

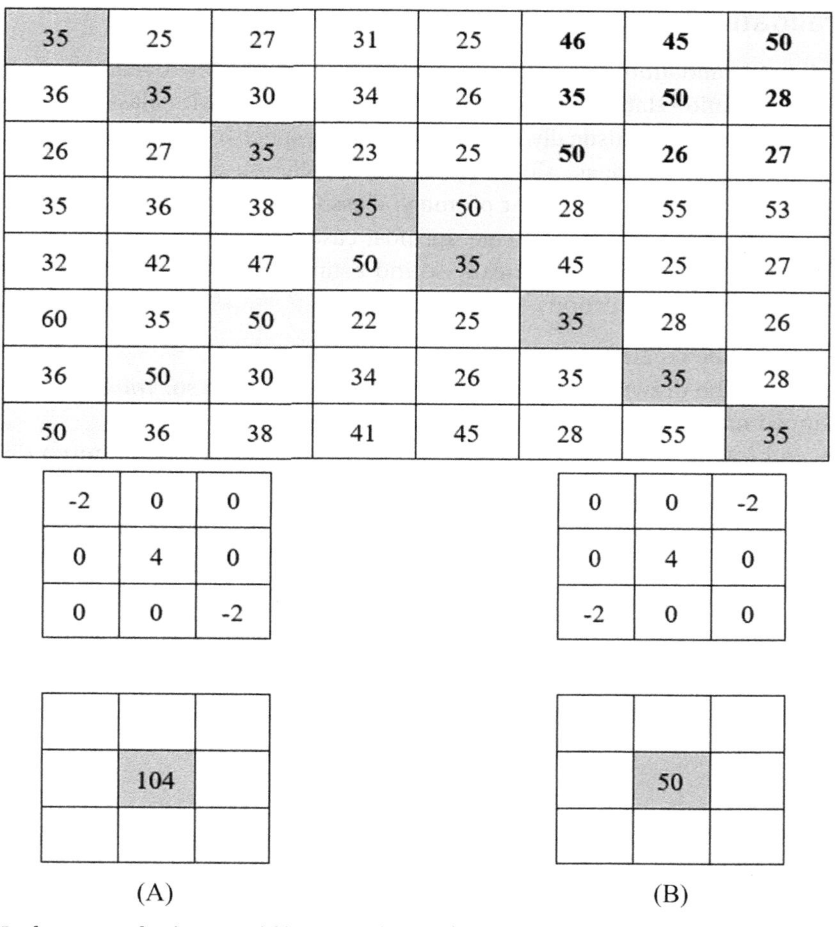

FIGURE 2.54 Performance of a directional filter on an image: (A) after applying the NE–SW filter and (B) after applying the NW–SE filter on the shaded part. *NE*, North-eastern; *NW*, north-western; *SE*, south-eastern; SW, south-western.

- the performance of the selected classification algorithm,
- the complexity of the classification scheme and the separability of the underlying classes.

For subsequent use of the classification map, the map accuracy and the sources of errors have to be known. In this regard, the quality of the classification map is evaluated through comparison to validation data representing the truth (also called ground truth data). The accuracy assessment aims to provide answers to the following questions:

- What is the overall accuracy (OA) of the classification map?
- What is the accuracy of single classes in the classification map?
- Which classes are misclassified and confused with each other?

2.10.2 Validation of the classified image

The availability of validation data is indispensable for assessing the accuracy of classification maps. Mostly, validation data are represented by point samples with class labels representing the truth. These are then statistically compared to the respective class labels of the classification map. In the classified image, all the pixels are usually not placed correctly in the groups they belong to. Examining the amount of image classification error helps the user to understand the accuracy of the classified image. In most cases, samples are generated and labeled by the map producer. To get a statistically sound estimate of map accuracy, attention must be paid to the way sample locations are selected:

- Samples should be created independently from training data.
- Samples should be drawn randomly (different strategies to do so, you may take a look at the advanced materials).
- Samples and training data should not be autocorrelated, that is, a minimum distance between points should be defined.

Different strategies exist regarding the labeling of the samples. Often, class labels are assigned through visual interpretation of the underlying satellite image or very high-resolution imagery, for example, from the Google Earth. Sometimes, samples are labeled based on field visits or based on available independent land cover information (Fig. 2.55).

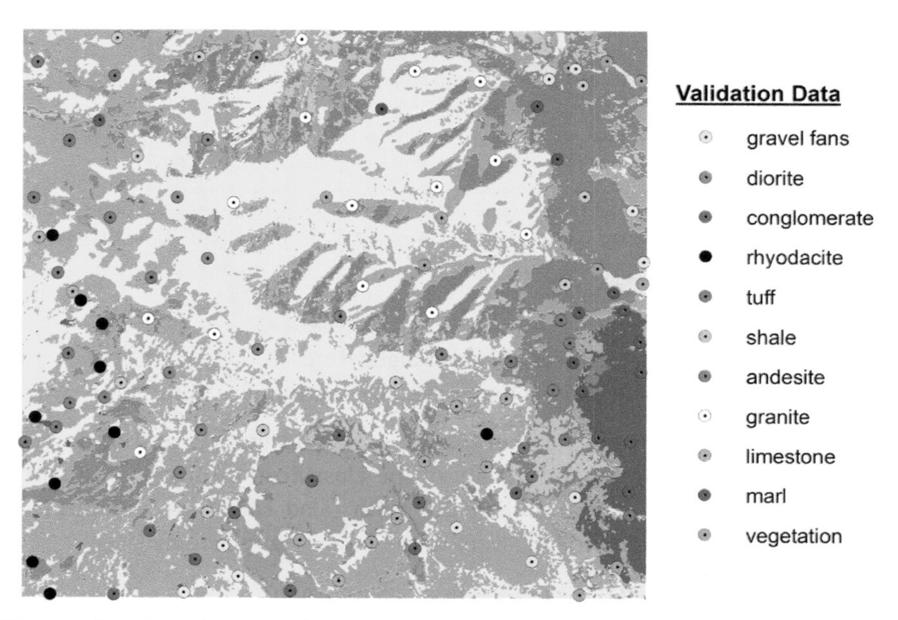

FIGURE 2.55 Examples of samples labeled based on field visits or based on available independent land cover information for the validation of the classified image.

2.10.3 Determining the number of samples

Statistical methods are usually used to determine the number of samples. In addition, involving the science of geostatistics can be one of the effective methods to estimate the number of samples (Jensen, 2016). Fitzpatrick-Lins (1981) proposed the following relationship to determine the number of samples for evaluating a classified map:

$$N = \frac{Z^2(p)(q)}{E^2}, \tag{2.33}$$

where p is the expected accuracy level for the map, $q = 100 - p$, E is the acceptable error level and $Z = 2$ are considered. For example, if an accuracy of 85% with a permissible error of 5% is considered, the number of samples needed to obtain valid results is equal to the following equation:

$$N = \frac{2^2(85)(15)}{5^2} = \text{aminimum of 204 points.} \tag{2.34}$$

According to Eq. (2.34), for 85% accuracy and 10% acceptable error, the number of required samples is reduced to 51 samples:

$$N = \frac{2^2(85)(15)}{10^2} = 51. \tag{2.35}$$

Therefore the lower the expected accuracy (p) and the higher the allowable error (E), the less the number of real ground data to evaluate the results of classification. These equations are presented in the evaluation studies of agricultural land classification maps. In the past and in the initial studies of evaluating the results of remote sensing, a lot of emphases have been placed on the accuracy of 85%. Foody (2008) stated that in other fields, using 85% accuracy can be considered a completely biased choice. Pontius and Millones (2011) suggested that there is no need to pay attention to a universal standard for accuracy assessment because this standard cannot be defined in the same way in all situations.

Unfortunately, it is not always possible to collect the number of calculated samples. As a result, a balance should be established between what is obtained through statistics and what can be achieved in practice. Congalton and Green (2009) suggested that as a general rule, we should have at least 50 samples in the error matrix for each class. If the area and the number of classes are very large (the area is more than 1 million hectares and the number of classes is more than 10 classes), the minimum sample required for each class should be increased to 75 or 100 samples (McCoy, 2005). Of course, the number of samples can be determined based on the importance of each class and the goals of the project. In general, the goal is to obtain a representative sample to be used in the error matrix.

2.10.4 Confusion matrix

Fig. 2.56 is used as an example herein to provide a step-by-step overview of the basic procedure and common error metrics for assessing the accuracy of classification maps. The left side illustrates the classification result of an image with a dimension of 10 × 11 pixels. The right side illustrates a set of 15 samples with labels representing the true classes. Please note that comparison between classification and validation is mostly on a pixel basis and samples are therefore rasterized to match with the image pixels of the classification map as demonstrated here.

From here onward, we focus further analysis on the samples by comparing predicted classes from the classification map with the true classes from the validation data (Fig. 2.57).

At first look, we see both agreements, but also confusion, between classification and validation. A simple summary is as follows:

Classification: Five dolomite samples, four limestone samples, three shale samples, and three evaporate samples.

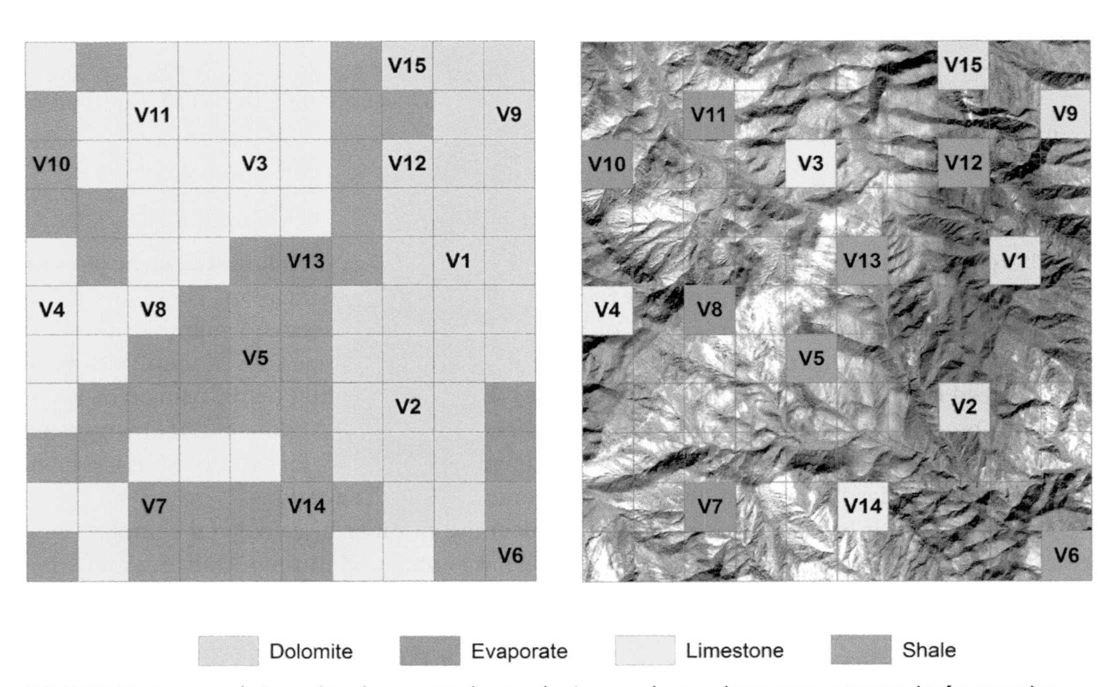

FIGURE 2.56 An example is used to show a step-by-step basic procedure and common error metrics for assessing the accuracy of classification maps in this subsection.

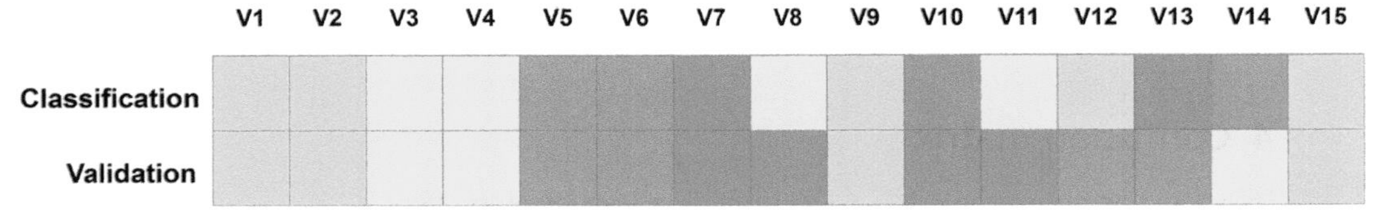

FIGURE 2.57 Comparison of classified pixels and validation pixels.

Validation: Four dolomite samples, three limestone samples, four shale samples, and four evaporate samples.

This summary provides an indication that the classification contains errors when compared to the truth. However, it does neither provide a statistical measure of the accuracy nor information on class confusion. We, therefore, extend our example to identify correctly and falsely classified samples (Fig. 2.58).

This leads to the following summary:

Correct: 10 out of 15 samples
False: 5 out of 15 samples

This summary allows us to calculate the percentage of correctly classified samples as a measure of accuracy but still does not provide information on class confusion. We, therefore, translate our comparison between classification and validation into a table layout, which is referred to as a confusion matrix (Fig. 2.59).

The confusion matrix is the foundation for statistical accuracy assessment and enables us to calculate a variety of accuracy metrics. The columns of the matrix contain the instances of the validation, and the rows of instances of the classification (Congalton, 1991; King and Clark, 2000). The diagonal of the matrix illustrates the correctly classified samples. Row and column sums are additionally presented.

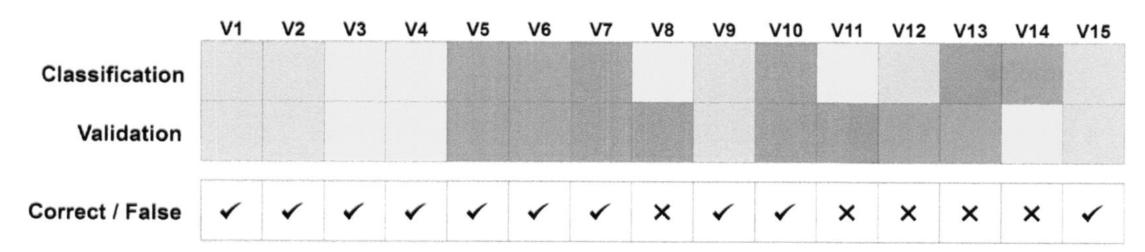

FIGURE 2.58 Comparison of classified pixels and validation pixels for identifying correctly and falsely classified samples.

		Validation				
		Dolomite	Limestone	Shale	Evaporation	**Sum**
Class.	Dolomite	4	0	1	0	**5**
	Limestone	0	2	0	2	**4**
	Shale	0	1	2	0	**3**
	Evaporation	0	0	1	2	**3**
	Sum	4	3	4	4	15

FIGURE 2.59 Generating a confusion matrix for the previous example.

2.10.5 Overall accuracy

The OA represents the proportion of correctly classified samples (Story and Congalton, 1986). The OA is calculated by dividing the number of correctly classified samples by the total number of samples (Fig. 2.60).

The OA is a general measure of map accuracy, however, does not tell us whether errors are evenly distributed for all classes. We do not know which classes were well classified or poorly mapped.

2.10.6 User's accuracy and commission error

The User's accuracy represents class-wise accuracies from the point of view of the map user. The User's Accuracy is calculated by dividing the number of correctly classified samples of class x by the number of samples classified as class *x*. It, therefore, provides the map user with a probability that a particular map location of class *x* is also the same class *x* in truth (Story and Congalton, 1986). The Commission Error is the complementary measure to the User's Accuracy and can be calculated by subtracting the User's Accuracy from 100%.

Based on our initial example as in Fig. 2.58, we only focus on the classification (map user perspective) and summarize the following:

User's Accuracy:

Dolomite (80%): Four correctly classified dolomite samples, five classified dolomite samples

		Validation				
		Dolomite	Limestone	Shale	Evaporation	**Sum**
Class.	Dolomite	4	0	1	0	**5**
	Limestone	0	2	0	2	**4**
	Shale	0	1	2	0	**3**
	Evaporation	0	0	1	2	**3**
	Sum	**4**	**3**	**4**	**4**	15

Overall Accuracy (OA)

$$OA = \frac{\text{Number of correctly classified samples}}{\text{Number of samples}}$$

$$OA = \frac{4+2+2+2}{15} = 0.66 = 66\%$$

FIGURE 2.60 The formulation for calculating the overall accuracy.

Limestone (50%): Two correctly classified limestone samples, four classified limestone samples
Shale (67%): Two correctly classified shale samples, three classified shale samples.
Evaporate (67%): Two correctly classified evaporate samples, three classified evaporate samples

Commission Error:

Dolomite (20%): One wrongly classified dolomite sample, five classified dolomite samples.
Limestone (50%): Two wrongly classified limestone samples, four classified limestone samples.
Shale (33%): One wrongly classified shale sample, three classified shale samples.
Evaporate (33%): One wrongly classified evaporate sample, three classified evaporate samples.

The User's Accuracy for each class is calculated by going through each row of the confusion matrix and by dividing the number of correctly classified samples through the row sum. The Commission Error is simply calculated as 100% of the User's Accuracy (Fig. 2.61).

		Validation						
		Dolomite	Limestone	Shale	Evaporation	**Sum**	**UA**	**Com. Err.**
Class.	Dolomite	4	0	1	0	5	80%	20%
	Limestone	0	**2**	0	2	**4**	**50%**	**50%**
	Shale	0	1	**2**	0	**3**	**33%**	**67%**
	Evaporation	0	0	1	**2**	**3**	**33%**	**67%**
	Sum	**4**	**3**	**4**	**4**	15		

User's Accuracy (UA) of class x

$$UA_x = \frac{\text{Number of correctly classified samples of } x}{\text{Sum of samples } classified\ as\ x}$$

Commission error (Com. Err.) of class x

$$Com.Err._x = 100\% - UA_x$$

$$UA_{Dolomite} = \frac{4}{5} = 0.8 = 80\% \qquad Com.Err._{Dolomite} = 100\% - 80\% = 20\%$$

FIGURE 2.61 The formulation for calculating User's Accuracy and Commission Error.

2.10.7 Producer's Accuracy and Omission Error

The Producer's Accuracy represents class-wise accuracies from the point of view of the map maker. The Producer's Accuracy is calculated by dividing the number of correctly classified samples of class x by the number of samples with the true labels of class x. It, therefore, provides a probability that a particular sample of class x is mapped as the same class x in the classification map (Story and Congalton, 1986). The Omission Error is the complementary measure to the Producer's Accuracy and can be calculated by subtracting the Producer's Accuracy from 100%.

Based on our initial example as in Fig. 2.58, we only focus on the validation (map maker perspective) and summarize the following:

Producer's Accuracy:

Dolomite (100%): Four correctly classified dolomite samples, four dolomite samples.
Limestone (67%): Two correctly classified limestone samples, three limestone samples.
Shale (50%): Two correctly classified shale samples, four shale samples.
Evaporate (50%): Two correctly classified evaporate samples, four evaporate samples.

Omission Error:

Dolomite (0%): 0 wrongly classified dolomite, four dolomite samples
Limestone (33%): One wrongly classified limestone sample, three limestone samples.
Shale (50%): Two wrongly classified shale samples, four shale samples.
Evaporate (50%): Two wrongly classified evaporate samples, four evaporate samples.

The Producer's Accuracy for each class is calculated by going through each column of the confusion matrix and by dividing the number of correctly classified samples by the column sum. The Omission Error is simply calculated as 100% Producer's Accuracy (Fig. 2.62).

2.10.8 Kappa coefficient

Kappa can be used as another measure of agreement or accuracy. Kappa values can range from +1 to −1. However, since there should be a positive correlation between the remotely sensed classification and the reference data, positive values are expected (Lillesand and Kiefer, 1994; Lunetta and Lyon, 2004). Landis and Koch (1977) divided the possible ranges for kappa into three groups: A value greater than 0.80 (i.e., 80%) represents strong agreement; a value between 0.40 and 0.80 (i.e., 40%–80%) represents moderate agreement; and a value below 0.40 (i.e., 40%) represents poor agreement. The kappa coefficient is calculated according to the following equation:

$$\hat{K} = \frac{N\sum_{i=1}^{r} x_{ii} - \sum_{i=1}^{r} (x_{i+} \times x_{+i})}{N^2 - \sum_{i=1}^{r} (x_{i+} \times x_{+i})} \tag{2.36}$$

where r is the number of rows in the matrix, x_{ii} is the number of observations in row i and column i (i.e., the ith diagonal element), x_{i+} and x_{+i} are the marginal totals of row i and

		Validation						
		Dolomite	Limestone	Shale	Evaporation	Sum	UA	Com. Err.
Class.	Dolomite	4	0	1	0	5	80%	20%
	Limestone	0	2	0	2	4	50%	50%
	Shale	0	1	2	0	3	33%	67%
	Evaporation	0	0	1	2	3	33%	67%
	Sum	4	3	4	4	15		
	PA	100%	67%	50%	50%			
	Om. Err.	0%	33%	50%	50%			

Producer's Accuracy (PA) of class x

$$UA_x = \frac{\text{Number of correctly classified samples of } x}{\text{Sum of samples } \mathit{with\ true\ label\ of\ x}}$$

Ommission error (Om. Err.) of class x

$$Om.Err._x = 100\% - PA_x$$

$$PA_{Dolomite} = \frac{4}{4} = 1 = 100\% \qquad Om.Err._{Dolomite} = 100\% - 100\% = 0\%$$

FIGURE 2.62 The formulation for calculating the Producer's Accuracy and Omission Error.

column i, respectively, and N is the total number of observations (Bishop et al., 2007). The kappa coefficient, unlike the OA parameter, includes all the pixels that are correctly or incorrectly classified in its calculations. For this reason, some researchers consider the validity of this coefficient to be much higher than the OA parameter. Fig. 2.63 shows an example of the calculation of the kappa coefficient.

2.11 Interpretation of remote sensing data for alteration mineral mapping

Porphyry copper deposits typically occur in association with hydrothermal alteration mineral zones such as phyllic, argillic, potassic, and propylitic (Fig. 2.64; Lowell and Guilbert, 1970). A core of quartz and potassium-bearing minerals is surrounded by multiple zones that contain clay and other hydroxyl minerals with diagnostic spectral absorption features in the SWIR portion of the electromagnetic spectrum. The hydrothermal alteration zones associated with porphyry copper deposit is illustrated in Fig. 2.64.

Furthermore, at the same time, an oxide zone with extensive iron oxide minerals is developed by virtue of supergene alteration processes over porphyry copper bodies. Iron oxides are one of the important mineral groups that are associated with hydrothermally altered

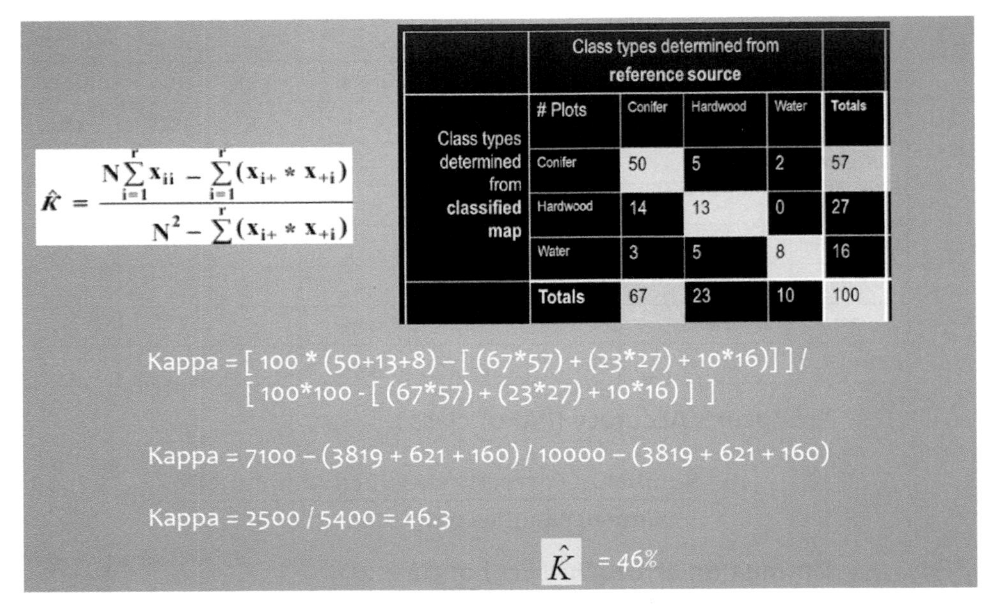

$$\hat{K} = \frac{N\sum_{i=1}^{r} x_{ii} - \sum_{i=1}^{r}(x_{i+} * x_{+i})}{N^2 - \sum_{i=1}^{r}(x_{i+} * x_{+i})}$$

	Class types determined from reference source				
Class types determined from classified map	# Plots	Conifer	Hardwood	Water	Totals
	Conifer	50	5	2	57
	Hardwood	14	13	0	27
	Water	3	5	8	16
	Totals	67	23	10	100

FIGURE 2.63 An example of calculating the kappa coefficient.

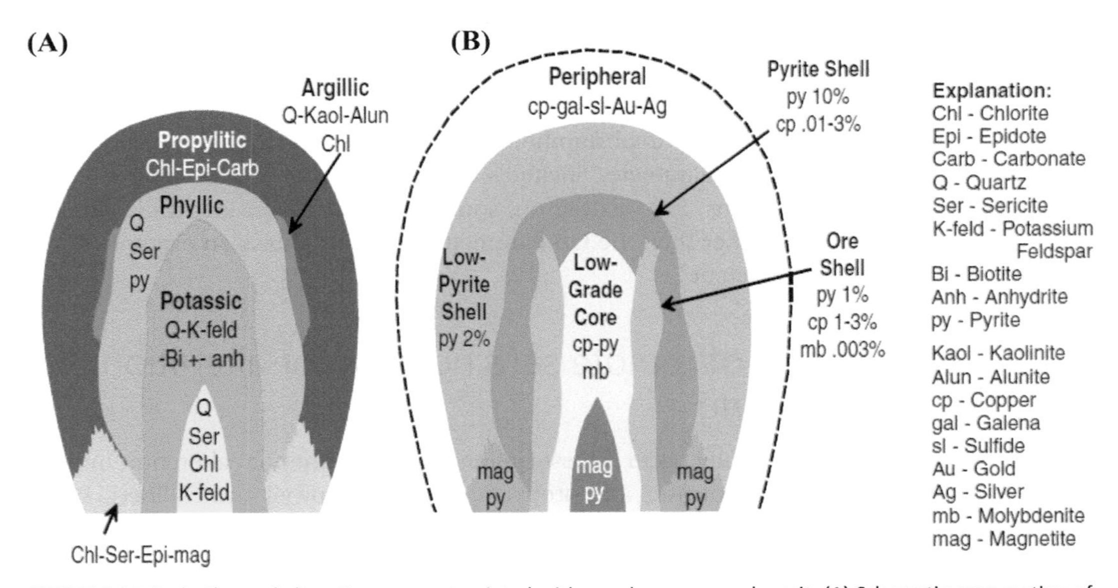

FIGURE 2.64 Hydrothermal alteration zones associated with porphyry copper deposit. (A) Schematic cross section of hydrothermal alteration mineral zones, which consist of propylitic, phyllic, argillic, and potassic alteration zones; and (B) schematic cross section of ores associated with each alteration zone. *Modified from Mars, J.C., Rowan, L.C., 2006. Regional mapping of phyllic- and argillic-altered rocks in the Zagros magmatic arc, Iran, using advanced spaceborne thermal emission and reflection radiometer (ASTER) data and logical operator algorithms. Geosphere 2 (3), 161–186. https://doi.org/10.1130/GES00044.1.*

rocks and are rendered to characteristic yellowish or reddish color to the altered rocks that are termed gossan (Sabins, 1999). Iron oxide minerals have low reflectance and high absorption properties in the VNIR region (Hunt, 1977; Hunt and Salisbury, 1974). Hydrothermal alteration minerals with diagnostic spectral absorption properties in the VNIR through SWIR can be identified by multispectral and hyperspectral remote sensing data as a tool for the initial stages of porphyry copper exploration.

As mentioned before, an ideal porphyry copper deposit is typically characterized by hydrothermal alteration mineral zones (Lowell and Guilbert, 1970). The core of quartz and potassium-bearing minerals is surrounded by multiple zones that contain minerals with diagnostic spectral absorption features in the SWIR region. Fig. 2.65 illustrates laboratory

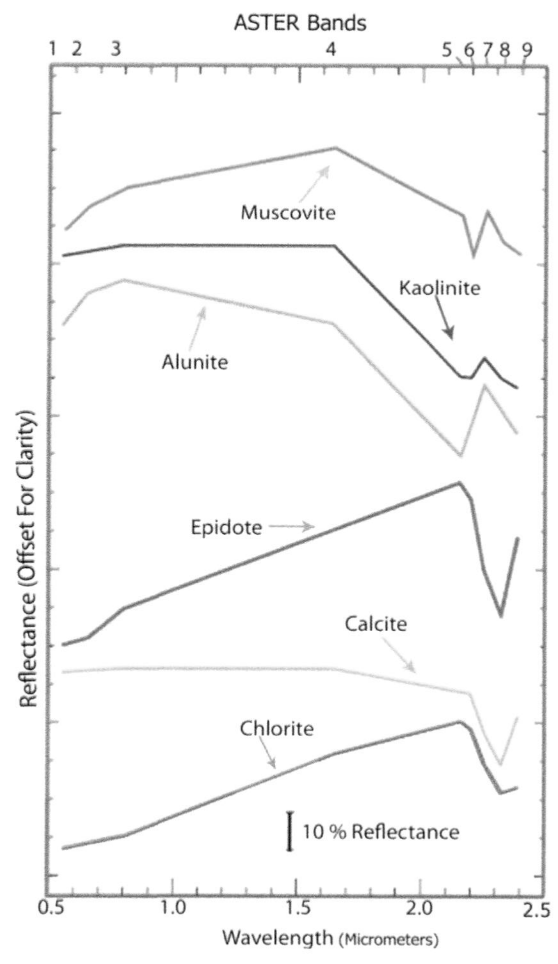

FIGURE 2.65 Laboratory spectra of muscovite, kaolinite, alunite, epidote, calcite, and chlorite resampled to ASTER bandpasses (Mars and Rowan, 2006). *ASTER*, Advanced Spaceborne Thermal Emission and Reflection Radiometer.

spectra of muscovite, kaolinite, alunite, epidote, calcite, and chlorite resampled to ASTER bandpasses. The broad phyllic zone is characterized by illite/muscovite (sericite) that yields an intense Al–OH absorption feature centered at 2.20 μm, coinciding with ASTER band 6 (Fig. 2.65). The narrower argillic zone includes kaolinite and alunite, which collectively displays a secondary Al–OH absorption feature at 2.17 μm that corresponds with ASTER band 5 (Fig. 2.65). The mineral assemblages of the outer propylitic zone include epidote, chlorite, and calcite that all exhibit absorption features situated in the 2.35 μm, which coincide with ASTER band 8 (Fig. 2.65) (Mars and Rowan, 2006; Rowan et al., 2006; Hecker et al., 2010).

The differentiation between the hydrothermal alteration zones and especially targeted identification of the phyllic zone is important in the exploration of porphyry copper mineralization because phyllic zone is an indicator of the high-economic potential for copper mineralization within the central shell of mineralization (Lowell and Guilbert, 1970). This alteration zone normally overprints the potassic and chlorite–sericite assemblages, which are the main and common ore contributors to the porphyry copper mineralization system (Dilles and Einaudi, 1992; Sillitoe, 2010). The Fe^{3+} absorption features of the supergene minerals, limonite, goethite, and hematite (limonite-rich rocks) are a potential indicator of supergene deposits (Rowan et al., 1974; Schmidt, 1976; Krohn et al., 1978; Raines, 1978). Multispectral and hyperspectral sensors have been used for mapping hydrothermally altered rocks associated with porphyry copper deposits. Although hyperspectral detectors have more limited coverage (swath width) compared to multispectral sensors, hyperspectral detectors (i.e., HyMap and AVIRIS) have much higher spectral resolution than multispectral sensors (i.e., ASTER and Landsat TM) and can map a detailed variety of hydrothermal minerals associated with porphyry copper deposits (Pour et al., 2021b).

Landsat TM and ETM^+ bands 1–4 span the Fe^{3+} absorption features of iron oxide/hydroxide minerals. Band 7 (centered at the 2.20 μm) spans the Al–OH, Mg–OH, and CO_3 absorption features associated with alunite, kaolinite, muscovite, epidote, chlorite, and calcite. TM and ETM^+ spectra of alunite, kaolinite, muscovite, and calcite show a lower reflectance (absorption features) in band 7 compared with band 5 spectral reflectance features due to Al–O–H and CO_3 absorption. Thus the TM and ETM^+ band ratio 5/7 typically has been used to map advanced argillically, sericitically, and propylitically altered rocks associated with porphyry copper deposits (Abrams et al., 1983; Ott et al., 2006). Nonetheless, because Landsat TM and ETM^+ band 7 is the only band that spans all of the argillic, sericitic, and propylitic SWIR absorption features, it is not capable of spectrally delineating the different types of altered rocks.

Iron oxide/hydroxide mineral groups can be detected using three VNIR spectral bands of ASTER. SWIR bands of ASTER have sufficient spectral resolution to illustrate different spectral signatures for advanced argillic (alunite–kaolinite), sericitic (muscovite), propylitic (epidote–chlorite–calcite), and supergene mineral assemblages. TIR bands of ASTER have sufficient spectral resolution to delineate the quartz reststrahlen feature (Rowan and Mars, 2003). ASTER mineral maps from previous studies illustrate that structurally undeformed porphyry copper deposits are typically characterized by circular to elliptical patterns of sericitically and advanced argillically-altered rocks (Mars and Rowan, 2006; Rowan et al., 2006; Di

Tommaso and Rubinstein, 2007; Yousefi et al., 2021, 2022). Spectral resolution of AVIRIS data is sufficient to map advanced argillic (alunite and kaolinite), sericitic (muscovite), and propylitic (epidote, chlorite, and calcite) alteration minerals with AL–O–H, Fe, Mg–O–H and CO_3 absorption features (Berger et al., 2003; Cunningham et al., 2005; Chen et al., 2007). High-altitude AVIRIS data have been used to map several concealed porphyry copper deposits in the Patagonia Mountains, Argentina (Berger et al., 2003).

2.12 Remote sensing structural analysis for mineral exploration

Geologic structures serve as reminders of significant Earth history events. Patterns in the arrangement of rock inside the Earth are known as geologic structures. They can be of enormous economic significance because of the fact that they frequently serve to concentrate deposits of significant resources, such as metals and petroleum. Then, several studies of ore deposits worldwide, such as porphyry ore deposits, volcanic massive sulfide (VMS) deposits, IOCG deposits, skarn deposits, and epithermal deposits, have revealed that they are controlled by specific structural/lithological settings (Bark and Weihed, 2012; Joly et al., 2010; Zoheir and Emam, 2012; Hor, 1998; Blenkinsop and Doyle, 2014; Dalstra, 2014; Peng et al., 2019; Zhou et al., 2021; Santosh and Groves, 2022; Kwak et al., 2022). Among the different types of structural controls on mineralization, the most common are (1) expansion and contraction along faults/shear zones (Micklethwaite et al., 2015; Kwak et al., 2022); (2) intersection of two synmineralization structures (Robert et al., 1995; Robert and Kelly, 1987; Robert and Poulsen, 2001; Bark and Weihed, 2012; Kwak et al., 2022); (3) intersection of faults/shear zones with highly competent and/or chemically reactive rocks (Robert and Poulsen, 2001); (4) faults/shear zones along lithological contacts between competent and less competent rocks (Bierlein et al., 2006; Micklethwaite et al., 2015; Zoheir et al., 2019a,b,c); (5) areas dipping parallel to a stretch lineation and (Alsop and Holdsworth, 2004; Alsop and Holdsworth, 2012; Alsop et al., 2021; Bons et al., 2012; Groves and Santosh, 2021); and (6) fold limbs and hinge zones (Robert and Poulsen, 2001). Therefore exploration and development of rock-hosted ore deposits at the regional, district, and orebody scales depend heavily on structural geology.

It is possible to analyze and categorize fluid conduits and create genetic models for ore production by determining the time of significant structural events in an ore region (Pour et al., 2016; Pour and Hashim, 2017; Funedda et al., 2018; Abd El-Wahed et al., 2016; El-Wahed et al., 2021; Cocco et al., 2022; Xie et al., 2022). The most useful applications of structural geology involve defining and measuring the various constituent parts of the orebody, which may then be directly used for ore reserve estimation, ground control, grade control, safety concerns, and mine planning. Long-term strategic planning, economic analysis, and property ownership are all directly applicable to the district- and regional-scale structural investigations. A thorough understanding of the structural regulation of the mineralization processes is necessary for the accurate identification, quantification, and utilization of ore deposits. This is especially true for ores housed in polydeformed basements, where the superposition of many deformation periods highlights the complexity of the structural context. The formation of hydrothermal ore deposits is highly dependent on structural controls (Funedda et al., 2018; Zoheir et al.,

2019a,b,c; Xie et al., 2022). In crustal regions impacted by mineralizing events, they specify the pattern, range, and modalities of fluid flow (Cathles, 1981; Ord et al., 2012; Lester et al., 2012; Ingebritsen and Appold, 2012; Cocco et al., 2022). The prevalence of structural controls, in which a frequently multiscale network of previously existing structural discontinuities in the host rocks forms a preestablished plumbing system for circulation and entrapment of hydrothermal fluids, which depends on the tectonic regime in which the mineralizing phenomena occur (Cocco et al., 2022).

Although it can have a significant influence on a successful exploration plan or resource recovery, the analysis of the role of structures in determining mineralization or deposit form and stability is frequently overlooked and is crucial for safe mine operation. By thoroughly comprehending the structural system, including the fault network, the morphology of the faults, the strain (and fractures) related to the faulting, and the development of structures through time, exploration for deeply buried or blind deposits may be enhanced and costs lowered. By having a greater understanding of the fracture distribution and the risk of structures failing under specific pressures, mining procedures like block caving may be better planned for. When there are numerous generations of mineralization, structural controls of ore deposits housed in polydeformed low-grade metamorphic basements are sometimes difficult to identify, primarily because the connections between structures and ore bodies are frequently unclear (Funedda et al., 2018). In fact, describing the deposit's properties, its emplacement style, its age, and, consequently, its origin, requires an accurate knowledge of the structural controls. Sometimes, the complexity of the tectonic structures prevents an understanding of the geometry of the ore bodies, which causes errors in mineral exploration, ore deposit appraisal, mine design, and even mineral exploitation. It can be useful to know if the structural controls on ore deposits are "active," or connected to tectonic structures that are gradually evolving with mineralization, as opposed to "passive," which is owing to tectonic structures created before mineralization (Funedda et al., 2018; Xie et al., 2022).

Remote sensing data are commonly used to identify geological structures associated with extensive shear belts, particularly in vast and rugged terrains. The innovative technologies in remote sensing promoted the application of multispectral and SAR imagery for detailed structural geology mapping (Pour et al., 2013, 2014, 2016; Pour and Hashim, 2017, 2017; Pour and Hashim, 2014a,b, 2017; Zoheir et al., 2019a,b,c). Remote sensing satellite imagery has a high capability of providing a synoptic view of geological structures, alteration zones, and lithological units in metallogenic provinces. Typically, the application of multisensor satellite imagery can be considered a cost-efficient exploration strategy for prospecting orogenic gold mineralization in transpression and transtension zones, which are located in harsh regions around the world (Zoheir and Emam, 2014; Pour and Hashim, 2014a,b; Adiri et al., 2017; Pour et al., 2018a,b; Zoheir et al., 2019a,b,c; Sheikhrahimi et al., 2019; El-Wahed et al., 2021).

2.13 Case studies

In this section, some of the most cited and recent studies that used remote sensing data as a tool for mapping hydrothermal alteration zones, lithological units, and structural features for ore minerals exploration were concisely reviewed and introduced to readers.

Cudahy et al. (2001) evaluated the capacity of VNIR and SWIR subsystems of Hyperion data for mineral mapping at Mount Fitton, South Australia. The Hyperion-derived mineral map indicated spatially coherent mineral distributions consistent with the geology map as well as superimposed alteration. The results showed the capability of Hyperion data and the spectral power for mineral mapping, especially in SWIR bands. Hewson et al. (2001) simulated the capability of the ASTER multispectral data for geologic and alteration mineral mapping in Mt Fitton, South Australia. This test site has been previously surveyed by visible-shortwave hyperspectral AMS (HyMap), Thermal Infrared Multispectral Scanner (TIMS) data, and several field campaigns collecting relevant spectral measurements. They applied decorrelation stretch on simulated ASTER bands 3-2-1 to delineate drainage and vegetation, and band 13-12-10 for the identification of quartz-rich areas. They also implemented the MTMF method on the simulated ASTER SWIR bands to obtain spectrally unmixed end-members related to the rich areas of hydrothermally alteration mineral assemblages. Their results showed good accuracy with field spectral measurements and compared well with HyMap and TIMS outputs that were collected previously for the study area.

Cudahy and Barry (2002) used MTMF method to Hyperion data that involved all available bands with particular attention to the SWIR region (2000–2400 nm) for hydrothermal alteration mapping at Panorama, Western Australia. They recognized two types of white mica (Al-rich and Al-poor), chlorite and pyrophyllite. The resultant Hyperion-derived mineral maps of white mica abundance and Al-chemistry were correlated well with the corresponding HyMap white mica maps and the published geologic maps. Kruse et al. (2003) compared the Airborne Visible/Infrared Imaging Spectrometer (AVIRIS) data with Hyperion for mineral mapping in Cuprite, Nevada, and northern Death Valley, south-central California/Nevada, USA. They used Hourglass hyperspectral imaging (HHI) analysis approach for processing the data. Visual comparison of the Hyperion and AVIRIS mineral maps for both case studies indicated that Hyperion generally identified similar minerals and produced similar mineral mapping results to AVIRIS. However, the lower SNR of the Hyperion data in the SWIR region has affected the ability to extract characteristic spectra and identify individual minerals. Results established that the Hyperion SWIR (2.0–2.4 μm) data can be used to produce useful mineralogic information. Hubbard et al. (2003) used ALI, Hyperion, and ASTER data for alteration mineral mapping in the Central Andes between Volcan Socompa and Salar de Liullaillaco located in the border region between Chile and Argentina. The comparison between results revealed that ALI has a sufficient spectral resolution in the VNIR wavelength range to discriminate several important ferric-iron oxide minerals. ASTER SWIR bands allow key distinctions to be mapped between various clay and sulfate mineral types. Hyperion is useful for calibrating ASTER and ALI data and can also be used for evaluating the mineral mapping results. They concluded that the integration of ALI, ASTER, and Hyperion imagery can be an effective technique for mapping a variety of minerals characteristic of hydrothermally altered rocks in remote areas of the earth, where available geologic and other ground truth information is limited.

Rowan et al. (2003) evaluated the capability of the ASTER data for mapping the hydrothermally altered rocks and the unaltered country rocks in the Cuprite mining district in

Nevada, USA. They used MF method for identifying the surface distribution of hydrothermal alteration minerals. Their results indicated that spectral reflectance differences in the nine bands of visible near-infrared through the SWIR (0.52–2.43 μm) can provide subtle spectral information for discriminating main hydrothermal alteration mineral zones. They identified a silicified zone, an opalized zone, an argillized zone, and the distribution of unaltered country rock units. Their results compared well with those obtained using Airborne Visible/Infrared Imaging Spectrometer (AVIRIS) hyperspectral data. Rowan and Mars (2003) used ASTER data for lithological mapping in Mountain Pass, California, USA. They applied RABD, MF, and SAM methods for differentiating calcitic, granodioritic, gneissic, granitic, and quartzos rock units. The results showed good similarities between the patterns of the identified rock units and the geologic map of the study area.

Ćrosta et al. (2003) used PCA on ASTER VNIR and SWIR bands to target key alteration minerals associated with epithermal gold deposits in Los Menucos, Patagonia, Argentina. PCA was applied to selected subsets of four ASTER bands according to the position of characteristic spectral absorption features of key hydrothermal alteration mineral end-members such as alunite, illite, smectite, and kaolinite in the VNIR and SWIR regions. Their results revealed that the PCA technique can extract detailed mineralogical spectral information from ASTER data by producing abundance images of selected minerals. Thus this technique can be used for identifying hydrothermal alteration minerals associated with precious and base–metal deposits. Volesky et al. (2003) distinguished the propylitic alteration zone and gossan associated with massive sulfide mineralization in host rocks by ASTER (4/2, 4/5, 5/6) band ratio images covering the Neoproterozoic Wadi Bidah shear zone, southwestern Saudi Arabia. Hewson et al. (2005) generated "seamless" regional-scale maps of Al–OH and Mg–OH/carbonate minerals and ferrous iron content from ASTER SWIR data, as well as a map of quartz content from ASTER TIR data for the Broken Hill, Curnamona province of Australia.

Rowan et al. (2005) evaluated ASTER band ratio and RABD, MF, and SAM methods for lithological mapping of the ultramafic complex in the Mordor Pound, NT, Australia. They identified subtle and major spectral features useful for distinguishing and classifying felsic and mafic–ultramafic rocks, alluvial–colluvial deposits, and quartzose to intermediate composition rocks from one another. The classification was based on spectral absorption features of Al–OH and ferric-iron mineralogical groups for felsic rock, ferrous-iron, and Fe, Mg–OH mineralogical absorption features for mafic–ultramafic rock in the ASTER VNIR + SWIR data. Additional Si–O spectral features were used to map more lithologic diversity within ultramafic complex and adjacent rocks such as mafic-gneisses, felsic–gneisses, intermediate composition rocks such as syenite and quartzite using ASTER TIR data. Ninomiya et al. (2005) applied the quartz index (QI), carbonate index (CI), and mafic index (MI) to ASTER-TIR "radiance-at-sensor" data for detecting the mineralogic or chemical composition of quartzose, carbonate, and silicate rocks in Mountain Yushishan, China and Mountain Fitton, Australia. These lithologic indices discriminated quartz, carbonate, and mafic–ultramafic rocks, which were compatible well with published geologic maps and field observation. They suggested that these lithologic indices can be one unified approach for lithological mapping of the Earth, especially in arid and semiarid regions.

Hubbard and Crowley (2005) used ALI, ASTER, and Hyperion data for mineral mapping in a volcanic terrane area of the Chilen–Bolivian Altiplano. ASTER and ALI channels were coregistered and jointed to produce a 13-channel reflectance cube spanning the visible to SWIR radiation (0.4–2.4 μm). MNF transformation, pixel purity index (PPI), and *n*-dimensional visualizer were applied to identify spectral end-members. SAM and LSU were applied to map altered rocks using extracted spectral end-members. Results showed that the Hyperion data were only marginally better for mineral mapping than the merged ALI + ASTER datasets. Galvão et al. (2005) used the SAM method on the ASTER SWIR bands to investigate the spectral discrimination of hydrothermally altered minerals in a tropical savannah environment in the northern portion of Gious state, central Brazil. Their results showed the efficacy of ASTER SWIR data in discriminating areas of altered minerals from the surrounding vegetated environment.

Gomez et al. (2005) evaluated the nine ASTER bands for geological mapping in the Western margin of the Kalahari Desert in Namibia. They applied PCA and supervised classification on visible near-infrared and SWIR ASTER bands to identify lithological units. The processing of ASTER data demonstrated the validation of the lithological boundaries defined on the previous geological maps and also provided the information for characterizing new lithological units, which were previously unrecognized. Kruse et al. (2006) used Hyperion and AVIRIS data for district-level mineral surveying in the Los Menucos District, Rio Negro, Argentina. Hourglass hyperspectral imaging (HIS) analysis approach has been applied to detect hydrothermal alteration minerals associated with the epithermal gold system in the study area. VNIR and SWIR bands of Hyperion and AVIRIS were analyzed to identify iron oxides, clay minerals, and carbonates. Hematite, goethite, dickite, alunite, pyrophyllite, muscovite/sericite, montmorillonite, calcite, and zeolites were identified in the study area using Hyperion and AVIRIS data. Field reconnaissance verification and spectral measurements showed the accuracy of hyperspectral mapping results.

Rowan et al. (2006) identified the distribution of hydrothermally altered rocks consisting of phyllic, argillic, and propylitic alteration zones based on spectral analysis of VNIR + SWIR ASTER bands. Additional hydrothermally silicified rocks were mapped using TIR ASTER bands covering the Reko Diq, Pakistan Cu–Au mineralized area. In a follow-up study, Mars and Rowan (2006) developed logical operator algorithms based on ASTER-defined band ratios for regional mapping of phyllic and argillic altered rocks in Zagros magmatic arc, Iran. The logical operator algorithms were used to illustrate distinctive patterns of argillic and phyllic alteration zones associated with Eocene to Miocene intrusive igneous rocks, as well as known and undiscovered porphyry copper deposits. Numerous high-potential areas of porphyry copper and epithermal or polymetallic vein-type mineralization were identified based on argillic and phyllic alteration patterns in the study area.

Ducart et al. (2006) applied MTMF method to ASTER SWIR data to provide regional and local information on the spatial distribution of hydrothermal alteration zones associated with epithermal gold mineralization at the Somún Curá Massif, Patagonia, Argentina. Three major areas of alteration were clearly recognized: Cerro La Mina, Cerro Abanico, and Aguada de Guerra. They identified alteration zones such as advanced argillic, argillic, and silicic with satisfactory correlation with their field spectroscopy data. Zhang and Pazner (2007) employed MF method to EO-1 Hyperion and ASTER data to extract abundance images for gold-associated lithological mapping in southeastern Chocolate Mountain, California, USA. The assessment of MF score index indicated that the ASTER data have good capability in the discrimination and classification of rock types. Although the Hyperion data can produce better accuracy than ASTER data, the lithologic information extracted from ASTER image data is mostly similar to Hyperion results. The better availability and vast spatial coverage of ASTER data make it more suitable for regional-scale lithological mapping. Di Tommaso and Rubinstein (2007) used band ratios, certain color band combinations, and the SAM method for mapping hydrothermal alteration minerals associated with Infiernillo porphyry copper deposit using ASTER data covering the San Rafael Massif, southern Mendoza Province, Argentina. They detected illite, kaolinite, sericite, and jarosite through spectral analysis of SWIR bands, and surface silica and potassic alteration using ASTER TIR bands.

Gad and Kusky (2007) resolved geological mapping problems by using new ASTER band ratio images 4/7, 4/6, and 4/10 for lithological mapping in the Arabian–Nubian Shield, the Neoproterozoic Wadi Kid area, Sinai, Egypt. These ASTER band ratios mapped the main rock units consisting of gneiss and migmatite, amphibolite, volcanogenic sediments with banded iron formation, metapelites, talc schist, metapsammites, metaacidic volcanics, meta-pyroclastics volcaniclastics, albitites, and granitic rocks. Zhang et al. (2007) evaluated ASTER surface reflectance (AST-07) data for gold-related lithologic mapping and alteration mineral detection in the south Chocolate Mountains area, CA, USA. They applied PCA transformation to mineralogic indices, including the OH-bearing altered mineral Index (OHI), kaolinite index (KLI), alunite index (ALI), and calcite index (CLI) for delineating alteration zones. The CEM technique (a subpixel unmixing algorithm) was used to detect alunite, kaolinite, muscovite, and montmorillonite fractional abundances using ASTER VNIR and SWIR surface reflectance data and reference spectra from the ASTER spectral library (Baldridge et al., 2009). Moghtaderi et al. (2007) used ASTER data for distinguishing sodic–calcic, potassic, and silicic–phyllic alteration patterns associated with hydrothermal iron oxide deposits in the Chadormalu paleocrater, Bafq region, Central Iran. The alteration minerals were identified by using false color composite (FCC), decorrelation-stretch, MNF transform, correlated filter, and mathematical evaluation method (MEM) to ASTER data.

Kruse and Perry (2007) integrated ASTER multispectral data with the Airborne Visible/Infrared Imaging Spectrometer (AVIRIS) and EO-1 Hyperion hyperspectral data to extend hyperspectral signatures to regional scales mineral mapping and environmental monitoring in northern death-valley, south-central California/Nevada, USA. The AIG-Developed hyperspectral analysis approach was applied to ASTER data. Their results indicated that the AIG methods are not only a way to analyze hyperspectral data but can achieve accurate results when selectively employed on ASTER multispectral data. Moreover, AIG methods can also provide a consistent way to extract spectral information from hyperspectral and multispectral data without prior knowledge or requiring ground observations. Rockwell and Hofstra (2008) used ASTER TIR emissivity data for identifying quartz and carbonate minerals in northern Nevada, USA. A QI and a CI were implemented using ASTER Level-2 surface emissivity

(Level-2B04) data for geologic mapping and mineral resource investigations. They concluded that mapping hydrothermal quartz and carbonate rocks at regional and local scales have considerable economical attention for ore deposit exploration because these rocks can host rock a wide range of metallic ore deposit types.

Gersman et al. (2008) identified hydrothermally altered rocks and a Percambrian metamorphic sequence at and around the Alid volcanic dome, at the northern Danakil Depression, Eritrea using Hyperion data. They discriminated the different types of rock groups by using unsupervised and supervised classification approaches. The ability of the Hyperion to detect ammonium spectral signatures was reported. The existence of ammonium in hydrothermally altered rocks within the Alid dome has been confirmed by previous studies. Amer et al. (2010) suggested new ASTER band ratios $(2+4)/3$, $(5+7)/6$, and $(7+9)/8$ for mapping ophiolitic rocks (serpentinites, metagabbros, and metabasalts) in the Central Eastern Desert of Egypt. PCA was also applied for discriminating between ophiolitic rocks and gray granite and pink granite. The achieved results from field works verified the accuracy and potential of these methods for lithological mapping in arid and semiarid regions. Aboelkhair et al. (2010) used ASTER Level-1B "radiance-at-sensor" and Level-2B04 "surface emissivity" TIR bands for mapping albite granite in the Central Eastern Desert of Egypt. Running band ratio, band combinations, and QI allowed the discrimination of albite granite from the other rock types in the study area.

Gabr et al. (2010) detected areas of high potential gold mineralization using ASTER surface reflectance (AST-07) data covering Abu Marawat, North-Eastern Desert of Egypt. Spectral discrimination between high-potential and low-potential areas of gold mineralization was recognized by using PCA-transformed mineral indices, color ratio images (e.g., 4/8, 4/2, 8/9 displayed as RGB), CEM, and SAM methods. The results of their field investigations proved the accuracy of their image-processing results. Mars and Rowan (2010) assessed two new ASTER SWIR surface reflectance data products, namely, RefL1b and AST-07XT for a spectroscopic mapping of rocks and minerals. Their results indicated that the new ASTER products are more capable than previous ASTER products for discriminating hydrothermal alteration minerals and mineral groups without the use of additional spectral data from the site for calibration. Mars and Rowan (2011) used ASTER Level-1B "radiance-at-sensor" data for lithologic mapping of the Khanneshin carbonatite volcano, southwest of Kandahar, Afghanistan. They used false-color composite image, band ratio, logical operator algorithms, and MF methods to VNIR-SWIR and TIR bands of ASTER. Quaternary carbonate rocks within the volcano were identified and discriminated from Neogene ferruginous polymitic and argillite rocks. Their results showed the distribution of calcitic and ankeritic carbonatites, agglomerates, contact metamorphosed rocks, argillic and sandstone, and iron-rich sandstone using VNIR-SWIR bands. Widespread silica and carbonate rocks, mafic-rich rock, and sediment were identified using TIR bands. Results provided an image-based map of rocks and minerals that are consistent with the available geologic map of the study area.

Bishop et al. (2011) used Hyperion and ASTER data for mineral mapping in the Pulang, Yunnan Province, China. They used ASTER data to locate target areas characterized by hydrothermal alteration minerals and Hyperion data for detailed mineral mapping. PCA and band ratioing methods were applied to ASTER data to detect target areas characterized by argillic alteration, iron oxides, and sulfate minerals. SAM and MTMF were implemented on Hyperion data to discriminate mineral species in the target areas. Iron oxide minerals consisting of hematite, goethite, limonite, and jarosite were detected using VNIR bands of Hyperion. Sericite, kaolinite, montmorillonite, muscovite, and illite were discriminated using SWIR bands of Hyperion. Results indicated that the combination of multispectral and hyperspectral data can be advantageous for mineral exploration in remote areas with limited or unavailable primary information. Ranjbar et al. (2011) evaluated the ASTER, ETM^+, and airborne magnetic-radiometric data for hydrothermal alteration mapping at the Sar Cheshmeh porphyry copper deposit, in southeastern Iran. They used PCA, band ratio, and the SAM methods to map hydrothermally altered rocks. The results showed that ASTER SWIR–derived images enhanced hydrothermally altered rocks using PCA (PCs 2 and 3) and band ratios (4/9 and 7/6) methods. SAM classification image detected sericite, chlorite, and calcite with a total accuracy of 71.3%. ETM^+ data were used to enhance iron oxide-rich areas using the PC_5 image. The potassic alteration was recognized well using airborne magnetic-radiometric data.

Haselwimmer et al. (2011) used ASTER data for lithological mapping in the Oscar Coast area, Graham Land, Antarctic Peninsula. MF method was applied to ASTER Level-1B and the standard TIR emissivity (AST05) data to discriminate the major lithologic groups within the study area as well as delineation of hydrothermal alteration zones. The results have shown the discrimination of most of the major lithologic units, and the delineation of propylitic and argillic alteration zones associated with volcanic rocks. The outcomes have enabled important revisions to the existing geological map of the study area. Bedini (2011) used ASTER and HyMap data for mineral and lithological mapping in the Kap Simpson complex, central East Greenland. MF algorithm was applied to map jarosite, ferric oxides, and Al–OH clays minerals using ASTER VNIR/SWIR data. Lithological units have been identified by applying a color composite of the ASTER TIR bands. The integration of the results with HyMap data produced useful information for mineral exploration activities in the Arctic regions of central East Greenland. Amer et al. (2012) used ASTER data for mapping lithology and gold-related alteration zones in the Um Rus area, Central Eastern Desert of Egypt. They applied PCA and band ratioing on VNIR + SWIR bands of ASTER to discriminate lithological units. SAM and Spectral Information Divergence (SID) classification methods were used to detect alteration minerals consisting of sericite, calcite, and clay minerals associated with mineralized granodiorite. Their field verification work indicated that the image-processing methods were capable of lithological and alteration mineral mapping. Zoheir and Emam (2012) used band ratioing, PCA, false-color composition (FCC), and frequency filtering (FFT-RWT) of ASTER and ETM^+ data to improve the visual interpretation for detailed mapping of the Gebel Egat area in South Estern Desert of Egypt. By compiling field, petrographic and spectral data, controls on gold mineralization have been assessed in terms of the association of gold lodes with particular lithological units and structures.

Pour and Hashim (2014a) used PALSAR data for structural geology mapping in the Bau gold mining district, Sarawak, Malaysia. Geological analyses combined with the PALSAR data were used to detect structural elements associated with gold mineralization. The PALSAR data were used to perform lithological–structural mapping of mineralized zones in the study area and surrounding terrain. Structural elements were detected along the SSW to NNE trend of the Tuban fault zone and Tai Parit fault that corresponds to the areas of occurrence of the gold mineralization in the Bau Limestone. Most of quartz–gold-bearing veins occur in high-angle faults, fractures, and joints within massive units of the Bau Limestone. The results show that four deformation events (D1–D4) in the structures of the Bau district and structurally controlled gold mineralization indicators, including faults, joints, and fractures, are detectable using PALSAR data at both regional and district scales. Pour et al. (2016) investigated the Bentong-Raub Suture Zone (BRSZ) of Peninsular Malaysia using PALSAR remote sensing data for sediment-hosted/orogenic gold mineral systems exploration. Major structural lineaments such as the Bentong-Raub Suture Zone (BRSZ) and Lebir Fault Zone, ductile deformation related to crustal shortening, brittle disjunctive structures (faults and fractures), and collisional mountain range (Main Range granites) were detected and mapped at regional scale using PALSAR ScanSAR data. Gold-mineralized trend lineaments are associated with the intersection of N–S, NE–SW, NNW–SSE, and ESE–WNW faults and curvilinear features in shearing and alteration zones. Compressional tectonic structures such as the NW–SE trending thrust, ENE–WSW-oriented faults in mylonite and phyllite, recumbent folds and asymmetric anticlines in argillite are high potential zones for gold prospecting in the Central Gold Belt of Peninsular Malaysia.

Ahmadirouhani et al. (2018) integrated the information extracted from SPOT-5 and ASTER data, field and mineralogy studies to map phyllic alteration zone associated with NW–SE structural fractures to explore Cu–Fe–Au vein-type mineralization in the Bajestan region, the Lut block, east Iran. The fractal pattern was calculated for the fractures map using the Box-Counting algorithm to the SPOT-5 data. Statistical parameters of fractures, such as density, intensity, and fractures' intersection were also determined. Band composition, specialized band ratio, and SAM classification methods were implemented in the ASTER dataset for detecting hydrothermal alteration zones, such as propylitic, phyllic, argillic, and gossan. Results indicate that the maximum value of the fractal dimension, intensity, density, and the intersection of the fractures are concentrated in the NW and SE parts of SPOT image maps. On the other hand, phyllic alteration zone containing sericite, alunite, kaolinite, and jarosite mineral assemblages was also identified in several zones of the NW and SE parts of the ASTER image maps. Integration of the results indicates the high potential zones for the occurrence of Cu–Fe–Au mineralization in the Bajestan region.

Pour et al. (2018b) used Landsat 8, PALSAR, and ASTER data for exploring zinc in the trough sequences and shelf-platform carbonate of the Franklinian Basin, North Greenland. A series of robust image-processing algorithms were implemented for detecting the spatial distribution of pixels/subpixels related to key alteration mineral assemblages and structural features that may represent potential undiscovered Zn–Pb deposits. Fusion of directed PCA (DPCA) and ICA was applied to some selected Landsat 8 mineral indices for mapping gossan, clay-rich zones, and dolomitization. Major lineaments, intersections, curvilinear structures, and sedimentary formations were traced by the application of feature-oriented principal components selection (FPCS) to cross-polarized backscatter PALSAR ratio images.

MTMF algorithm was applied to ASTER VNIR/SWIR bands for subpixel detection and classification of hematite, goethite, jarosite, alunite, gypsum, chalcedony, kaolinite, muscovite, chlorite, epidote, calcite, and dolomite in the prospective targets. Several high-potential zones characterized by distinct alteration mineral assemblages and structural fabrics were identified.

Sheikhrahimi et al. (2019) mapped hydrothermal alteration zones and lineaments associated with orogenic gold mineralization in the Sanandaj-Sirjan Zone, Iran using ASTER data. Image transformation techniques such as specialized band ratioing and PCA are used to delineate lithological units and alteration minerals. Supervised classification techniques, namely, SAM and spectral information divergence (SID) are applied to detect subtle differences between indicator alteration minerals associated with ground-truth gold locations in the area. The directional filtering technique is applied to help in tracing the strike of the different linear structures. The structural intersections that are associated with zones rich in jarosite, epidote, chlorite, hematite, and muscovite are considered the high-potential zones in the study area.

Pour et al. (2019a) used Landsat 8, ASTER, and WorldView-3 data for hydrothermal alteration mapping and prospecting copper–gold mineralization in the Northeastern Inglefield Mobile Belt (IMB), Northwest Greenland. DPCA technique was applied to map iron oxide/hydroxide, Al/Fe–OH, Mg–Fe–OH minerals, silicification (Si–OH), and SiO_2 mineral groups using specialized band ratios of the multispectral datasets. For extracting reference spectra directly from the Landsat 8, ASTER, and WorldView-3 data s to generate fraction images of end-member minerals, the automated spectral hourglass (ASH) approach was implemented. LSU algorithm was thereafter used to produce a mineral map of fractional images. Furthermore, the ACE algorithm was applied to the VINR + SWIR bands of ASTER using laboratory reflectance spectra extracted from the USGS spectral library for verifying the presence of mineral spectral signatures. Results indicate that the boundaries between the Franklinian sedimentary successions and the Etah metamorphic and metaigneous complex, the orthogneiss in the northeastern part of the Cu–Au mineralization belt adjacent to Dallas Bugt, and the southern part of the Cu–Au mineralization belt nearby Marshall Bugt show a high content of iron oxides/hydroxides and Si–OH/SiO_2 mineral groups, which warrant high potential for Cu–Au prospecting.

Pour et al. (2019b) used ASTER data to detect listvenite occurrences and alteration mineral assemblages in the poorly exposed damage zones of the boundaries between the Wilson, Bowers, and Robertson Bay terranes in Northern Victoria Land (NVL), Antarctica. Spectral information for detecting alteration mineral assemblages and listvenites were extracted at pixel and subpixel levels using PCA/ICA fusion technique, LSU, and CEM algorithms. Mineralogical assemblages containing Fe^{2+}, Fe^{3+}, Fe–OH, Al–OH, Mg–OH, and CO_3 spectral absorption features were detected. Silicate lithological groups were mapped and discriminated using PCA/ICA fusion to TIR bands of ASTER. Fraction images of prospective alteration minerals, including goethite, hematite, jarosite, biotite, kaolinite, muscovite, antigorite, serpentine, talc, actinolite, chlorite, epidote, calcite, dolomite, and siderite and possible zones encompassing listvenite occurrences were produced. Several potential zones for listvenite occurrences were identified, typically in association with mafic metavolcanic rocks (Glasgow Volcanics) in the Bowers Mountains.

Pour et al. (2019c) investigated the applications of Landsat 8 and ASTER data to extract geological information for lithological and alteration mineral mapping in poorly exposed lithologies located in inaccessible regions of the northeastern Graham Land, Antarctic Peninsula (AP). In this study, a two-stage methodology was implemented to differentiate pixel and subpixel targets in the satellite images. In the first stage, the CR spectral mapping tool and ICA techniques were applied to Landsat 8 and ASTER spectral bands to map the pixels related to poorly exposed lithological units. The second step was accomplished based on the application of target detection algorithms such as CEM, orthogonal subspace projection (OSP), and ACE to SWIR bands of ASTER for detecting spectral features attributed to alteration mineral assemblages at the subpixel level. Pixels composed of distinctive absorption features of alteration mineral assemblages and Si—O bond emission minima features were detected by applying the CR mapping tool to reflective and thermal bands of Landsat 8 and ASTER. Anomaly pixels related to spectral features of Al—O—H, Fe, Mg—O—H, and CO_3 groups as well as lithological attributions from felsic to mafic rocks were detected by the implementation of the ICA technique to reflective and thermal bands of Landsat 8 and ASTER. ICA method provided image maps of alteration mineral assemblages and lithological units (mafic to felsic trend) for poorly mapped and/or unmapped regions. A fractional abundance of alteration minerals, such as muscovite, kaolinite, illite, montmorillonite, epidote, chlorite, and biotite, were detected in poorly exposed lithologies using target detection algorithms. Several prospecting areas for Cu, Mo, Au, and Ag mineralization related to propyllitically and argillically altered units of Andean Intrusive Suite (AIS) were identified in the southern sector of the study region.

Zoheir et al. (2019a) used Sentinel-1, PALSAR, ASTER, and Sentinel-2 datasets to explicate the regional structural control of gold mineralization in the Barramiya—Mueilha area in the Central Eastern Desert of Egypt. Feature-oriented Principal Components Selection (FPCS) applied to polarized backscatter ratio images of Sentinel-1 and PALSAR datasets show appreciable capability in tracing along the strike of regional structures and identification of potential dilation loci. The PCA, band combination, and band ratioing techniques were applied to the ASTER and Sentinel-2 datasets for lithological and hydrothermal alteration mapping. Results demonstrated that gold occurrences are confined to major zones of fold superimposition and transpression along flexural planes in the foliated ophiolite-island arc belts. The radar and multispectral satellite data abetted a better understanding of the structural framework and unraveled settings of the scattered gold occurrences in the study area.

Zoheir et al. (2019b) used ASTER, Sentinel-2, Sentinel-1, and PALSAR remote sensing data for interpreting the setting and controls of gold-bearing quartz veins in the central part of the greater Wadi Allaqi district in the South Eastern Desert of Egypt. The ASTER VNIR + SWIR band combinations and band rationing techniques were used to enhance the spectral differences between bands and to enable detailed lithological mapping. The Sentinel-2 data promote lithological mapping of the gold mine areas. Band combinations and ratios of Sentinel-2 were used for discriminating the rock units and delineating the mineralized alteration zones. The Sentinel-1 and PALSAR data have significantly enhanced structural mapping and lineaments extraction. Zoheir et al. (2019c) used Landsat 8, ASTER, PALSAR, and Sentinel-1B data coupled with field and microscopic investigations to unravel the setting and controls of gold mineralization in the Wadi Beitan–Wadi Rahaba area in the South Eastern Desert of Egypt. Band rationing, RABD, PCA, ICA, directional filtering, and automated and semiautomated lineament extraction techniques were implemented. The results reveal no particular spatial association between gold occurrences and certain lithological units but show a preferential distribution of gold–quartz veins in zones of chlorite–epidote alteration overlapping with high-density intersections of lineaments.

Takodjou Wambo et al. (2020) used Landsat 8 and ASTER remote sensing data to identify and map high potential zones of gold mineralization in the Ngoura-Colomines goldfield, eastern Cameroon a subtropical region. The PCA, ICA, and specialized spectral band ratios were used to extract spectral information related to vegetation, iron oxide/hydroxide minerals, Al–OH, Fe–Mg–OH, carbonate group minerals, and silicification. The LSU algorithm was implemented to ASTER VNIR + SWIR bands for detailed discrimination of hematite, jarosite, kaolinite, muscovite, chlorite, and epidote at the district scale. The ASH technique was employed to extract reference spectra directly from the ASTER bands for producing fraction images of end-members using the LSU. Several hydrothermal alteration zones of iron oxide/hydroxide, clay, carbonate minerals, and silicification zones were identified, and high-potential prospects were also delineated, including the Ngoura-Colomines prospects and the newly discovered Yangamo-Ndatanga and Tapare-Tapondo prospects.

Sekandari et al. (2020) integrated Landsat 8, Sentinel-2, ASTER, and WorldView-3 datasets were used for prospecting Zn–Pb mineralization in the central part of the Kashmar–Kerman Tectonic Zone (KKTZ), the Central Iranian Terrane (CIT). Band ratios and PCA techniques were adopted and implemented. Fuzzy logic modeling was applied to integrate the thematic layers produced by image processing techniques for generating mineral prospectivity maps of the study area. The most favorable/prospective zones for hydrothermal ore mineralizations and carbonate-hosted Pb–Zn mineralization in the study region were particularly mapped and indicated. Ngassam Mbianya et al. (2021) used VNIR and SWIR bands of Landsats 8 and 7 (ETM +) images and field data for lineaments and hydrothermal alterations mapping in tropical mining goldfield of Ketté in Cameroon. PCA, MNF transformation, and band ratios were applied to map alteration minerals. The result revealed NE–SW to ENE–WSW main trends. Gold mineralization occurrences are spatially associated with WNW–ESE to NW–SE lineaments/faults and show a strong correlation with medium-to-high lineament density zones. Hydrothermal alteration minerals are spatially associated closely with gold occurrences and known mining sites that are structurally controlled by the NE–SW to ENE–WSW shear zone.

Ishagh et al. (2021) used Landsat 8 and ASTER data to map lithological units and alteration minerals, aiming at uranium exploration in the Bir Lemjed area of the eastern Paleoproterozoic domain, the Reguibat Shield, north Mauritania. Band ratios, PCA, MNF, and maximum likelihood classification were exclusively applied to Landsat 8 data for mapping lithological units and alteration zones in the study area. Besides, specialized band ratios and LSU algorithm were executed to ASTER VNIR + SWIR bands to detect alteration minerals, particularly. Some zones contain a high surface abundance of iron oxides, and Mg, Fe–OH, and CO_3 minerals were identified, which are typically associated with calcrete deposits in the study area. Abd El-Wahed et al. (2021) processed Landsat 8 OLI, ASTER, and PALSAR remote sensing data using band combinations, BM, PCA, decorrelation stretch and mineralogical indices for identification of shear-related gold ores in the Wadi Hodein Shear Belt, South Eastern Desert of Egypt. Gold-mineralized shear zones cut heterogeneously deformed ophiolites and metavolcaniclastic rocks and attenuate in and around granodioritic intrusions. Shallow NNW- or SSE-plunging mineral and stretching lineations on steeply dipping shear planes depict a considerably simple shear component. The results of image-processing complying with field observations and structural analyses suggest that the coincidence of shear zones, hydrothermal alteration, and crosscutting dikes in the study area could be considered a model criterion in the exploration of new gold targets.

Guha et al. (2021) evaluated the potential of Airborne Hyperspectral AVIRIS-NG Data for the exploration of base-metal deposits in Bhilwara, Rajasthan. The spectral bands of AVIRIS-NG data were processed for delineating the surface signatures associated with the base–metal mineralization. The FCCs of different RBD images were used for delineating alteration minerals. These minerals are typically associated with the surface alteration resulting from the ore-bearing fluid migration or associated with the redox-controlled supergene enrichments of the ore deposit. The results show that the AVIRIS-NG image can delineate surface signatures of mineralization in 1:10,000 to 1:15,000 scales to narrow down the targets for detailed exploration.

Pour et al. (2021a,b) evaluated the capability of ASTER data for mapping and discrimination of phyllosilicate mineral groups in the Antarctic environment of northern Victoria Land. The MTMF and CEM algorithms were used to detect the subpixel abundance of Al-rich, Fe^{3+}-rich, Fe^{2+}-rich, and Mg-rich phyllosilicates using VNIR, SWIR, and TIR bands of ASTER. Results indicate that Al-rich phyllosilicates are strongly detected in the exposed outcrops of the Granite Harbour granitoids, Wilson Metamorphic Complex, and the Beacon Supergroup. The presence of the smectite mineral group derived from the Jurassic basaltic rocks (Ferrar Dolerite and Kirkpatrick Basalts) by weathering and decomposition processes implicates Fe^{3+}- and Fe^{2+}-rich phyllosilicates. Biotite (Fe^{2+}-rich phyllosilicate) is detected associated with the Granite Harbour granitoids, Wilson Metamorphic Complex, and Melbourne Volcanics. Mg-rich phyllosilicates are mostly mapped in the scree, glacial drift, moraine, and crevasse fields derived from weathering and decomposition of the Kirkpatrick Basalt and Ferrar Dolerite. Chlorite (Mg-rich phyllosilicate) was generally mapped in the exposures of Granite Harbour granodiorite and granite and partially identified in the Ferrar Dolerite, the Kirkpatrick Basalt, the Priestley Formation, and Priestley Schist and the scree, glacial drift, and moraine. Ground truth with detailed geological data, petrographic study, and XRD analysis verified the remote sensing results. Consequently, an ASTER image map of phyllosilicate minerals is generated for the Mesa Range, Campbell, and Priestley Glaciers, northern Victoria Land of Antarctica.

Sekandari et al. (2022) used ASTER and WorldView-3 datasets for mapping lithological units and hydrothermal alteration zones associated with Pb–Zn mineralization in the

Kerman–Kashmar Tectonic Zone (KKTZ), Iran. The VNIR, SWIR, and TIR bands of ASTER were used to map iron oxide/hydroxides, Al–OH minerals, Fe, Mg–OH minerals, quartz, and carbonate minerals. The VNIR bands of WV-3 were used to discriminate Fe^{3+} and Fe^{2+} absorption intensities. Selective PCA (SPCA), SAM, LSU, and Automatic Lineament Extraction techniques were implemented. Lithological units were discriminated based on Al/Fe–OH, Fe^{2+}/Fe^{3+}, and Mg–Fe–OH/CO_3 absorption properties. The spatial distribution of hematite, goethite, jarosite, gypsum, calcite, dolomite, kaolinite, and muscovite were comprehensively detected. Some prospective zones were identified in the intersection of N–S, NW–SE, and NE–SW trending fault systems, gossan, argillic/phyllic, and dolomitic units.

Yousefi et al. (2022) evaluated Hyperion data to map alteration zones of porphyry copper systems using the Dirichlet process Stick-Breaking model-based clustering algorithm. The DPSB clustering algorithm was implemented and subsequently compared with the *k*-means algorithm, CLARA clustering, hierarchical clustering, Gaussian finite mixture model (GFMM), Gaussian model for high-dimensional (GMHD), and spectral clustering as well as spectral angle mapping (SAM). Results derived from the DPSB model-based clustering algorithm show 88.6% accuracy in distinguishing propylitic, argillic, advanced argillic, propylitic–argillic, and phyllic alteration zones. Maleki et al. (2022) integrated ASTER remote sensing data with geological, structural, geophysical, and geochemical datasets for gold prospectivity mapping in the Godarsorkh area, Central Iran. The evidential belief functions (EBFs) algorithm was executed to fuse the evidential layers. Results show that the orogenic gold mineralization occurred in shear zones, which are mainly controlled by tectonic structures. Therefore analyzing the structural features can be considered a key feature for identifying potential orogenic gold mineralization zones in the study area.

Shirazi et al. (2022) developed the neuro-fuzzy-AHP (NFAHP) technique for fusing ASTER alteration mineral image maps, lithological map, geochronological map, structural map, and geochemical map to identify high-potential zones of VMS copper mineralization in the Sahlabad mining area, east Iran. Argillic, phyllic, propylitic, and gossan alteration zones were identified in the study area using band ratio and SPCA methods implemented to ASTER VNIR and SWIR bands. For each of the copper deposits, old mines, and mineralization indices in the study area, information related to exploration factors such as ore mineralization, host-rock lithology, alterations, geochronological, geochemistry, and distance from high-intensity lineament factor communities were investigated. Subsequently, the predictive power of these factors in identifying copper occurrences was evaluated using Back propagation neural network (BPNN) technique. The BPNN results demonstrated that using the exploration factors, copper mineralizations in the Sahlabad mining area could be identified with high accuracy. Lastly, using the fuzzy-analytic hierarchy process (fuzzy-AHP) method, information layers were weighted and fused. As a result, a potential map of copper mineralization was generated, which pinpointed several high-potential zones in the study area.

Aali et al. (2022) fused Sentinel-2 and Landsat 7, aerial magnetic geophysical data, and geological data for identifying polymetallic mineralization potential zones in the Chakchak region, Yazd province, Iran. Hydrothermal alteration mineral zones and surface and deep intrusive masses, hidden faults and lineaments, and lithological units were detected using remote sensing, aerial magnetic, and geological data, respectively. The exploratory/information layers were fused using fuzzy logic modeling and the multi-class index overlap method. Subsequently, mineral potential maps were generated for the study area. Some high-potential zones of polymetallic mineralization were identified and verified through a detailed field campaign and drilling programs in the Chakchak region.

References

Aali, A.A., Shirazy, A., Shirazi, A., Pour, A.B., Hezarkhani, A., Maghsoudi, A., et al., 2022. Fusion of remote sensing, magnetometric, and geological data to identify polymetallic mineral potential zones in Chakchak region, Yazd. Iran. Remote Sens. 14 (23), 6018. Available from: https://doi.org/10.3390/rs14236018.

Abd El-Wahed, M.A., Harraz, H., El-Behairy, M.H., 2016. Transpressional imbricate thrust zones controlling gold mineralization in the Central Eastern Desert of Egypt. Ore Geol. Rev. 78, 424–446. Available from: https://doi.org/10.1016/j.oregeorev.2016.03.022; http://www.sciencedirect.com/science/journal/01691368.

Abdelkader, M.A., Watanabe, Y., Shebl, A., El-Dokouny, H.A., Dawoud, M., Csámer, Á., 2022. Effective delineation of rare metal-bearing granites from remote sensing data using machine learning methods: a case study from the Umm Naggat Area, Central Eastern Desert, Egypt. Ore Geol. Rev. 150. Available from: https://doi.org/10.1016/j.oregeorev.2022.105184; http://www.sciencedirect.com/science/journal/01691368.

Abdelsalam, M.G., Stern, R.J., Berhane, W.G., 2000. Mapping gossans in arid regions with Landsat TM and SIR-C images: the Beddaho Alteration Zone in northern Eritrea. J. Afr. Earth Sci. 30 (4), 903–916. Available from: https://doi.org/10.1016/S0899-5362(00)00059-2.

Aboelkhair, H., Ninomiya, Y., Watanabe, Y., Sato, I., 2010. Processing and interpretation of ASTER TIR data for mapping of rare-metal-enriched albite granitoids in the Central Eastern Desert of Egypt. J. Afr. Earth Sci. 58 (1), 141–151. Available from: https://doi.org/10.1016/j.jafrearsci.2010.01.007.

Abrams, M., 2000. The Advanced Spaceborne Thermal Emission and Reflection Radiometer (ASTER): data products for the high spatial resolution imager on NASA's Terra platform. Int. J. Remote Sens. 21 (5), 847–859. Available from: https://doi.org/10.1080/014311600210326.

Abrams, M.J., Brown, D., 1984. The American Association of Petroleum Geologists, The Joint NASA–Geosat Test Case Project. Final Report, pp. 4–73.

Abrams, M., Hook, S.J., 1995. Simulated aster data for geologic studies. IEEE Trans. Geosci. Remote Sens. 33 (3), 692–699. Available from: https://doi.org/10.1109/36.387584.

Abrams, M., Hook, S., Ramachandran, B., 2004. ASTER User Handbook, Version 2. Jet Propulsion Laboratory, California Institute of Technology. Available from: http://asterweb.jpl.nasa.gov/content/03_data/04_Documents/aster_guide_v2.pdf.

Abrams, M., Yamaguchi, Y., 2019. Twenty years of ASTER contributions to lithologic mapping and mineral exploration. Remote Sens. 11, 1394.

Abrams, M.J., Brown, D., Lepley, L., Sadowski, R., 1983. Remote sensing for porphyry copper deposits in southern Arizona. Econ. Geol. 78 (4), 591–604. Available from: https://doi.org/10.2113/gsecongeo.78.4.591.

Adams, J.B., Smith, M.O., Gillespie, A.R., 1993. Remote Geochemical Analysis: Elemental and Mineralogical Composition 145–166. Cambridge University Press Imaging Spectroscopy: Interpretation based on Spectral Mixture Analysis.

Adams, J.B., Sabol, D.E., Kapos, V., Almeida Filho, R., Roberts, D.A., Smith, M.O., et al., 1995. Classification of multispectral images based on fractions of endmembers: application to land-cover change in the Brazilian Amazon. Remote Sens. Environ. 52 (2), 137–154. Available from: https://doi.org/10.1016/0034-4257(94)00098-8.

Adão, T., Hruška, J., Pádua, L., Bessa, J., Peres, E., Morais, R., et al., 2017. Hyperspectral imaging: a review on UAV-based sensors, data processing and applications for agriculture and forestry. Remote Sens. 9 (11). Available from: https://doi.org/10.3390/rs9111110; http://www.mdpi.com/2072-4292/9/11/1110/pdf.

Adiri, Z., El Harti, A., Jellouli, A., Lhissou, R., Maacha, L., Azmi, M., et al., 2017. Comparison of Landsat-8, ASTER and Sentinel 1 satellite remote sensing data in automatic lineaments extraction: A case study of Sidi Flah-Bouskour inlier, Moroccan Anti Atlas. Adv. Space Research. 60 (11), 2355–2367. Available from: https://doi.org/10.1016/j.asr.2017.09.006; http://www.journals.elsevier.com/advances-in-space-research/.

Ahmadirouhani, R., Rahimi, B., Karimpour, M.H., Malekzadeh Shafaroudi, A., Afshar Najafi, S., Pour, A.B., 2017. Fracture mapping of lineaments and recognizing their tectonic significance using SPOT-5 satellite data: A case study from the Bajestan area, Lut Block, east of Iran. J. Afr. Earth Sciences. 134, 600–612. Available from: https://doi.org/10.1016/j.jafrearsci.2017.07.027; http://www.sciencedirect.com/science/journal/1464343X.

Ahmadirouhani, R., Karimpour, M.H., Rahimi, B., Malekzadeh-Shafaroudi, A., Pour, A.B., Pradhan, B., 2018. Integration of SPOT-5 and ASTER satellite data for structural tracing and hydrothermal alteration mineral mapping: implications for Cu–Au prospecting. Int. J. Image Data Fusion. 9 (3), 237–262. Available from: https://doi.org/10.1080/19479832.2018.1469548; http://www.tandfonline.com/toc/tidf20/current.

Alimuddin, I., Sumantyo, J.T.S., Kuze, H., 2012. Assessment of pan-sharpening methods applied to image fusion of remotely sensed multi-band data. Int. J. Appl. Earth Obs. Geoinf. 18, 165–175.

Alsop, G.I., Holdsworth, R.E., 2004. Shear zone folds: Records of flow perturbation or structural inheritance? Geol. Soc. Spec. Publ. 224, 177–199. Available from: https://doi.org/10.1144/gsl.sp.2004.224.01.12; http://sp.lyellcollection.org/.

Alsop, G.I., Holdsworth, R.E., 2012. The three dimensional shape and localisation of deformation within multilayer sheath folds. J. Struct. Geol. 44, 110–128. Available from: https://doi.org/10.1016/j.jsg.2012.08.015.

Alsop, G.I., Weinberger, R., Marco, S., Levi, T., 2021. Detachment fold duplexes within gravity-driven fold and thrust systems. J. Struct. Geology. 142. Available from: https://doi.org/10.1016/j.jsg.2020.104207; http://www.sciencedirect.com/science/journal/01918141.

Alvey, B., Zare, A., Cook, M., Ho, D.K.C., 2016. Adaptive coherence estimator (ACE) for explosive hazard detection using wideband electromagnetic induction (WEMI). In: Proceedings of SPIE – The International Society for Optical Engineering, SPIE United States, vol. 9823, 1996756X. Available from: https://doi.org/10.1117/12.2223347; http://spie.org/x1848.xml.

Amer, R., Kusky, T., Ghulam, A., 2010. Lithological mapping in the Central Eastern Desert of Egypt using ASTER data. J. Afr. Earth Sci. 56 (2–3), 75–82. Available from: https://doi.org/10.1016/j.jafrearsci.2009.06.004.

Amer, R., Kusky, T., El Mezayen, A., 2012. Remote sensing detection of gold related alteration zones in Um Rus area, central eastern desert of Egypt. Adv. Space Res. 49 (1), 121–134. Available from: https://doi.org/10.1016/j.asr.2011.09.024; http://www.journals.elsevier.com/advances-in-space-research/.

Anderson, K., Gaston, K.J., 2013. Lightweight unmanned aerial vehicles will revolutionize spatial ecology. Front. Ecol. Environ. 11 (3), 138–146. Available from: https://doi.org/10.1890/120150; http://www.esajournals.org/doi/pdf/10.1890/120150. United Kingdom.

Anderson, G.P., Felde, G.W., Hoke, M.L., Ratkowski, A.J., Cooley, T., Chetwynd, J.H., et al., 2002. MODTRAN4-based atmospheric correction algorithm: FLAASH (fast line-of-sight atmospheric analysis of spectral hypercubes). In: Proceedings of SPIE – The International Society for Optical Engineering, December 12, 2002, United States, vol. 4725, pp. 65–71. Available from: https://doi.org/10.1117/12.478737.

Anguita, D., Ghio, A., Ridella, S., Sterpi, D., 2009. K-Fold Cross Validation for Error Rate Estimate in Support Vector Machines. DMIN 291–297.

Assiri, A.M., 2016. Remote Sensing Applications for Carbonatite Assessment and Mapping Using VNIR and SWIR Bands at Aluyaynah.

Attema, E., Bargellini, P., Edwards, P., Levrini, G., Lokas, S., Moeller, L., et al., 2007. The radar mission for GMES operational land and sea services. Eur. Space Agency Bull., 131, 10–17.

Baldridge, A.M., Hook, S.J., Grove, C.I., Rivera, G., 2009. The ASTER spectral library version 2.0. Remote Sens. Environ. 113, 711–715.

Bark, G., Weihed, P., 2012. Geodynamic settings for Paleoproterozoic gold mineralization in the Svecofennian domain: a tectonic model for the Fäboliden orogenic gold deposit, northern Sweden. Ore Geol. Rev. 48, 403–412. Available from: https://doi.org/10.1016/j.oregeorev.2012.05.007.

Barry, P.S., Pearlman, J., 2001. The EO-1 Misson: Hyperion data. National Aeronautics and Space Administration (NASA), pp. 35–40.

Barry, P., Pearlman, J.S., Jarecke, P., Folkman, M., 2002. Hyperion data collection: performance assessment and science application. IEEE Trans. Geosci. Remote Sens. 3, 1439–1499.

Baugh, W.M., Kruse, F.A., Atkinson, W.W., 1998. Quantitative geochemical mapping of ammonium minerals in the southern Cedar Mountains, Nevada, using the airborne visible/infrared imaging spectrometer (AVIRIS). Remote Sens. Environ. 65 (3), 292–308. Available from: https://doi.org/10.1016/S0034-4257(98)00039-X.

Beck, R., 2003. EO-1 User Guide v. 2.3. Department of Geography, University of Cincinnati.

Bedini, E., 2009. Mapping lithology of the Sarfartoq carbonatite complex, southern West Greenland, using HyMap imaging spectrometer data. Remote Sens. Environ. 113 (6), 1208–1219. Available from: https://doi.org/10.1016/j.rse.2009.02.007.

Bedini, E., 2011. Mineral mapping in the Kap Simpson complex, central East Greenland, using HyMap and ASTER remote sensing data. Adv. Space Res. 47 (1), 60–73. Available from: https://doi.org/10.1016/j.asr.2010.08.021; http://www.journals.elsevier.com/advances-in-space-research/.

Bedini, E., 2012. Mapping alteration minerals at Malmbjerg molybdenum deposit, central East Greenland, by Kohonen self-organizing maps and matched filter analysis of HyMap data. Int. J. Remote Sens. 33 (4), 939–961. Available from: https://doi.org/10.1080/01431161.2010.542202; https://www.tandfonline.com/loi/tres20.

Bedini, E., Chen, J., 2020. Application of PRISMA satellite hyperspectral imagery to mineral alteration mapping at Cuprite, Nevada, USA. J. Hyperspectral Remote Sens. 10 (2), 87. Available from: https://doi.org/10.29150/jhrs.v10.2.p87-94.

Bedini, E., Chen, J., 2022. Prospection for economic mineralization using PRISMA satellite hyperspectral remote sensing imagery: an example from central East Greenland. J. Hyperspectral Remote Sens. 12 (3), 124. Available from: https://doi.org/10.29150/2237-2202.2022.253484.

Bekkari, A., El, M., Idbraim, S., Mammass, D., Elhassouny, A., Ducrot, D., 2013. SVM classification of urban high-resolution imagery using composite kernels and contour information. Int. J. Adv. Comput. Sci. Appl. 4 (7). Available from: https://doi.org/10.14569/IJACSA.2013.040718.

Ben-Dor, E., Kruse, F.A., 1994. The relationship between the size of spatial subsets of GER 63 channel scanner data and the quality of the internal average relative reflectance (IARR) atmospheric correction technique. Int. J. Remote Sens. 15 (3), 683–690. Available from: https://doi.org/10.1080/01431169408954107.

Benhalouche, F.Z., Benabbou, O., Karoui, M.S., Kebir, L.W., Bennia, A., Deville, Y., 2022. Minerals detection and mapping in the Southwestern Algeria Gara-Djebilet Region with a multistage informed NMF-based unmixing approach using prisma remote sensing hyperspectral data. In: International Geoscience and Remote Sensing Symposium (IGARSS), January 1, 2022, Institute of Electrical and Electronics Engineers Inc. Algeria 2022, pp. 6422–6425. Available from: https://doi.org/10.1109/IGARSS46834.2022.9884746 9781665427920.

Berger, B.R., King, T.V.V., Morath, L.C., Phillips, J.D., 2003. Utility of high-altitude infrared spectral data in mineral exploration: application to Northern Patagonia Mountains, Arizona. Econ. Geol. 98 (5), 1003–1018. Available from: https://doi.org/10.2113/gsecongeo.98.5.1003.

Bhardwaj, A., Sam, L., Akanksha, Martín-Torres, F.J., Kumar, R., 2016. UAVs as remote sensing platform in glaciology: present applications and future prospects. Remote Sens. Environ. 175, 196–204. Available from: https://doi.org/10.1016/j.rse.2015.12.029; http://www.elsevier.com/inca/publications/store/5/0/5/7/3/3.

Bidon, S., Besson, O., Tourneret, J.Y., 2008. The adaptive coherence estimator is the generalized likelihood ratio test for a class of heterogeneous environments. IEEE Signal. Process. Lett. 15, 281–284. Available from: https://doi.org/10.1109/LSP.2007.916044.

Bierlein, F.P., Murphy, F.C., Weinberg, R.F., Lees, T., 2006. Distribution of orogenic gold deposits in relation to fault zones and gravity gradients: Targeting tools applied to the Eastern Goldfields, Yilgarn Craton, Western Australia. Mineralium Deposita 41 (2), 107–126. Available from: https://doi.org/10.1007/s00126-005-0044-4.

Biggar, S.F., Thome, K.J., McCorkel, J.T., D'Amico, J.M., 2005. Vicarious calibration of the ASTER SWIR sensor including crosstalk correction. In: Proceedings of SPIE – The International Society for Optical Engineering, December 12, 2005, United States, vol. 588. pp. 1–8. Available from: https://doi.org/10.1117/12.620090 0277786X.

Bishop, J.L., Murad, E., 2005. The visible and infrared spectral properties of jarosite and alunite. Am. Mineralogist. 90 (7), 1100–1107. Available from: https://doi.org/10.2138/am.2005.1700; https://www.degruyter.com/view/j/ammin.

Bishop, Y.M., Fienberg, S.E., Holland, P.W., 2007. Discrete Multivariate Analysis: Theory and Practice. Springer Science & Business Media.

Bishop, J.L., Lane, M.D., Dyar, M.D., Brown, A.J., 2008. Reflectance and emission spectroscopy study of four groups of phyllosilicates: smectites, kaolinite-serpentines, chlorites and micas. Clay Miner. 43 (1), 35–54. Available from: https://doi.org/10.1180/claymin.2008.043.1.03.

Bishop, C.A., Liu, J.G., Mason, P.J., 2011. Hyperspectral remote sensing for mineral exploration in Pulang, Yunnan province, China. Int. J. Remote Sens. 32 (9), 2409–2426. Available from: https://doi.org/10.1080/01431161003698336; https://www.tandfonline.com/loi/tres20.

Blenkinsop, T.G., Doyle, M.G., 2014. Structural controls on gold mineralization on the margin of the Yilgarn craton, Albany-Fraser orogen: the Tropicana deposit, Western Australia. J. Struct. Geol. 67, 189–204. Available from: https://doi.org/10.1016/j.jsg.2014.01.013; http://www.sciencedirect.com/science/journal/01918141.

Boardman, J.W., 1989. Inversion of imaging spectrometry data using singular value decomposition. Digest – Int. Geosci. Remote Sens. Symp. (IGARSS) 4, 2069–2072. Available from: https://doi.org/10.1109/igarss.1989.577779.

Boardman, J.W., 1992. Sedimentary Facies Analysis using Imaging Spectrometry: A Geophysical Inverse Problem: Unpublished Ph.

Boardman, J.W., 1998. Leveraging the High Dimensionality of AVIRIS Data for Improved sub-Pixel Target Unmixing and Rejection of False Positives: Mixture Tuned Matched Filtering Summaries of the Seventh JPL Airborne Geoscience Workshop. JPL Publication, pp. 55–56.

Boardman, J.W., Kruse, F.A., 1994. Automated spectral analysis: a geological example using AVIRIS data. In: 1994 Proceedings, ERIM Tenth Thematic Conference on Geologic Remote Sensing.

Boardman, J.W., Kruse, F.A., 2011. Analysis of imaging spectrometer data using N-dimensional geometry and a mixture-tuned matched filtering approach. IEEE Trans. Geosci. Remote Sens. 49 (11), 4138–4152. Available from: https://doi.org/10.1109/TGRS.2011.2161585.

Bolouki, S.M., Ramazi, H.R., Maghsoudi, A., Pour, A.B., Sohrabi, G., 2020. A remote sensing-based application of Bayesian networks for epithermal gold potential mapping in Ahar-Arasbaran area, NW Iran. Remote Sens. 12 (1). Available from: https://doi.org/10.3390/rs12010105; http://www.mdpi.com/journal/remotesensing.

Bons, P.D., Elburg, M.A., Gomez-Rivas, E., 2012. A review of the formation of tectonic veins and their microstructures. J. Struct. Geol. 43, 33–62. Available from: https://doi.org/10.1016/j.jsg.2012.07.005.

Campbel, J.B., 2007. Introduction to Remote Sensing. The Guilford Press.

Camps-Valls, G., Bruzzone, L., 2005. Kernel-based methods for hyperspectral image classification. IEEE Trans. Geosci. Remote Sens. 43 (6), 1351–1362. Available from: https://doi.org/10.1109/TGRS.2005.846154.

Cao, P., Zhu, Y., Zhao, W., Liu, S., Gao, H., 2019. Chromaticity measurement based on the image method and its application in water quality detection. Water. 11 (11), 2339. Available from: https://doi.org/10.3390/w11112339.

Cathles, L.M., 1981. Fluid flow and genesis of hydrothermal ore deposits. In: Economic Geology, Seventy-Fifth Anniversary Volume (1905–1980), pp. 424–457.

Chang, C.I., 2003. Hyperspectral Imaging: Techniques for Spectral Detection and Classification, vol. 1.

Chang, C.-I., Heinz, D.C., 2000. Constrained subpixel target detection for remotely sensed imagery. IEEE Trans. Geosci. Remote Sens. 38 (3), 1144–1159. Available from: https://doi.org/10.1109/36.843007.

Chang, C.I., Liu, J.M., Chieu, B.C., Ren, H., Wang, C.M., Lo, C.S., et al., 2000. Generalized constrained energy minimization approach to subpixel target detection for multispectral imagery. Opt. Eng. 39 (5), 1275–1281. Available from: https://doi.org/10.1117/1.602486.

Chavez, P.S., 1988. An improved dark-object subtraction technique for atmospheric scattering correction of multispectral data. Remote Sens. Environ. 24 (3), 459–479. Available from: https://doi.org/10.1016/0034-4257(88)90019-3.

Chavez, P., Guptill, S.C., Bowell, J.A., 1984. Image processing techniques for thematic mapper data. In: Technical Papers of the American Society of Photogrammetry, Annual Meeting. American Society of Photogrammetry Undefined, January 1, 1984, vol. 2, pp. 728–743.

Chen, X., Warner, T.A., Campagna, D.J., 2007. Integrating visible, near-infrared and short-wave infrared hyperspectral and multispectral thermal imagery for geological mapping at Cuprite, Nevada. Remote Sens. Environ. 110 (3), 344–356. Available from: https://doi.org/10.1016/j.rse.2007.03.015.

Clark, R.N., 1999. Remote Sensing for the Earth Sciences: Manual of Remote Sensing, 3. John Wiley Sons, pp. 3–58.

Clark, R.N., Swayze, G.A., Gallagher, A., Gorelick, N., Kruse, F.A., 1991. Mapping with imaging spectrometer data using the complete band shape least-squares algorithm simultaneously fit to multiple spectral features from multiple materials. In: Proceedings, 3rd Airborne Visible/Infrared Imaging Spectrometer (AVIRIS) Workshop, pp. 2–3.

Clark, R.N., Swayze, G., Boardman, J., Kruse, F., 1993a. Comparison of three methods for material identification and mapping. In: JPL, Summaries of the 4th Annual JPL Airborne Geoscience Workshop, JPL Publication, vol. 1, pp. 31–33.

Clark, R.N., Swayze, G.A., Gallagher, A., King, T.V.V., Calvin, W.M., 1993b. The U.S. Geological Survey, Digital Spectral Library.

Clark, R.N., King, T.V.V., Klejwa, M., Swayze, G.A., 1990. High spectral resolution reflectance spectroscopy of minerals. J. Geophys. Res. 95, 12653–12680.

Cloutis, E.A., 1996. Review article hyperspectral geological remote sensing: Evaluation of analytical techniques. Int. J. Remote Sens. 17 (12), 2215–2242. Available from: https://doi.org/10.1080/01431169608948770.

Cloutis, E.A., Hawthorne, F.C., Mertzman, S.A., Krenn, K., Craig, M.A., Marcino, D., et al., 2006. Detection and discrimination of sulfate minerals using reflectance spectroscopy. Icarus. 184 (1), 121–157. Available from: https://doi.org/10.1016/j.icarus.2006.04.003.

Cocco, F., Attardi, A., Deidda, M.L., Fancello, D., Funedda, A., Naitza, S., 2022. Passive structural control on skarn mineralization localization: a case study from the Variscan Rosas shear zone (SW Sardinia, Italy). Minerals. 12 (2). Available from: https://doi.org/10.3390/min12020272; https://www.mdpi.com/2075-163X/12/2/272/pdf.

Cocks, T., Jenssen, R., Stewart, A., Wilson, I., Shields, T., 1998. The HyMap airborne hyperspectral sensor: the system, calibration and performance. In: 1998 Proceedings of the 1st EARSeL Workshop on Imaging Spectroscopy, pp. 6–8.

Colomina, I., Molina, P., 2014. Unmanned aerial systems for photogrammetry and remote sensing: a review. ISPRS J. Photogrammetry Remote Sens. 92, 79–97. Available from: https://doi.org/10.1016/j.isprsjprs.2014.02.013; http://www.elsevier.com/inca/publications/store/5/0/3/3/4/0.

Comon, P., 1994. Independent component analysis, a new concept? Signal. Process. 36 (3), 287–314. Available from: https://doi.org/10.1016/0165-1684(94)90029-9.

Congalton, R.G., 1991. A review of assessing the accuracy of classifications of remotely sensed data. Remote Sens. Environ. 37 (1), 35–46. Available from: https://doi.org/10.1016/0034-4257(91)90048-B.

Congalton, R.G., Green, K., 2009. Assessing the Accuracy of Remotely Sensed Data: Principles and Practices, p. 183.

Cracknell, A.P., 2017. UAVs: regulations and law enforcement. Int. J. Remote Sens. 38 (8–10), 3054–3067. Available from: https://doi.org/10.1080/01431161.2017.1302115; https://www.tandfonline.com/loi/tres20.

Crósta, A.P., De Souza Filho, C.R., Azevedo, F., Brodie, C., 2003. Targeting key alteration minerals in epithermal deposits in Patagonia, Argentina, using ASTER imagery and principal component analysis. Int. J. Remote Sens. 24 (21), 4233–4240. Available from: https://doi.org/10.1080/0143116031000152291; https://www.tandfonline.com/loi/tres20.

Crowley, J.K., 1991. Visible and near-infrared (0.4-2.5 μm) reflectance spectra of playa evaporite minerals. J. Geophys. Res. 96 (10), 16–240. Available from: https://doi.org/10.1029/91jb01714.

Cudahy, T.J., Barry, P.S., 2002. Earth magmatic-seawater hydrothermal alteration revealed through satellite-borne hyperion imagery at Panorama, Western Australia. In: International Geoscience and Remote Sensing Symposium (IGARSS), January 1, 2002, Australia, vol. 1, pp. 590–592.

Cudahy, T.J., Hewson, R., Huntington, J.F., Quigley, M.A., Barry, P.S., 2001. The performance of the satellite-borne hyperion hyperspectral VNIR-SWIR imaging system for mineral mapping at Mount Fitton, South Australia. In: International Geoscience and Remote Sensing Symposium (IGARSS), December 12, 2001, Australia, vol. 1, pp. 314–316.

Cunningham, C.G., Rye, R.O., Rockwell, B.W., Kunk, M.J., Councell, T.B., 2005. Supergene destruction of a hydrothermal replacement alunite deposit at Big Rock Candy Mountain, Utah: Mineralogy, spectroscopic remote sensing, stable-isotope, and argon-age evidences. Chem. Geol. 215 (1–4), 317–337. Available from: https://doi.org/10.1016/j.chemgeo.2004.06.055.

Dagras, C.H., Duran, M., Zarrouati, O., Fratter, C., 1995. The SPOT-5 mission. Acta Astronautica. 35 (9-11), 651–660. Available from: https://doi.org/10.1016/0094-5765(95)00016-S.

Dalstra, H.J., 2014. Structural evolution of the Mount Wall region in the Hamersley province, Western Australia and its control on hydrothermal alteration and formation of high-grade iron deposits. J. Struct. Geol. 67, 268–292. Available from: https://doi.org/10.1016/j.jsg.2014.03.005; http://www.sciencedirect.com/science/journal/01918141.

Davis, J.C., Sampson, R.J., 1986. Statistics and Data Analysis in Geology, Vol. 646. Wiley, New York.

Di Tommaso, I., Rubinstein, N., 2007. Hydrothermal alteration mapping using ASTER data in the Infiernillo porphyry deposit, Argentina. Ore Geol. Rev. 32 (1-2), 275–290. Available from: https://doi.org/10.1016/j.oregeorev.2006.05.004.

Ding, C., Liu, X., Liu, W., Liu, M., Li, Y., 2014. Mafic-ultramafic and quartz-rich rock indices deduced from ASTER thermal infrared data using a linear approximation to the Planck function. Ore Geol. Rev. 60, 161–173. Available from: https://doi.org/10.1016/j.oregeorev.2014.01.005.

Dilles, J.H., Einaudi, M.T., 1992. Wall-rock alteration and hydrothermal flow paths about the Ann-Mason porphyry copper deposit, Nevada; a 6-km vertical reconstruction. Econ. Geol. 87 (8), 1963–2001. Available from: https://doi.org/10.2113/gsecongeo.87.8.1963.

Ding, C., Li, X., Liu, X., Zhao, L., 2015. Quartzose-mafic spectral feature space model: a methodology for extracting felsic rocks with ASTER thermal infrared radiance data. Ore Geol. Rev. 66, 283–292. Available from: https://doi.org/10.1016/j.oregeorev.2014.11.006; http://www.sciencedirect.com/science/journal/01691368.

Du, Q., Younan, N.H., King, R., Shah, V.P., 2007. On the performance evaluation of pan-sharpening techniques. IEEE Geosci. Remote Sens. Lett. 4 (4), 518–522. Available from: https://doi.org/10.1109/LGRS.2007.896328.

Ducart, D.F., Crosta, A.P., Filho, C.R.S., Coniglio, J., 2006. Alteration mineralogy at the Cerro La Mina Epithermal Prospect, Patagonia, Argentina: field mapping, short-wave infrared spectroscopy, and ASTER images. Econ. Geol. 101 (5), 981–996. Available from: https://doi.org/10.2113/gsecongeo.101.5.981.

Earth Remote Sensing Data Analysis Center (ERSDAC), 2006. PALSAR User's Guide, first ed.

Eisenbeiß, H., 2009. UAV photogrammetry. Institute of Photogrammetry and Remote Sensing. E.T.H. Zürich. Available from: https://doi.org/10.3929/ethz-a-005939264.

El-Wahed, M.A., Zoheir, B., Pour, A.B., Kamh, S., 2021. Shear-related gold ores in the Wadi Hodein shear belt, South Eastern Desert of Egypt: Analysis of remote sensing, field and structural data. Minerals. 11 (5). Available from: https://doi.org/10.3390/min11050474; https://www.mdpi.com/2075-163X/11/5/474/pdf.

ENVI Tutorials, 2008. Research Systems, Inc.

Fitzpatrick-Lins, K., 1981. Comparison of sampling procedures and data analysis for a land- use and land-cover map. Photogrammetric Eng. Remote Sens. 47 (3), 343–351.

Folkman, M., Pearlman, J., Liao, L., Jarecke, P., 2001. EO-1/Hyperion hyperspectral imager design, development, characterization, and calibration. Proc. SPIE- Int. Soc. Opt. Eng. 4151, 40–51. Available from: https://doi.org/10.1117/12.417022.

Fombuena, A., 2017. Unmanned aerial vehicles and spatial thinking: boarding education with geotechnology and drones. IEEE Geosci. Remote Sens. Mag. 5 (3), 8–18. Available from: https://doi.org/10.1109/MGRS.2017.2710054; http://ieeexplore.ieee.org/xpl/RecentIssue.jsp?punumber = 6245518.

Foody, G., 2008. Harshness in image classification accuracy assessment. Int. J. Remote Sens. 29 (11), 3137–3158. Available from: https://doi.org/10.1080/01431160701442120; https://www.tandfonline.com/loi/tres20.

Franceschetti, G., Lanari, R., 1999. Synthetic Aperture Radar Processing. CRC Press.

Fujisada, H., 1995. Design and performance of ASTER instrument. In: Proceedings of SPIE – The International Society for Optical Engineering, December 12, 1995, Japan, vol. 2583, pp. 16–25. Available from: https://doi.org/10.1117/12.228565.

Funedda, A., Naitza, S., Buttau, C., Cocco, F., Dini, A., 2018. Structural controls of ore mineralization in a polydeformed basement: field examples from the Variscan Baccu Locci Shear Zone (SE Sardinia, Italy). Minerals. 8 (10), 456. Available from: https://doi.org/10.3390/min8100456.

Gabr, S., Ghulam, A., Kusky, T., 2010. Detecting areas of high-potential gold mineralization using ASTER data. Ore Geol. Rev. 38 (1–2), 59–69. Available from: https://doi.org/10.1016/j.oregeorev.2010.05.007.

Gad, S., Kusky, T., 2007. ASTER spectral ratioing for lithological mapping in the Arabian-Nubian shield, the Neoproterozoic Wadi Kid area, Sinai, Egypt. Gondwana Res. 11 (3), 326–335. Available from: https://doi.org/10.1016/j.gr.2006.02.010; http://www.sciencedirect.com/science/journal/1342937X.

Gaffey, S.J., 1986. Spectral reflectance of carbonate minerals in the visible and near infrared (0.35-2.55 microns): calcite, aragonite, and dolomite. Am. Mineralogist. 71 (1-2), 151–162.

Galvão, L.S., Almeida-Filho, R., Vitorello, I., 2005. Spectral discrimination of hydrothermally altered materials using ASTER short-wave infrared bands: evaluation in a tropical savannah environment. Int. J. Appl. Earth Observ. Geoinform. 7 (2), 107–114. Available from: https://doi.org/10.1016/j.jag.2004.12.003; http://www.elsevier.com/locate/jag.

Gasmi, A., Gomez, C., Chehbouni, A., Dhiba, D., El Gharous, M., 2022. Using PRISMA hyperspectral satellite imagery and GIS approaches for soil fertility mapping (FertiMap) in Northern Morocco. Remote Sens. 14 (16), 4080. Available from: https://doi.org/10.3390/rs14164080.

Gelautz, M., Frick, H., Raggam, J., Burgstaller, J., Leberl, F., 1998. SAR image simulation and analysis of alpine terrain. ISPRS J. Photogrammetry Remote Sens. 53 (1), 17–38. Available from: https://doi.org/10.1016/S0924-2716(97)00028-2.

Gersman, R., Ben-Dor, E., Beyth, M., Avigad, D., Abraha, M., Kibreab, A., 2008. Mapping of hydrothermally altered rocks by the EO-1 Hyperion sensor, Northern Danakil Depression, Eritrea. Int. J. Remote Sens. 29 (13), 3911–3936. Available from: https://doi.org/10.1080/01431160701874587; https://www.tandfonline.com/loi/tres20.

Giardino, C., Bresciani, M., Braga, F., Fabbretto, A., Ghirardi, N., Pepe, M., et al., 2020. First evaluation of PRISMA level 1 data for water applications. Sensors. 20 (16), 4553. Available from: https://doi.org/10.3390/s20164553.

Gillespie, A., Rokugawa, S., Matsunaga, T., Steven Cothern, J., Hook, S., Kahle, A.B., 1998. A temperature and emissivity separation algorithm for advanced spaceborne thermal emission and reflection radiometer (ASTER) images. IEEE Trans. Geosci. Remote Sens. 36 (4), 1113–1126. Available from: https://doi.org/10.1109/36.700995.

Gillespie, A., Abrams, M., Yamaguchi, Y., 2005. Scientific results from ASTER. Remote Sens. Environ. 99, 1.

Gillingham, S.S., Flood, N., Gill, T.K., Mitchell, R.M., 2012. Limitations of the dense dark vegetation method for aerosol retrieval under Australian conditions. Remote Sens. Lett. 3 (1), 67–76. Available from: https://doi.org/10.1080/01431161.2010.533298.

Girard, M.-C., Girard, C., Courault, D., Gilliot, J.-M., Loubersac, L., Meyer-Roux, J., et al., 2018. Processing of Remote Sensing Data. Routledge.

Gomez, C., Purdie, H., 2016. UAV- based photogrammetry and geocomputing for hazards and disaster risk monitoring – a review. Geoenviron. Disasters. 3 (1). Available from: https://doi.org/10.1186/s40677-016-0060-y; https://link.springer.com/journal/40677.

Gomez, C., Delacourt, C., Allemand, P., Ledru, P., Wackerle, R., 2005. Using ASTER remote sensing data set for geological mapping, in Namibia. Phys. Chem. Earth. 30 (1-3), 97–108. Available from: https://doi.org/10.1016/j.pce.2004.08.042.

Gonzalez-Aguilera, D., Rodriguez-Gonzalvez, P., 2017. Drones—an open access journal. Drones 1 (1), 1–5. Available from: https://doi.org/10.3390/drones1010001; https://www.mdpi.com/2504-446X/1/1/1/pdf.

Goodenough, D.G., Dyk, A., Niemann, K.O., Pearlman, J.S., Chen, H., Han, T., et al., 2003. Processing Hyperion and ALI for forest classification. IEEE Trans. Geosci. Remote Sens. 41 (6), 1321–1331. Available from: https://doi.org/10.1109/TGRS.2003.813214.

Graham, R.W., 1988. Small format aerial surveys from light and microlight aircraft. Photogrammetric Rec. 12 (71), 561–573. Available from: https://doi.org/10.1111/j.1477-9730.1988.tb00605.x.

Green, A.A., Berman, M., Switzer, P., Craig, M.D., 1988. A transformation for ordering multispectral data in terms of image quality with implications for noise removal. IEEE Trans. Geosci. Remote Sens. 26 (1), 65–74. Available from: https://doi.org/10.1109/36.3001.

Green, R.O., Eastwood, M.L., Sarture, C.M., Chrien, T.G., Aronsson, M., Chippendale, B.J., et al., 1998. Imaging spectroscopy and the Airborne Visible/Infrared Imaging Spectrometer (AVIRIS). Remote Sens. Environ. 65 (3), 227–248. Available from: https://doi.org/10.1016/S0034-4257(98)00064-9.

Green, R.O., Pavri, B.E., Chrien, T.G., 2003. On-orbit radiometric and spectral calibration characteristics of EO-1 hyperion derived with an underflight of AVIRIS and In situ measurements at Salar de Arizaro, Argentina. IEEE Trans. Geosci. Remote Sens. 41 (6), 1194–1203. Available from: https://doi.org/10.1109/TGRS.2003.813204.

Groves, D.I., Santosh, M., 2021. Craton and thick lithosphere margins: the sites of giant mineral deposits and mineral provinces. Gondwana Res. 100, 195–222. Available from: https://doi.org/10.1016/j.gr.2020.06.008; http://www.sciencedirect.com/science/journal/1342937X.

Gruen, A., 2012. Development and status of image matching in photogrammetry. Photogrammetric Rec. 27 (137), 36–57. Available from: https://doi.org/10.1111/j.1477-9730.2011.00671.x.

Guha, A., K., V.K., 2016. New ASTER derived thermal indices to delineate mineralogy of different granitoids of an Archaean Craton and analysis of their potentials with reference to Ninomiya's indices for delineating quartz and mafic minerals of granitoids—an analysis in Dharwar Craton, India. Ore Geol. Rev. 74, 76–87. Available from: https://doi.org/10.1016/j.oregeorev.2015.10.033.

Guha, A., Kumar, K.V., 2016. Integrated approach of using ASTER-derived emissivity and pixel temperature for delineating different granitoids – a case study in parts of Dharwar Craton, India. Geocarto Int. 31 (8), 860–869. Available from: https://doi.org/10.1080/10106049.2015.1086904; http://www.tandfonline.com/toc/tgei20/current.

Guha, A., Ghosh, U.K., Sinha, J., Pour, A.B., Bhaisal, R., Chatterjee, S., et al., 2021. Potentials of airborne hyperspectral aviris-ng data in the exploration of base metal deposit—a study in the parts of Bhilwara, Rajasthan. Remote Sens. 13 (11). Available from: https://doi.org/10.3390/rs13112101; https://www.mdpi.com/2072-4292/13/11/2101/pdf.

Gupta, R.P., 2018. Remote Sensing Geology. Springer.

Hagen, N., Kudenov, M.W., 2013. Review of snapshot spectral imaging technologies. Opt. Eng. 52 (9), 090901. Available from: https://doi.org/10.1117/1.OE.52.9.090901.

Hall, D.K., Riggs, G.A., Salomonson, V.V., DiGirolamo, N.E., Bayr, K.J., 2002. MODIS snow-cover products. Remote Sens. Environ. 83 (1-2), 181–194. Available from: https://doi.org/10.1016/S0034-4257(02)00095-0.

Hardin, P., Jensen, R., 2011. Small-scale unmanned aerial vehicles in environmental remote sensing: challenges and opportunities. GIScience Remote Sens. 48 (1), 99–111. Available from: https://doi.org/10.2747/1548-1603.48.1.99.

Harris, J.R., Rogge, D., Hitchcock, R., Ijewliw, O., Wright, D., 2005. Mapping lithology in Canada's Arctic: application of hyperspectral data using the minimum noise fraction transformation and matched filtering. Can. J. Earth Sci. 42 (12), 2173–2193. Available from: https://doi.org/10.1139/e05-064.

Harris Geospatial Solutions, 2017. Continuum Removal. https://www.harrisgeospatial.com/docs/ContinuumRemoval.html (accessed 10.06.2017).

Harsanyi, J.C., 1993. Detection and Classification of Subpixel Spectral Signatures in Hyperspectral Image Sequences.

Harsanyi, J.C., Farrand, W.H., Chang, C.I., 1994. Detection of subpixel signatures in hyperspectral image sequences. In: 1994 Proc. American Society for Photogrammetry and Remote Sensing (ASPRS) Annual Meeting, pp. 236–247.

Hartmann, W., Havlena, M., Schindler, K., 2016. Recent developments in large-scale tie-point matching. ISPRS J. Photogrammetry Remote Sens. 115, 47–62. Available from: https://doi.org/10.1016/j.isprsjprs.2015.09.005; http://www.elsevier.com/inca/publications/store/5/0/3/3/4/0.

Haselwimmer, C.E., Riley, T.R., Liu, J.G., 2011. Lithologic mapping in the Oscar II Coast area, Graham Land, Antarctic Peninsula using ASTER data. Int. J. Remote Sens. 32 (7), 2013–2035. Available from: https://doi.org/10.1080/01431161003645824; https://www.tandfonline.com/loi/tres20.

Hearn, D.R., Digenis, C.J., Lencioni, D.E., Mendenhall, J.A., Evans, J.B., Welsh, R.D., 2001. EO-1 Advanced Land Imager overview and spatial performance. In: International Geoscience and Remote Sensing Symposium (IGARSS), December 12, 2001, United States vol. 2, pp. 897–900.

Hecker, C., der Meijde, Mv, van der Meer, F.D., 2010. Thermal infrared spectroscopy on feldspars – successes, limitations and their implications for remote sensing. Earth-Sci. Rev. 103 (1–2), 60–70. Available from: https://doi.org/10.1016/j.earscirev.2010.07.005.

Heincke, B., Jackisch, R., Saartenoja, A., Salmirinne, H., Rapp, S., Zimmermann, R., et al., 2019. Developing multi-sensor drones for geological mapping and mineral exploration: setup and first results from the MULSEDRO project. Geol. Surv. Den. Greenl. Bull. 43. Available from: https://doi.org/10.34194/GEUSB-201943-03-02; https://geusbulletin.org/index.php/geusb.

Heller Pearlshtien, D., Pignatti, S., Greisman-Ran, U., Ben-Dor, E., 2021. PRISMA sensor evaluation: a case study of mineral mapping performance over Makhtesh Ramon, Israel. Int. J. Remote Sens. 42 (15), 5882–5914. Available from: https://doi.org/10.1080/01431161.2021.1931541.

Hewson, R.D., Cudahy, T.J., Huntington, J.F., 2001. Geologic and alteration mapping at Mt fitton, South Australia, using ASTER satellite-borne data. In: International Geoscience and Remote Sensing Symposium (IGARSS), December 12, 2001, Australia, vol. 2, pp. 724–726.

Hewson, R.D., Cudahy, T.J., Mizuhiko, S., Ueda, K., Mauger, A.J., 2005. Seamless geological map generation using ASTER in the Broken Hill-Curnamona province of Australia. Remote Sens. Environ. 99 (1–2), 159–172. Available from: https://doi.org/10.1016/j.rse.2005.04.025.

Honarmand, M., Ranjbar, H., Shahabpour, J., 2012. Application of principal component analysis and spectral angle mapper in the mapping of hydrothermal alteration in the Jebal-Barez Area, Southeastern Iran. Resour. Geol. 62 (2), 119–139. Available from: https://doi.org/10.1111/j.1751-3928.2012.00184.x.

Hook, S.J., Gabell, A.R., Green, A.A., Kealy, P.S., 1992. A comparison of techniques for extracting emissivity information from thermal infrared data for geologic studies. Remote Sens. Environ. 42 (2), 123–135. Available from: https://doi.org/10.1016/0034-4257(92)90096-3.

Hook, S.J., Abbott, E.A., Grove, C., Kahle, A.B., Palluconi, F., 1999. Use of Multispectral Thermal Infrared Data in Geological Studies, p. 3.

Hor, 1998. Structural Setting of Hypogene Iron Ore along the Southwest Margin of the Proterozoic Hamersley Basin, Western Australia. Internal Hamersley Iron Pty Ltd Report.

Hosseini, S., Lashkaripour, G.R., Moghadas, N.H., Ghafoori, M., Pour, A.B., 2019. Lineament mapping and fractal analysis using SPOT-ASTER satellite imagery for evaluating the severity of slope weathering process. Adv. Space Res. 63 (2), 871–885. Available from: https://doi.org/10.1016/j.asr.2018.10.005; http://www.journals.elsevier.com/advances-in-space-research/.

Hu, J., Lanzon, A., 2018. An innovative tri-rotor drone and associated distributed aerial drone swarm control. Robot. Autonomous Syst. 103, 162–174. Available from: https://doi.org/10.1016/j.robot.2018.02.019.

Hubbard, B.E., Crowley, J.K., 2005. Mineral mapping on the Chilean-Bolivian Altiplano using co-orbital ALI, ASTER and Hyperion imagery: data dimensionality issues and solutions. Remote Sens. Enviro. 99 (1–2), 173–186. Available from: https://doi.org/10.1016/j.rse.2005.04.027.

Hubbard, B.E., Crowley, J.K., Zimbelman, D.R., 2003. Comparative alteration mineral mapping using visible to shortwave infrared (0.4-2.4 μm) Hyperion, ALI, and ASTER imagery. IEEE Trans. Geosci. Remote Sens. 41 (6), 1401–1410. Available from: https://doi.org/10.1109/TGRS.2003.812906.

Hunt, G.R., Salisbury, J.W., 1976. Mid-infrared spectral behavior of metamorphic rocks. Technical Report AFRCL-TR-76-0003. US Air Force Cambridge Research Laboratory, Cambridge, MA.

Hunt, G.R., Salisbury, J.W., 1974. Mid-infrared spectral behavior of igneous rocks. Technical Report AFRCL-TR-75-0356. US Air Force Cambridge Research Laboratory, Cambridge, MA.

Hunt, G.R., 1977. Spectral signatures of particulate minerals in the visible and near infrared. Geophysics. 42 (3), 501–513. Available from: https://doi.org/10.1190/1.1440721.

Hunt, G.R., Ashley, R.P., 1979. Spectra of altered rocks in the visible and near infrared. Econ. Geol. 74 (7), 1613–1629. Available from: https://doi.org/10.2113/gsecongeo.74.7.1613.

Hunt, G.R., Salisbury, J.W., Lenhoff, C.J., 1971a. Visible and near-infrared spectra of minerals and rocks: III. Oxides hydroxides. Mod. Geol. 2, 195–205.

Hunt, G.R., Salisbury, J.W., Lenhoff, C.J., 1971b. Visible and near-infrared spectra of minerals and rocks: IV. Sulphides sulphates. Mod. Geol. 3, 1–14.

Huntington, J.F., 1996. The role of remote sensing in finding hydrothermal mineral deposits on earth. CIBA Foundation Symposia, vol. 202, pp. 214–235.

Huo, H., Ni, Z., Jiang, X., Zhou, P., Liu, L., 2014. Mineral mapping and ore prospecting with HyMap data over eastern Tien Shan, Xinjiang Uyghur autonomous region. Remote Sens. 6 (12), 11829–11851. Available from: https://doi.org/10.3390/rs61211829; http://www.mdpi.com/2072-4292/6/12/11829/pdf.

Huurneman, G.C., Bakker, W.H., Janssen, L.L.F., 2009. Principles of Remote Sensing: An Introductory Textbook ITC. The Netherlands, p. 591.

Hyvärinen, A., Oja, E., 1997. A fast fixed-point algorithm for independent component analysis. Neural Comput. 9 (7), 1483–1492. Available from: https://doi.org/10.1162/neco.1997.9.7.1483; http://www.mit-pressjournals.org/loi/neco.

Hyvärinen, A., Oja, E., 2000. Independent component analysis: algorithms and applications. Neural Net. 13 (4–5), 411–430. Available from: https://doi.org/10.1016/S0893-6080(00)00026-5; http://www.elsevier.com/locate/neunet.

Hyvärinen, A., Karhunen, J., Oja, E., 2004. Independent Component Analysis, p. 46.

Igarashi, T., 2001. ALOS mission requirement and sensor specifications. Adv. Space Res. 28 (1), 127–131. Available from: https://doi.org/10.1016/S0273-1177(01)00316-7; http://www.journals.elsevier.com/advances-in-space-research/.

Ingebritsen, S.E., Appold, M.S., 2012. The physical hydrogeology of ore deposits. Econ. Geol. 107 (4), 559–584. Available from: https://doi.org/10.2113/econgeo.107.4.559; http://economicgeology.org/content/107/4/559.full.pdf + html. United States.

Irons, J.R., Dwyer, J.L., Barsi, J.A., 2012. The next Landsat satellite: the Landsat Data Continuity Mission. Remote Sens. Environ. 122, 11–21. Available from: https://doi.org/10.1016/j.rse.2011.08.026.

Ishagh, M.M., Pour, A.B., Benali, H., Idriss, A.M., Reyoug, S.S., Muslim, A.M., et al., 2021. Lithological and alteration mapping using Landsat 8 and ASTER satellite data in the Reguibat Shield (West African Craton), North of Mauritania: implications for uranium exploration. Arab. J. Geosci. 14 (23). Available from: https://doi.org/10.1007/s12517-021-08846-x.

Iwasaki, A., Tonooka, H., 2005. Validation of a crosstalk correction algorithm for ASTER/SWIR. IEEE Trans. Geosci. Remote Sens. 43 (12), 2747–2751. Available from: https://doi.org/10.1109/TGRS.2005.855066.

Jackisch, R., 2020. Drone-based surveys of mineral deposits. Nat. Rev. Earth Environ. 1 (4), 187. Available from: https://doi.org/10.1038/s43017-020-0042-1; http://nature.com/natrevearthenviron/.

Jackisch, R., Madriz, Y., Zimmermann, R., Pirttijärvi, M., Saartenoja, A., Heincke, B.H., et al., 2019. Drone-borne hyperspectral and magnetic data integration: Otanmäki Fe-Ti-V deposit in Finland. Remote Sens. 11 (18). Available from: https://doi.org/10.3390/rs11182084; https://res.mdpi.com/d_attachment/remotesensing/remotesensing-11-02084/article_deploy/remotesensing-11-02084-v2.pdf.

Jensen, J.R., 2005. Introductory Digital Image Processing a Remote Sensing Perspective, No. 621, p. 3678.

James, K., Vergo, C., Vergo, N., 1988. Near-infrared reflectance spectra of mixtures of kaolin-group minerals: use in clay mineral studies. Clays Clay Miner. 36 (4), 310–316.

Jensen, J.R., 2009. Remote Sensing of the Environment: An Earth Resource Perspective 2/e. Pearson Education, India.

Jensen, J.R., 2016. Introductory Digital Image Processing: A Remote Sensing Perspective. Pearson Education.

Jiao, X., Chang, C.I., 2008. Kernel-based Constrained Energy Minimization (K-CEM). In: Proceedings of SPIE – The International Society for Optical Engineering, June 6, 2008, United States, vol. 6966. Available from: https://doi.org/10.1117/12.782221 0277786X.

Johnson, S., 2003. Constrained energy minimization and the target-constrained interference-minimized filter. Opt. Eng. 42 (6), 1850–1854. Available from: https://doi.org/10.1117/1.1571062.

Joly, A., Miller, J., McCuaig, T.C., 2010. Archean polyphase deformation in the Lake Johnston Greenstone Belt area: Implications for the understanding of ore systems of the Yilgarn Craton. Precambrian Res. 177 (1-2), 181–198. Available from: https://doi.org/10.1016/j.precamres.2009.11.010.

Kanlinowski, A., Oliver, S., 2004. ASTER Mineral Index Processing. Remote Sensing Application Geoscience Australia. Australian Government Geoscience Website.

Karpouzli, E., Malthus, T., 2003. The empirical line method for the atmospheric correction of IKONOS imagery. Int. J. Remote Sens. 24 (5), 1143–1150. Available from: https://doi.org/10.1080/0143116021000026779; https://www.tandfonline.com/loi/tres20.

Khan, S.D., Mahmood, K., Casey, J.F., 2007. Mapping of Muslim Bagh ophiolite complex (Pakistan) using new remote sensing, and field data. J. Asian Earth Sci. 30 (2), 333–343. Available from: https://doi.org/10.1016/j.jseaes.2006.11.001.

King, T.V.V., Clark, R.N., 2000. Verification of Remotely Sensed Data. Springer Science and Business Media LLC, pp. 59–62. Available from: https://doi.org/10.1007/978-3-642-56978-4_5.

Kirsch, M., Lorenz, S., Zimmermann, R., Tusa, L., Möckel, R., Hödl, P., et al., 2018. Integration of terrestrial and drone-borne hyperspectral and photogrammetric sensing methods for exploration mapping and mining monitoring. Remote Sens. 10 (9), 1366. Available from: https://doi.org/10.3390/rs10091366.

Kraut, S., Scharf, L.L., McWhorter, L.T., 2001. Adaptive subspace detectors. IEEE Trans. Signal. Process. 49 (1), 1–16. Available from: https://doi.org/10.1109/78.890324.

Krohn, M.D., Abrams, M.J., Rowan, L.C., 1978. Discrimination of Hydrothermal Altered Rocks along the Battle Mountain. Mineral Belt using Landsat Images: U.S. Geological Survey Open-File Report, pp. 78–585.

Kruse, F.A., 1993. Artificial Intelligence for Geologic Mapping with Imaging Spectrometers Center for the Study of Earth from Space (CSES).

Kruse, F.A., Perry, S.L., 2007 Regional mineral mapping by extending hyperspectral signatures using multispectral data. In: IEEE Aerospace Conference Proceedings, September 9, 2007, undefined. Available from: https://doi.org/10.1109/AERO.2007.353059 1095323X.

Kruse, F.A., Perry, S.L., 2013. Mineral mapping using simulated worldview-3 short-wave-infrared imagery. Remote Sens. 5 (6), 2688–2703. Available from: https://doi.org/10.3390/rs5062688; http://www.mdpi.com/2072-4292/5/6/2688/pdf. United States.

Kruse, F.A., Lefkoff, A.B., Boardman, J.W., Heidebrecht, K.B., Shapiro, A.T., Barloon, P.J., et al., 1993. The spectral image processing system (SIPS)-interactive visualization and analysis of imaging spectrometer data. Remote Sens. Environ. 44 (2-3), 145–163. Available from: https://doi.org/10.1016/0034-4257(93)90013-N.

Kruse, F.A., Boardman, J.W., Huntington, J.F., 2003. Comparison of airborne hyperspectral data and EO-1 Hyperion for mineral mapping. IEEE Trans. Geosci. Remote Sens. 41 (6), 1388–1400. Available from: https://doi.org/10.1109/TGRS.2003.812908.

Kruse, F.A., Perry, S.L., Caballero, A., 2006. District-level mineral survey using airborne hyperspectral data, Los Menucos, Argentina. Ann. Geophys. 49 (1), 83–92.

Kruse, F.A., Baugh, W.M., Perry, S.L., 2015. Validation of DigitalGlobe WorldView-3 Earth imaging satellite shortwave infrared bands for mineral mapping. J. Appl. Remote Sens. 9 (1). Available from: https://doi.org/10.1117/1.JRS.9.096044; http://www.spie.org/x3636.xml.

Kuester, M., 2016. Radiometric use of WV-3 imagery. Technical Note.

Kuester, M.A., Ochoa, M., Dayer, A., Levin, J., Aaron, D., Helder, D.L., et al., 2015. Absolute Radiometric Calibration of the DigitalGlobe Fleet and Updates on the New WV-3 Sensor Suite.

Kurz, T.H., Buckley, S.J., 2016. A review of hyperspectral imaging in close range applications. Int. Arch. Photogrammetry Remote Sens. Spat. Inf. Sci. XLI-B5, 865–870. Available from: https://doi.org/10.5194/isprs-archives-xli-b5-865-2016.

Kwak, Y., Park, S.I., Park, C., 2022. Structural controls on crustal fluid redistribution and hydrothermal gold deposits: a review on the suction pump and fault valve models. Econ. Environ. Geol. 55 (2), 183–195. Available from: https://doi.org/10.9719/EEG.2022.55.2.183; https://www.kseeg.org/journal/view.html?uid = 2117&page = &sort = &scale = 10&all_k = &s_t = &s_a = &s_k = &s_v = 55&s_n = 2&spage = &pn = search&year = &vmd = Full.

Landis, J.R., Koch, G.G., 1977. The measurement of observer agreement for categorical data. Biometrics. 33 (1), 159–174. Available from: https://doi.org/10.2307/2529310.

Leary, D., 2017. Drones on ice: an assessment of the legal implications of the use of unmanned aerial vehicles in scientific research and by the tourist industry in Antarctica. Polar Record. 53 (4), 343–357. Available from: https://doi.org/10.1017/S0032247417000262; http://uk.cambridge.org/journals/journal_catalogue.asp?historylinks = ALPHA&mnemonic = POL.

Legg, C.A., 1994. Remote Sensing and Geographical Information Systems: Geological Mapping, Mineral Exploration and Mining. Ellis Horwood.

Lencioni, D.E., Digenis, C.J., Bicknell, W.E., Hearn, D.R., Mendenhall, J.A., 1999. Design and performance of the EO1 advanced land imager. In: SPIE Conference on Sensors, Systems, and Next Generation Satellites III.

Lester, D.R., Ord, A., Hobbs, B.E., 2012. The mechanics of hydrothermal systems: II. Fluid mixing and chemical reactions. Ore Geol. Rev. 49, 45–71. Available from: https://doi.org/10.1016/j.oregeorev.2012.08.002.

Liao, L., Jarecke, P., Gleichauf, D., Hedman, T., 2000. Performance characterization of the Hyperion imaging spectrometer instrument. In: Proceedings of SPIE – The International Society for Optical Engineering, December 12, 2000, United States, vol. 4135, pp. 264–275. Available from: https://doi.org/10.1117/12.494253 0277786X.

Liao, X., Zhou, C., Su, F., Lu, H., Yue, H., Gou, J., 2016. The mass innovation era of UAV remote sensing. J. Geo-Inf. Sci. 18 (11), 1439–1448 Available from: https://doi.org/10.3724/SP.J.1047.2016.01439.

Lillesand, T.M., Kiefer, R.W., 1994. Remote Sensing and Image Interpretation, third ed.

Lillesand, T.M., Kiefer, R.W., Chipman, J.W., 2011. Remote Sensing and Image Interpretation. Wiley India Pvt. Limited.

Lillesand, T., Kiefer, R.W., Chipman, J., 2015. Remote Sensing and Image Interpretation. John Wiley & Sons.

Liu, J.G., Mason, P.J., 2016. Image Processing and GIS for Remote Sensing: Techniques and Applications. Wiley. Available from: https://doi.org/10.1002/9781118724194.

Lobell, D.B., Asner, G.P., 2003. Comparison of earth observing-1 ALI and Landsat ETM + for crop identification and yield prediction in Mexico. IEEE Trans. Geosci. Remote Sens. 41 (6), 1277–1282. Available from: https://doi.org/10.1109/TGRS.2003.812909.

Logan, L.M., Hunt, G.R., Salisbury, J.W., Balsamo, S.R., 1973. Compositional implications of Christiansen frequency maximums for infrared remote sensing applications. J. Geophys. Res. 78 (23), 4983–5003. Available from: https://doi.org/10.1029/jb078i023p04983.

Loizzo, R., Guarini, R., Longo, F., Scopa, T., Formaro, R., Facchinetti, C., et al., 2018. Prisma: The Italian hyperspectral mission. In: International Geoscience and Remote Sensing Symposium (IGARSS), October 31, 2018, Institute of Electrical and Electronics Engineers Inc., Italy, pp. 175–178. Available from: https://doi.org/10.1109/IGARSS.2018.8518512. 9781538671504.

Lowell, J.D., Guilbert, J.M., 1970. Lateral and vertical alteration-mineralization zoning in porphyry ore deposits. Econ. Geol. 65 (4), 373–408. Available from: https://doi.org/10.2113/gsecongeo.65.4.373; http://economicgeology.org/content/65/4/373.full.pdf + html.

Lunetta, R.S., Lyon, 2004. Remote Sensing and GIS Accuracy Assessment. CRC Press.

Lyon, R.J.P., 1972. Infrared spectral emittance in geological mapping: airborne spectrometer data from Pisgah Crater, California. Science. 175 (4025), 983–986. Available from: https://doi.org/10.1126/science.175.4025.983.

Mars, J.L., 2014. Regional mapping of hydrothermally altered igneous rocks along the Urumieh-Dokhtar, Chagai, and Alborz Belts of western Asia using Advanced Spaceborne Thermal Emission and Reflection Radiometer (ASTER) data and Interactive Data Language (IDL) logical operators: a tool for porphyry copper exploration and assessment: Chapter.

Maleki, M., Niroomand, S., Rajabpour, S., Pour, A.B., Ebrahimpour, S., 2022. Targeting local orogenic gold mineralization zones using data-driven evidential belief functions: the Godarsorkh area, Central Iran. All Earth 34 (1), 259–278. Available from: https://doi.org/10.1080/27669645.2022.2129132.

Mars, J.C., 2018. Mineral and lithologic mapping capability of worldview 3 data at Mountain Pass, California, using true- and false-color composite images, band ratios, and logical operator algorithms. Econ. Geol. 113 (7), 1587–1601. Available from: https://doi.org/10.5382/econgeo.2018.4604; https://watermark.silverchair.com.

Mars, J.C., Rowan, L.C., 2006. Regional mapping of phyllic- and argillic-altered rocks in the Zagros magmatic arc, Iran, using advanced spaceborne thermal emission and reflection radiometer (ASTER) data and logical operator algorithms. Geosphere 2 (3), 161–186. Available from: https://doi.org/10.1130/GES00044.1.

Mars, J.C., Rowan, L.C., 2010. Spectral assessment of new ASTER SWIR surface reflectance data products for spectroscopic mapping of rocks and minerals. Remote Sens. Environ. 114 (9), 2011–2025. Available from: https://doi.org/10.1016/j.rse.2010.04.008.

Mars, J.C., Rowan, L.C., 2011. ASTER spectral analysis and lithologic mapping of the Khanneshin carbonatite volcano, Afghanistan. Geosphere 7 (1), 276–289. Available from: https://doi.org/10.1130/GES00630.1; http://geosphere.gsapubs.org/content/7/1/276.full.pdf. United States.

Mather, P.M., Koch, M., 2011. Computer Processing of Remotely-Sensed Images: An Introduction. John Wiley & Sons, Ltd.

McCoy, R., 2005. Field Methods in Remote Sensing, p. 159.

Melgani, F., Bruzzone, L., 2004. Classification of hyperspectral remote sensing images with support vector machines. IEEE Trans. Geosci. Remote Sens. 42 (8), 1778–1790. Available from: https://doi.org/10.1109/TGRS.2004.831865.

Mendenhall, J.A., Ryanhoward, D.P., Willard, B.C., 2000. Earth Observing-1 Advanced Land Imager: Instrument and Flight Operations Overview.

Metternicht, G.I., Zinck, J.A., 1998. Evaluating the information content of JERS-1 SAR and Landsat TM data for discrimination of soil erosion features. ISPRS J. Photogrammetry Remote Sens. 53 (3), 143–153. Available from: https://doi.org/10.1016/S0924-2716(98)00004-5.

Micklethwaite, S., Ford, A., Witt, W., Sheldon, H.A., 2015. The where and how of faults, fluids and permeability – insights from fault stepovers, scaling properties and gold mineralisation. Geofluids 15 (1-2), 240–251. Available from: https://doi.org/10.1111/gfl.12102; https://www.hindawi.com/journals/geofluids/.

Moghtaderi, A., Moore, F., Mohammadzadeh, A., 2007. The application of advanced space-borne thermal emission and reflection (ASTER) radiometer data in the detection of alteration in the Chadormalu paleocrater, Bafq region, Central Iran. J. Asian Earth Sci. 30 (2), 238–252. Available from: https://doi.org/10.1016/j.jseaes.2006.09.004.

Moradpour, H., Rostami Paydar, G., Pour, A.B., Valizadeh Kamran, K., Feizizadeh, B., Muslim, A.M., et al., 2022. Landsat-7 and ASTER remote sensing satellite imagery for identification of iron skarn mineralization in metamorphic regions. Geocarto Int. 37 (7), 1971–1998. Available from: https://doi.org/10.1080/10106049.2020.1810327; http://www.tandfonline.com/toc/tgei20/current.

Morris, R.V., Lauer, H.V., Lawson, C.A., Gibson, E.K., Nace, G.A., Stewart, C., 1985. Spectral and other physicochemical properties of submicron powders of hematite (α-Fe_2O_3), maghemite (γ-Fe_2O_3), magnetite (Fe_3O_4), goethite (α-FeOOH), and lepidocrocite (γ-FeOOH). J. Geophys. Res. 90 (B4), 3126. Available from: https://doi.org/10.1029/jb090ib04p03126.

Ngassam Mbianya, G., Ngnotue, T., Takodjou Wambo, J.D., Ganno, S., Pour, A.B., Ayonta Kenne, P., et al., 2021. Remote sensing satellite-based structural/alteration mapping for gold exploration in the Ketté goldfield, Eastern Cameroon. J. Afr. Earth Sci. 184. Available from: https://doi.org/10.1016/j.jafrearsci.2021.104386; http://www.sciencedirect.com/science/journal/1464343X.

Ninomiya, Y., 1995. Quantitative estimation of SiO_2 content in igneous rocks using thermal infrared spectra with a neural network approach. IEEE Trans. Geosci. Remote Sens. 33 (3), 684–691. Available from: https://doi.org/10.1109/36.387583.

Ninomiya, Y., 2003a. A stabilized vegetation index and several mineralogic indices defined for ASTER VNIR and SWIR data. In: International Geoscience and Remote Sensing Symposium (IGARSS), November 11 2003, Japan, vol. 3, pp. 1552–1554.

Ninomiya, Y., 2003b. Advanced remote lithologic mapping in ophiolite zone with ASTER multispectral thermal infrared data. In: International Geoscience and Remote Sensing Symposium (IGARSS), November 11 2003, Japan, vol. 3, pp. 1561–1563.

Ninomiya, Y., Fu, B., 2019. Thermal infrared multispectral remote sensing of lithology and mineralogy based on spectral properties of materials. Ore Geol. Rev. 108, 54–72. Available from: https://doi.org/10.1016/j.oregeorev.2018.03.012; http://www.sciencedirect.com/science/journal/01691368.

Ninomiya, Y., Fu, B., Cudahy, T.J., 2005. Detecting lithology with Advanced Spaceborne Thermal Emission and Reflection Radiometer (ASTER) multispectral thermal infrared \radiance-at-sensor\ data. Remote Sens. Environ. 99 (1-2), 127–139. Available from: https://doi.org/10.1016/j.rse.2005.06.009; http://www.elsevier.com/inca/publications/store/5/0/5/7/3/3.

Ord, A., Hobbs, B.E., Lester, D.R., 2012. The mechanics of hydrothermal systems: I. Ore systems as chemical reactors. Ore Geol. Rev. 49, 1–44. Available from: https://doi.org/10.1016/j.oregeorev.2012.08.003.

Ott, N., Kollersberger, T., Tassara, A., 2006. GIS analyses and favorability mapping of optimized satellite data in northern Chile to improve exploration for copper mineral deposits. Geosphere 2 (4), 236–252. Available from: https://doi.org/10.1130/GES00017.1.

Ouchi, K., 1988. Synthetic aperture radar imagery of range traveling ocean waves. IEEE Trans. Geosci. Remote Sens. 26 (1), 30–37. Available from: https://doi.org/10.1109/36.2997.

Pádua, L., Vanko, J., Hruška, J., Adão, T., Sousa, J.J., Peres, E., et al., 2017. UAS, sensors, and data processing in agroforestry: a review towards practical applications. Int. J. Remote Sensing. 38 (8–10), 2349–2391. Available from: https://doi.org/10.1080/01431161.2017.1297548; https://www.tandfonline.com/loi/tres20.

Paganelli, F., Grunsky, E.C., Richards, J.P., Pryde, R., 2003. Use of RADARSAT-1 principal component imagery for structural mapping: a case study in the Buffalo Head Hills area, northern central Alberta, Canada. Can. J. Remote Sens. 29 (1), 111–140. Available from: https://doi.org/10.5589/m02-084.

Pajares, G., 2015. Overview and current status of remote sensing applications based on unmanned aerial vehicles (UAVs). Photogrammetric Eng. Remote Sens. 81 (4), 281–329. Available from: https://doi.org/10.14358/PERS.81.4.281; http://www.asprs.org/a/publications/pers/2015journals/PERS_April_2015/HTML/files/assets/common/downloads/PE&RS%20April%202015.pdf.

Pal, S.K., Majumdar, T.J., Bhattacharya, A.K., 2007. ERS-2 SAR and IRS-1C LISS III data fusion: a PCA approach to improve remote sensing based geological interpretation. ISPRS J. Photogrammetry Remote Sens. 61 (5), 281–297. Available from: https://doi.org/10.1016/j.isprsjprs.2006.10.001.

Park, S., Choi, Y., 2020. Applications of unmanned aerial vehicles in mining from exploration to reclamation: a review. Minerals. 10 (8), 663. Available from: https://doi.org/10.3390/min10080663.

Pearlman, J.S., Barry, P.S., Segal, C.C., Shepanski, J., Beiso, D., Carman, S.L., 2003. Hyperion, a space-based imaging spectrometer. IEEE Trans. Geosci. Remote Sens. 41 (6), 1160–1173. Available from: https://doi.org/10.1109/TGRS.2003.815018.

Peng, Z., Wang, C., Zhang, L., Zhu, M., Tong, X., 2019. Geochemistry of metamorphosed volcanic rocks in the Neoarchean Qingyuan greenstone belt, North China Craton: Implications for geodynamic evolution and VMS mineralization. Precambrian Res. 326, 196–221. Available from: https://doi.org/10.1016/j.precamres.2017.12.033; http://www.sciencedirect.com/science/journal/03019268.

Pengra, B.W., Johnston, C.A., Loveland, T.R., 2007. Mapping an invasive plant, *Phragmites australis*, in coastal wetlands using the EO-1 Hyperion hyperspectral sensor. Remote Sens. Environ. 108 (1), 74–81. Available from: https://doi.org/10.1016/j.rse.2006.11.002.

Pepe, M., Pompilio, L., Gioli, B., Busetto, L., Boschetti, M., 2020. Detection and classification of non-photosynthetic vegetation from PRISMA hyperspectral data in croplands. Remote Sens. 12 (23), 3903. Available from: https://doi.org/10.3390/rs12233903.

Perko, K.L., Huggins, R.W., Heisen, P.T., Miller, G.E., McMeen, D.J., Dod, T., et al., 2001. EO-1 Technology Validation Report X-Band Phase Array Antenna.

Pontius, R.G., Millones, M., 2011. Death to Kappa: birth of quantity disagreement and allocation disagreement for accuracy assessment. Int. J. Remote Sens. 32 (15), 4407–4429. Available from: https://doi.org/10.1080/01431161.2011.552923; https://www.tandfonline.com/loi/tres20.

Pour, A.B., Hashim, M., 2011a. Identification of hydrothermal alteration minerals for exploring of porphyry copper deposit using ASTER data, SE Iran. J. Asian Earth Sci. 42 (6), 1309–1323. Available from: https://doi.org/10.1016/j.jseaes.2011.07.017.

Pour, A.B., Hashim, M., 2011b. The Earth Observing-1 (EO-1) satellite data for geological mapping, southeastern segment of the Central Iranian Volcanic Belt, Iran. Int. J. Phys. Sci. 6 (33), 7638–7650. Available from: https://doi.org/10.5897/IJPS11.910Malaysia; http://www.academicjournals.org/ijps/PDF/pdf2011/9Dec/Pour%20and%20Hashim.pdf.

Pour, A.B., Hashim, M., 2012a. Identifying areas of high economic-potential copper mineralization using ASTER data in the Urumieh-Dokhtar Volcanic Belt, Iran. Adv. Space Res. 49 (4), 753–769. Available from: https://doi.org/10.1016/j.asr.2011.11.028; http://www.journals.elsevier.com/advances-in-space-research/.

Pour, A.B., Hashim, M., 2012b. The application of ASTER remote sensing data to porphyry copper and epithermal gold deposits. Ore Geol. Rev. 44, 1–9. Available from: https://doi.org/10.1016/j.oregeorev.2011.09.009.

Pour, A.B., Hashim, M., 2014a. ASTER, ALI and Hyperion sensors data for lithological mapping and ore minerals exploration. SpringerPlus 3 (1), 1–19. Available from: https://doi.org/10.1186/2193-1801-3-130; http://www.springerplus.com/archive.

Pour, A.B., Hashim, M., 2014b. Structural geology mapping using PALSAR data in the Bau gold mining district, Sarawak, Malaysia. Adv. Space Res. 54 (4), 644–654. Available from: https://doi.org/10.1016/j.asr.2014.02.012; http://www.journals.elsevier.com/advances-in-space-research/.

Pour, A.B., Hashim, M., 2015a. Evaluation of Earth Observing-1 (EO1) data for lithological and hydrothermal alteration mapping: a case study from Urumieh-Dokhtar Volcanic Belt, SE Iran. J. Indian. Soc. Remote Sens. 43 (3), 583–597. Available from: https://doi.org/10.1007/s12524-014-0444-y; http://www.springerlink.com/content/0255-660X.

Pour, A.B., Hashim, M., 2015b. Hydrothermal alteration mapping from Landsat-8 data, Sar Cheshmeh copper mining district, south-eastern Islamic Republic of Iran. J. Taibah Univ. Sci. 9 (2), 155–166. Available from: https://doi.org/10.1016/j.jtusci.2014.11.008; https://www.tandfonline.com/journals/tusc20.

Pour, A.B., Hashim, M., 2015c. Structural mapping using PALSAR data in the Central Gold Belt, Peninsular Malaysia. Ore Geol. Rev. 64 (1), 13–22. Available from: https://doi.org/10.1016/j.oregeorev.2014.06.011; http://www.sciencedirect.com/science/journal/01691368.

Pour, A.B., Hashim, M., 2017. Application of Landsat-8 and ALOS-2 data for structural and landslide hazard mapping in Kelantan, Malaysia. Nat. Hazards Earth Syst. Sci. 17 (7), 1285–1303. Available from: https://doi.org/10.5194/nhess-17-1285-2017; http://www.nat-hazards-earth-syst-sci.net/volumes_and_issues.html.

Pour, A.B., Hashim, M., van Genderen, J., 2013. Detection of hydrothermal alteration zones in a tropical region using satellite remote sensing data: Bau goldfield, Sarawak, Malaysia. Ore Geol. Rev. 54, 181–196. Available from: https://doi.org/10.1016/j.oregeorev.2013.03.010.

Pour, A.B., Hashim, M., Marghany, M., 2014. Exploration of gold mineralization in a tropical region using Earth Observing-1 (EO1) and JERS-1 SAR data: a case study from Bau gold field, Sarawak, Malaysia. Arab. J. Geosci. 7 (6), 2393–2406. Available from: https://doi.org/10.1007/s12517-013-0969-3; http://www.springer.com/geosciences/journal/12517?cm_mmc = AD-_-enews-_-PSE1892-_-0.

Pour, A.B., Hashim, M., Makoundi, C., Zaw, K., 2016. Structural mapping of the Bentong-Raub suture zone using PALSAR remote sensing data, peninsular Malaysia: implications for sediment-hosted/orogenic gold mineral systems exploration. Resour. Geol. 66 (4), 368–385. Available from: https://doi.org/10.1111/rge.12105; http://onlinelibrary.wiley.com/journal/10.1111/(ISSN)1751-3928.

Pour, A.B., Hashim, M., Park, Y., 2017. Alteration mineral mapping in inaccessible regions using target detection algorithms to ASTER data. Accepted for Oral presentation. International Conference on Space Science and Communication: Space Science for Sustainability, IconSpace 2017, Kuala Lumpur; Malaysia, 3 May 2017 through 5 May 2017. J. Phys.: Conf. Ser 852, 012022. Available from: https://doi.org/10.1088/1742-6596/852/1/012022.

Pour, A.B., Park, T.Y.S., Park, Y., Hong, J.K., Zoheir, B., Pradhan, B., et al., 2018a. Application of multi-sensor satellite data for exploration of Zn-Pb sulfide mineralization in the Franklinian Basin, North Greenland.

Remote Sens. 10 (8). Available from: https://doi.org/10.3390/rs10081186; https://res.mdpi.com/remotesensing/remotesensing-10-01186/article_deploy/remotesensing-10-01186.pdf?filename = &attachment = 1.

Pour, A.B., Park, Y., Park, T.Y.S., Hong, J.K., Hashim, M., Woo, J., et al., 2018b. Regional geology mapping using satellite-based remote sensing approach in Northern Victoria Land, Antarctica. Polar Sci. 16, 23–46. Available from: https://doi.org/10.1016/j.polar.2018.02.004; http://www.elsevier.com.

Pour, A.B., Park, T.Y.S., Park, Y., Hong, J.K., Muslim, A.M., Läufer, A., et al., 2019a. Landsat-8, advanced spaceborne thermal emission and reflection radiometer, and WorldView-3 multispectral satellite imagery for prospecting copper-gold mineralization in the northeastern Inglefield Mobile Belt (IMB), northwest Greenland. Remote Sens. 11 (20). Available from: https://doi.org/10.3390/rs11202430; https://res.mdpi.com/d_attachment/remotesensing/remotesensing-11-02430/article_deploy/remotesensing-11-02430-v2.pdf.

Pour, A.B., Hashim, M., Hong, J.K., Park, Y., 2019b. Lithological and alteration mineral mapping in poorly exposed lithologies using Landsat-8 and ASTER satellite data: North-eastern Graham Land, Antarctic Peninsula. Ore Geol. Rev. 108, 112–133. Available from: https://doi.org/10.1016/j.oregeorev.2017.07.018; http://www.sciencedirect.com/science/journal/01691368.

Pour, A.B., Park, Y., Crispini, L., Läufer, A., Hong, J.K., Park, T.Y.S., et al., 2019c. Mapping listvenite occurrences in the damage zones of Northern Victoria Land, Antarctica using ASTER satellite remote sensing data. Remote Sens. 11 (12). Available from: https://doi.org/10.3390/rs11121408; https://res.mdpi.com/remotesensing/remotesensing-11-01408/article_deploy/remotesensing-11-01408.pdf?filename = &attachment = 1.

Pour, A.B., Sekandari, M., Rahmani, O., Crispini, L., Läufer, A., Park, Y., et al., 2021a. Identification of phyllosilicates in the antarctic environment using aster satellite data: case study from the Mesa Range, Campbell and Priestley Glaciers, Northern Victoria Land. Remote Sens. 13 (1), 1–37. Available from: https://doi.org/10.3390/rs13010038; https://www.mdpi.com/2072-4292/13/1/38/pdf.

Pour, A.B., Zoheir, B., Pradhan, B., Hashim, M., 2021b. Editorial for the special issue: multispectral and hyperspectral remote sensing data for mineral exploration and environmental monitoring of mined areas. Remote Sens. 13 (3), 1–6. Available from: https://doi.org/10.3390/rs13030519; https://www.mdpi.com/2072-4292/13/3/519.

Pournamdari, M., Hashim, M., Pour, A.B., 2014. Application of ASTER and Landsat TM data for geological mapping of esfandagheh ophiolite complex, southern Iran. Resour. Geol. 64 (3), 233–246. Available from: https://doi.org/10.1111/rge.12038; http://onlinelibrary.wiley.com/journal/10.1111/(ISSN)1751-3928. Malaysia.

Price, J.C., 1999. Combining multispectral data of differing spatial resolution. IEEE Trans. Geosci. Remote Sens. 37 (3 I), 1199–1203. Available from: https://doi.org/10.1109/36.763272.

Pu, R., 2017. Hyperspectral Remote Sensing: Fundamentals and Practices. CRC Press, United States, pp. 1–466. Available from: http://www.tandfebooks.com/doi/book/10.1201/9781315120607; https://doi.org/10.1201/9781315120607.

Pushparaj, J., Hegde, A.V., 2017. Evaluation of pan-sharpening methods for spatial and spectral quality. Appl. Geomat. 9 (1), 1–12. Available from: https://doi.org/10.1007/s12518-016-0179-2; http://www.springerlink.com/content/1866-9298/.

Raines, G.L., 1978. Porphyry copper exploration model for Northern Sonora, Mexico. J. Res. US Geol. Surv. 6 (1), 51–58.

Rajendran, S., Nasir, S., 2017. Characterization of ASTER spectral bands for mapping of alteration zones of volcanogenic massive sulphide deposits. Ore Geol. Rev. 88, 317–335. Available from: https://doi.org/10.1016/j.oregeorev.2017.04.016; http://www.sciencedirect.com/science/journal/01691368.

Ramakrishnan, D., Bharti, R., Singh, K.D., Nithya, M., 2013. Thermal inertia mapping and its application in mineral exploration: results from mamandur polymetal prospect, India. Geophys. J. Int. 195 (1), 357–368. Available from: https://doi.org/10.1093/gji/ggt237.

Ranjbar, H., Honarmand, M., Moezifar, Z., 2004. Application of the Crosta technique for porphyry copperalteration mapping, using ETM + data in the southern part of the Iranian volcanic sedimentary belt. J. Asian Earth Sci. 24 (2), 237–243. Available from: https://doi.org/10.1016/j.jseaes.2003.11.001.

Ranjbar, H., Masoumi, F., Carranza, E.J.M., 2011. Evaluation of geophysics and spaceborne multispectral data for alteration mapping in the Sar Cheshmeh mining area, Iran. Int. J. Remote Sens. 32 (12), 3309–3327. Available from: https://doi.org/10.1080/01431161003745665; https://www.tandfonline.com/loi/tres20.

Remondino, F., Spera, M.G., Nocerino, E., Menna, F., Nex, F., 2014. State of the art in high density image matching. Photogrammetric Rec. 29 (146), 144–166. Available from: https://doi.org/10.1111/phor.12063; http://www.blackwellpublishing.com/journals/PhotRec.

Robert, F., Kelly, W.C., 1987. Ore-forming fluids in archean gold-bearing quartz veins at the sigma mine, Abitibi greenstone belt, Quebec, Canada. Econ. Geol. 82 (6), 1464–1482. Available from: https://doi.org/10.2113/gsecongeo.82.6.1464; http://economicgeology.org/content/82/6/1464.full.pdf + html. United States.

Robert, F., Boullier, A.M., Firdaous, K., 1995. Gold-quartz veins in metamorphic terranes and their bearing on the role of fluids in faulting. J. Geophys. Res. 100 (7), 12–879.

Robert, F., Poulsen, K.H., 2001. Vein formation and deformation in greenstone gold deposits. Soc. Econ. Geol. Rev. 14, 111–155.

Rockwell, B.W., Hofstra, A.H., 2008. Identification of quartz and carbonate minerals across northern Nevada using aster thermal infrared emissivity data-implications for geologic mapping and mineral resource investigations in well-studied and frontier areas. Geosphere 4 (1), 218–246. Available from: https://doi.org/10.1130/GES00126.1; http://geosphere.gsapubs.org/content/4/1/218.full.pdf#page = 1&view = FitH.

Rosenqvist, A., 1996. The global rain forest mapping project by JERS-1 SAR. Inter. Arch. Photo Remote Sens. 31 (B7), 594–598.

Rosenqvist, A., Shimada, M., Chapman, B., McDonald, K., De Grandi, G., Jonsson, H., et al. 2004 An overview of the JERS-1 SAR global boreal forest mapping (GBFM) project. In: International Geoscience and Remote Sensing Symposium (IGARSS), December 12, 2004, Japan, vol. 2, pp. 1033–1036.

Rosenqvist, A., Shimada, M., Chapman, B., Freeman, A., De Grandi, G., Saatchi, S., et al., 2010. The Global Rain Forest Mapping project – a review. Int. J. Remote Sens. 21 (6-7), 1375–1387. Available from: https://doi.org/10.1080/014311600210227.

Rowan, L.C., Mars, J.C., 2003. Lithologic mapping in the Mountain Pass, California area using Advanced Spaceborne Thermal Emission and Reflection Radiometer (ASTER) data. Remote Sens. Environ. 84 (3), 350–366. Available from: https://doi.org/10.1016/S0034-4257(02)00127-X.

Rowan, L.C., Wetlaufer, P.H., Goetz, A.F.H., Billingsley, F.C., Stewart, J.H., 1974. Discrimination of rock types and detection of hydrothermally altered areas in south-central Nevada by the use of computer-enhanced ERTS images. U. S. Geol. Surv. Prof. Pap. 883.

Rowan, L.C., Crowley, J.K., Schmidt, R.G., Ager, C.M., Mars, J.C., 2000. Mapping hydrothermally altered rocks by analyzing hyperspectral image (AVIRIS) data of forested areas in the Southeastern United States. J. Geochem. Explor 68 (3), 145–166. Available from: https://doi.org/10.1016/S0375-6742(99)00081-3.

Rowan, L.C., Hook, S.J., Abrams, M.J., Mars, J.C., 2003. Mapping hydrothermally altered rocks at Cuprite, Nevada, using the advanced spaceborne thermal emission and reflection radiometer (Aster), a new satellite-imaging system. Econ. Geol. 98 (5), 1019–1027. Available from: https://doi.org/10.2113/gsecongeo.98.5.1019.

Rowan, L.C., Mars, J.C., Simpson, C.J., 2005. Lithologic mapping of the Mordor, NT, Australia ultramafic complex by using the Advanced Spaceborne Thermal Emission and Reflection Radiometer (ASTER). Remote Sens. Environ. 99 (1-2), 105–126. Available from: https://doi.org/10.1016/j.rse.2004.11.021.

Rowan, L.C., Schmidt, R.G., Mars, J.C., 2006. Distribution of hydrothermally altered rocks in the Reko Diq, Pakistan mineralized area based on spectral analysis of ASTER data. Remote Sens. Environ. 104 (1), 74–87. Available from: https://doi.org/10.1016/j.rse.2006.05.014.

Roy, D.P., Wulder, M.A., Loveland, T.R., Woodcock, C.E., Allen, R.G., Anderson, M.C., et al., 2014. Landsat-8: science and product vision for terrestrial global change research. Remote Sens. Environ. 145, 154–172. Available from: https://doi.org/10.1016/j.rse.2014.02.001.

Sabine, C., Realmuto, V.J., Taranik, J.V., 1994. Quantitative estimation of granitoid composition from thermal infrared multispectral scanner (TIMS) data, Desolation Wilderness, northern Sierra Nevada, California. J. Geophys. Res. 99 (3), 4261–4271. Available from: https://doi.org/10.1029/93JB03127.

Sabins, F.F., 1987. Remote Sensing – Principles and Interpretation. WH Freeman and Company.

Sabins, F.F., 1999. Remote sensing for mineral exploration. Ore Geol. Rev. 14 (3–4), 157–183. Available from: https://doi.org/10.1016/S0169-1368(99)00007-4.

Safari, M., Maghsoudi, A., Pour, A.B., 2018. Application of Landsat-8 and ASTER satellite remote sensing data for porphyry copper exploration: a case study from Shahr-e-Babak, Kerman, south of Iran. Geocarto Int. 33 (11), 1186–1201. Available from: https://doi.org/10.1080/10106049.2017.1334834; http://www.tandfonline.com/toc/tgei20/current.

Salisbury, J.W., D'Aria, D.M., 1992. Emissivity of terrestrial materials in the 8-14 μm atmospheric window. Remote Sens. Environ. 42 (2), 83–106. Available from: https://doi.org/10.1016/0034-4257(92)90092-X.

Salisbury, J.W., Walter, L.S., 1989. Thermal infrared (2.5-13.5 μm) spectroscopic remote sensing of igneous rock types on particulate planetary surfaces. J. Geophys. Res. 94 (7), 9192–9202. Available from: https://doi.org/10.1029/JB094iB07p09192.

Salles, Rd.R., de Souza Filho, C.R., Cudahy, T., Vicente, L.E., Monteiro, L.V.S., 2017. Hyperspectral remote sensing applied to uranium exploration: a case study at the Mary Kathleen metamorphic-hydrothermal U-REE deposit, NW, Queensland, Australia. J. Geochem. Explor. 179, 36–50. Available from: https://doi.org/10.1016/j.gexplo.2016.07.002; http://www.sciencedirect.com/science/journal/03756742.

Santosh, M., Groves, D.I., 2022. Global metallogeny in relation to secular evolution of the Earth and supercontinent cycles. Gondwana Res. 107, 395–422. Available from: https://doi.org/10.1016/j.gr.2022.04.007; http://www.sciencedirect.com/science/journal/1342937X.

Scheidt, S., Ramsey, M., Lancaster, N., 2008. Radiometric normalization and image mosaic generation of ASTER thermal infrared data: an application to extensive sand sheets and dune fields. Remote Sens. Environ. 112 (3), 920–933. Available from: https://doi.org/10.1016/j.rse.2007.06.020.

Schmidt, R.G., 1976. Exploration for porphyry copper deposits in Pakistan using digital processing of Landsat-1 data. J. Res. US Geol. Surv. 4 (1), 27–34.

Seibert, J.A., Boone, J.M., Lindfors, K.K., 1998. Flat-field correction technique for digital detectors. Proceedings of SPIE – The International Society for Optical Engineering, December 12, 1998, United States, vol. 3336, pp. 348–354. Available from: https://doi.org/10.1117/12.317034.

Sekandari, M., Masoumi, I., Pour, A.B., Muslim, A.M., Rahmani, O., Hashim, M., et al., 2020. Application of Landsat-8, Sentinel-2, ASTER and Worldview-3 spectral imagery for exploration of carbonate-hosted Pb-Zn deposits in the Central Iranian Terrane (CIT). Remote Sens. 12 (8). Available from: https://doi.org/10.3390/RS12081239; https://www.mdpi.com/2072-4292/12/8/1239.

Sekandari, M., Masoumi, I., Pour, A.B., Muslim, A.M., Hossain, M.S., Misra, A., 2022. ASTER and WorldView-3 satellite data for mapping lithology and alteration minerals associated with Pb-Zn mineralization. Geocarto Int. 37 (6), 1782–1812. Available from: https://doi.org/10.1080/10106049.2020.1790676; http://www.tandfonline.com/toc/tgei20/current.

Shahbazi, M., Théau, J., Ménard, P., 2014. Recent applications of unmanned aerial imagery in natural resource management. GIScience Remote Sens. 51 (4), 339–365. Available from: https://doi.org/10.1080/15481603.2014.926650; http://www.tandfonline.com/loi/tgrs20?open = 50&repitition = 0#vol_50.

Shahriari, H., Ranjbar, H., Honarmand, M., 2013. Image segmentation for hydrothermal alteration mapping using PCA and concentration-area fractal model. Nat. Resour. Res. 22 (3), 191–206. Available from: https://doi.org/10.1007/s11053-013-9211-y.

Sheikhrahimi, A., Pour, A.B., Pradhan, B., Zoheir, B., 2019. Mapping hydrothermal alteration zones and lineaments associated with orogenic gold mineralization using ASTER data: A case study from the Sanandaj-Sirjan Zone, Iran. Adv. Space Res. 63 (10), 3315–3332. Available from: https://doi.org/10.1016/j.asr.2019.01.035; http://www.journals.elsevier.com/advances-in-space-research/.

Sherman, D.M., Waite, T.D., 1985. Electronic spectra of Fe3 + oxides and oxide hydroxides in the near IR to near UV. Am. Mineralogist. 70 (11-12), 1262–1269.

Shirazi, A., Hezarkhani, A., Pour, A.B., Shirazy, A., Hashim, M., 2022. Neuro-Fuzzy-AHP (NFAHP) technique for copper exploration using Advanced Spaceborne Thermal Emission and Reflection Radiometer (ASTER) and geological datasets in the Sahlabad mining area, east Iran. Remote Sens. 14 (22), 5562. Available from: https://doi.org/10.3390/rs14215562.

Shirmard, H., Farahbakhsh, E., Müller, R.D., Chandra, R., 2022. A review of machine learning in processing remote sensing data for mineral exploration. Remote Sens. Environ. 268, 112750. Available from: https://doi.org/10.1016/j.rse.2021.112750.

Sillitoe, R.H., 2010. Porphyry copper systems. Econ. Geol. 105 (1), 3–41. Available from: https://doi.org/10.2113/gsecongeo.105.1.3; http://economicgeology.org/cgi/reprint/105/1/3. United Kingdom.

Singhroy, V.H., 2001. Geological applications of RADARSAT-1: A review. In: International Geoscience and Remote Sensing Symposium (IGARSS), December 12, 2001, Canada, vol. 1, pp. 468–470.

Spatz, D.M., Wilson, R.T., 1995. Remote sensing characteristics of porphyry copper systems, western America Cordillera. p. 20.

Stöcker, C., Bennett, R., Nex, F., Gerke, M., Zevenbergen, J., 2017. Review of the current state of UAV regulations. Remote Sens. 9 (5). Available from: https://doi.org/10.3390/rs9050459; http://www.mdpi.com/2072-4292/9/5/459/pdf.

Stone, J.V., 2004. Independent Component Analysis. John Wiley, New York.

Story, M., Congalton, R.G., 1986. Accuracy assessment: a user's perspective. Photogrammetric Eng. Remote Sens. 52 (3), 397–399.

Sun, Y., Tian, S., Di, B., 2017. Extracting mineral alteration information using WorldView-3 data. Geosci. Front. 8 (5), 1051–1062. Available from: https://doi.org/10.1016/j.gsf.2016.10.008; https://www.sciencedirect.com/journal/geoscience-frontiers.

Takodjou Wambo, J.D., Pour, A.B., Ganno, S., Asimow, P.D., Zoheir, B., Salles, Rd.R., et al., 2020. Identifying high potential zones of gold mineralization in a sub-tropical region using Landsat-8 and ASTER remote sensing data: A case study of the Ngoura-Colomines goldfield, eastern Cameroon. Ore Geol. Rev. 122. Available from: https://doi.org/10.1016/j.oregeorev.2020.103530; http://www.sciencedirect.com/science/journal/01691368.

Thome, K., Palluconi, F., Takashima, T., Masuda, K., 1998. Atmospheric correction of ASTER. IEEE Trans. Geosci. Remote Sens. 36 (4), 1199–1211. Available from: https://doi.org/10.1109/36.701026.

Thome, K.J., Biggar, S.F., Wisniewski, W., 2003. Cross comparison of EO-1 sensors and other earth resources sensors to Landsat-7 ETM + using Railroad Valley Playa. IEEE Trans. Geosci. Remote Sens. 41 (6), 1180–1188. Available from: https://doi.org/10.1109/TGRS.2003.813210.

Tonooka, H., Iwasaki, A., 2004. Improvement of ASTER/SWIR crosstalk correction. In: Proceedings of SPIE – The International Society for Optical Engineering, May 5 2004, Japan, vol. 5234, pp. 168–179. Available from: https://doi.org/10.1117/12.511811.

Torresan, C., Berton, A., Carotenuto, F., Di Gennaro, S.F., Gioli, B., Matese, A., et al., 2017. Forestry applications of UAVs in Europe: a review. Int. J. Remote Sens. 38 (8-10), 2427–2447. Available from: https://doi.org/10.1080/01431161.2016.1252477; https://www.tandfonline.com/loi/tres20.

Tözün, K.A., Özyavaş, A., 2020. New logical operator algorithms for mapping of hydrothermally altered rocks using ASTER data: a case study from central Turkey. Ore Geol. Rev. 122. Available from: https://doi.org/10.1016/j.oregeorev.2020.103533; http://www.sciencedirect.com/science/journal/01691368.

Traore, M., Takodjou Wambo, J.D., Ndepete, C.P., Tekin, S., Pour, A.B., Muslim, A.M., 2020. Lithological and alteration mineral mapping for alluvial gold exploration in the south east of Birao area, Central African Republic using Landsat-8 Operational Land Imager (OLI) data. J. Afr. Earth Sci. 170. Available from: https://doi.org/10.1016/j.jafrearsci.2020.103933; http://www.sciencedirect.com/science/journal/1464343X.

Ungar, S.G., 2002. Overview of the earth observing one (EO-1) mission. IEEE Trans. Geosci. Remote Sens. 568–571.

Ungar, S.G., Pearlman, J.S., Mendenhall, J.A., Reuter, D., 2003. Overview of the Earth Observing One (EO-1) mission. IEEE Trans. Geosci. Remote Sens. 41 (6), 1149–1159. Available from: https://doi.org/10.1109/TGRS.2003.815999.

van der Meer, F., 2004. Analysis of spectral absorption features in hyperspectral imagery. Int. J. Appl. Earth Observ. Geoinform. 5 (1), 55–68. Available from: https://doi.org/10.1016/j.jag.2003.09.001; http://www.elsevier.com/locate/jag.

van der Meer, F., de Jong, S., 2003. Spectral Mapping Methods: Many Problems, Some Solutions φ. 2003 3rd EARSeL Workshop on Imaging Spectroscopy. pp. 146–162.

van der Meer, F.D., van der Werff, H.M.A., van Ruitenbeek, F.J.A., Hecker, C.A., Bakker, W.H., Noomen, M.F., et al., 2012. Multi- and hyperspectral geologic remote sensing: a review. Int. J. Appl. Earth Observ. Geoinform. 14 (1), 112–128. Available from: https://doi.org/10.1016/j.jag.2011.08.002; http://www.elsevier.com/locate/jag.

van der Meer, F.D., van der Werff, H.M.A., van Ruitenbeek, F.J.A., 2014. Potential of ESA's Sentinel-2 for geological applications. Remote Sens. Environ. 148, 124–133.

Van der Werff, H., Van der Meer, F., 2016. Sentinel-2A MSI and Landsat 8 OLI provide data continuity for geological remote sensing. Remote Sens. 8 (11), 883.

Vangi, E., D'Amico, G., Francini, S., Giannetti, F., Lasserre, B., Marchetti, M., et al., 2021. The new hyperspectral satellite PRISMA: imagery for forest types discrimination. Sensors. 21 (4), 1182. Available from: https://doi.org/10.3390/s21041182.

Verdel, C.S., Knepper, D., Livo, K.E., McLemore, V.T., Penn, B., Keller, R., 2001. Mapping minerals at the copper flat porphyry using AVIRIS data. In: 2001 Proceedings of the 10th JPL Airborne Earth Sciences Workshop, pp. 427–433.

Vapnik, V.N., 1998. Statistical Learning Theory. Wiley, New York.

Verhoeven, G.J.J., Loenders, J., Vermeulen, F., Docter, R., 2009. Helikite aerial photography – a versatile means of unmanned, radio controlled, low-altitude aerial archaeology. Archaeol. Prospect. 16 (2), 125–138. Available from: https://doi.org/10.1002/arp.353Belgium; http://www3.interscience.wiley.com/cgi-bin/fulltext/122342612/PDFSTART.

Vincent, R.K., 1997. Fundamentals of Geological and Environmental Remote Sensing. Prentice Hall.

Volesky, J.C., Stern, R.J., Johnson, P.R., 2003. Geological control of massive sulfide mineralization in the Neoproterozoic Wadi Bidah shear zone, southwestern Saudi Arabia, inferences from orbital remote sensing and field studies. Precambrian Res. 123 (2-4), 235–247. Available from: https://doi.org/10.1016/S0301-9268(03)00070-6; http://www.sciencedirect.com/science/journal/03019268.

Vrabel, J., 1996. Multispectral imagery band sharpening study. Photogrammetric Eng. Remote Sens. 62 (9), 1075–1083.

Waldhoff, G., Bubenzer, O., Bolten, A., Koppe, W., Bareth, G., 2008. Spectral analysis of aster, hyperion, and Quickbird data for geomorphological and geological research in Egypt (Dakhla Oasis, western desert). In: International Archives of the Photogrammetry, Remote Sensing and Spatial Information Sciences, January 1, 2008 – International Society for Photogrammetry and Remote Sensing Germany, vol. 37, ISPRS Archives 16821750, pp. 1201–1206. http://www.isprs.org/proceedings/XXXVIII/4-W15/.

Walter, C., Braun, A., Fotopoulos, G., 2020. High-resolution unmanned aerial vehicle aeromagnetic surveys for mineral exploration targets. Geophys. Prospect. 68 (1), 334–349. Available from: https://doi.org/10.1111/1365-2478.12914; http://onlinelibrary.wiley.com/journal/10.1111/(ISSN)1365-2478.

Warren, S.G., 1982. Optical properties of snow. Rev. Geophys 20 (1), 67–89. Available from: https://doi.org/10.1029/RG020i001p00067.

Watts, A.C., Ambrosia, V.G., Hinkley, E.A., 2012. Unmanned aircraft systems in remote sensing and scientific research: Classification and considerations of use. Remote Sens. 4 (6), 1671–1692. Available from: https://doi.org/10.3390/rs4061671; http://www.mdpi.com/2072-4292/4/6/1671/pdf. United States.

Wise, S., 2002. GIS Basics. Taylor & Francis, London.

Woodhouse, I.H., 2006. Introduction to Microwave Remote Sensing. Taylor & Francis Group.

Wulder, M.A., White, J.C., Goward, S.N., Masek, J.G., Irons, J.R., Herold, M., et al., 2008. Landsat continuity: issues and opportunities for land cover monitoring. Remote Sens. Environ. 112 (3), 955–969. Available from: https://doi.org/10.1016/j.rse.2007.07.004.

Xiang, T.Z., Xia, G.S., Zhang, L., 2019. Mini-unmanned aerial vehicle-based remote sensing: techniques, applications, and prospects. IEEE Geosci. Remote Sens. Mag. 7 (3), 29–63. Available from: https://doi.org/10.1109/MGRS.2019.2918840; http://ieeexplore.ieee.org/xpl/RecentIssue.jsp?punumber = 6245518.

Xie, Y., Gao, H., Kong, H., Zheng, H., 2022. Structural controls on mineralization within the Huanggou gold deposit in the Southern Mesozoic Xuefengshan Orogen, South China. Minerals. 12 (6), 751. Available from: https://doi.org/10.3390/min12060751.

Yajima, T., Yamaguchi, Y., 2013. Geological mapping of the Francistown area in northeastern Botswana by surface temperature and spectral emissivity information derived from Advanced Spaceborne Thermal Emission and Reflection Radiometer (ASTER) thermal infrared data. Ore Geol. Rev. 53, 134–144. Available from: https://doi.org/10.1016/j.oregeorev.2013.01.005.

Yamaguchi, Y., Fujisada, H., Kahle, A.B., Tsu, H., Kato, M., Watanabe, H., et al., 2001. ASTER instrument performance, operation status, and application to Earth sciences. In: International Geoscience and Remote Sensing Symposium (IGARSS), December 12, 2001, Japan, vol. 3, pp. 1215–1216.

Yamaguchi, Y.I., Fujisada, H., Kudoh, M., Kawakami, T., Tsu, H., Kahle, A.B., Pniel, M., 1999. ASTER instrument characterization and operation scenario. Adv. Space Res. 23 (8), 1415–1424.

Yao, H., Qin, R., Chen, X., 2019. Unmanned aerial vehicle for remote sensing applications – a review. Remote Sens. 11 (12), 1443. Available from: https://doi.org/10.3390/rs11121443; https://res.mdpi.com/remotesensing/remotesensing-11-01443/article_deploy/remotesensing-11-01443-v2.pdf?filename = &attachment = 1.

Yousefi, M., Tabatabaei, S.H., Rikhtehgaran, R., Pour, A.B., Pradhan, B., 2021. Application of dirichlet process and support vector machine techniques for mapping alteration zones associated with porphyry copper deposit using aster remote sensing imagery. Minerals. 11 (11). Available from: https://doi.org/10.3390/min11111235; https://www.mdpi.com/2075-163X/11/11/1235/pdf.

Yousefi, M., Tabatabaei, S.H., Rikhtehgaran, R., Pour, A.B., Pradhan, B., 2022. Detection of alteration zones using the Dirichlet process Stick-Breaking model-based clustering algorithm to hyperion data: the case study of Kuh-Panj porphyry copper deposits, Southern Iran. Geocarto Int. Available from: https://doi.org/10.1080/10106049.2022.2025917; http://www.tandfonline.com/toc/tgei20/current.

Zhang, X., Pazner, M., 2007. Comparison of lithologic mapping with ASTER, Hyperion, and ETM data in the southeastern Chocolate Mountains, USA. Photogrammetric Eng. Remote Sens. 73 (5), 555–561. Available from: https://doi.org/10.14358/PERS.73.5.555; http://www.asprs.org/Photogrammetric-Engineering-and-Remote-Sensing/PE-RS-Journals.html.

Zhang, X., Pazner, M., Duke, N., 2007. Lithologic and mineral information extraction for gold exploration using ASTER data in the south Chocolate Mountains (California). ISPRS J. Photogrammetry Remote Sens. 62 (4), 271–282. Available from: https://doi.org/10.1016/j.isprsjprs.2007.04.004.

Zhou, Y., Xu, D., Dong, G., Chi, G., Deng, T., Cai, J., et al., 2021. The role of structural reactivation for gold mineralization in northeastern Hunan Province, South China. J. Struct. Geol. 145, 104306. Available from: https://doi.org/10.1016/j.jsg.2021.104306.

Zoheir, B., Emam, A., 2012. Integrating geologic and satellite imagery data for high-resolution mapping and gold exploration targets in the South Eastern Desert, Egypt. J. Afr. Earth Sci. 66–67, 22–34. Available from: https://doi.org/10.1016/j.jafrearsci.2012.02.007.

Zoheir, B., Emam, A., 2014. Field and ASTER imagery data for the setting of gold mineralization in Western Allaqi-Heiani belt, Egypt: a case study from the Haimur deposit. J. Afr. Earth Sci. 99 (1), 150–164. Available from: https://doi.org/10.1016/j.jafrearsci.2013.06.006; http://www.sciencedirect.com/science/journal/1464343X.

Zoheir, B., Emam, A., Abd El-Wahed, M., Soliman, N., 2019a. Gold endowment in the evolution of the Allaqi-Heiani suture, Egypt: a synthesis of geological, structural, and space-borne imagery data. Ore Geol. Rev. 110, 102938. Available from: https://doi.org/10.1016/j.oregeorev.2019.102938.

Zoheir, B., Emam, A., Abdel-Wahed, M., Soliman, N., 2019b. Multispectral and radar data for the setting of gold mineralization in the South Eastern Desert, Egypt. Remote Sens. 11 (12), 1450. Available from: https://doi.org/10.3390/rs11121450.

Zoheir, B., El-Wahed, M.A., Pour, A.B., Abdelnasser, A., 2019c. Orogenic gold in transpression and transtension zones: field and remote sensing studies of the Barramiya-Mueilha sector, Egypt. Remote Sens. 11 (18). Available from: https://doi.org/10.3390/rs11182122; https://res.mdpi.com/d_attachment/remotesensing/remotesensing-11-02122/article_deploy/remotesensing-11-02122.pdf.

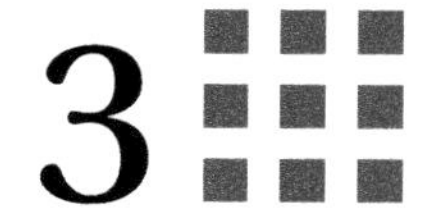

The geographical information system toolbox for mineral exploration

Amin Beiranvand Pour[1], Renguang Zuo[2], Jeff Harris[3]

[1]INSTITUTE OF OCEANOGRAPHY AND ENVIRONMENT (INOS), UNIVERSITY MALAYSIA TERENGGANU (UMT), KUALA NERUS, TERENGGANU, MALAYSIA [2]STATE KEY LABORATORY OF GEOLOGICAL PROCESSES AND MINERAL RESOURCES, CHINA UNIVERSITY OF GEOSCIENCES, WUHAN, P.R. CHINA [3]METAL EARTH RESEARCH CENTER, HARQUAIL SCHOOL OF MINES, LAURENTIAN UNIVERSITY, SUDBURY, ON, CANADA

3.1 The geographical information system toolbox for mineral exploration

3.1.1 Introduction

Despite numerous applications of geographical information system (GIS) in earth sciences and mineral exploration, ranging from visualization to manipulation of geoscientific databases, this chapter focuses on predicting the location of both known and unknown mineral deposits using a GIS. In mineral exploration, different datasets gathered from diverse sources (i.e., geochemical, geophysical, geological, and remotely sensed datasets) should be managed and integrated into predictive models that show the probability of discovering new mineral deposits of a certain type. This process is often referred to as mineral prospectivity modeling/mapping (MPM). Since laying the foundations of MPM by Bonham-Carter and Agterberg (1990), there have been many studies and/or reviews explaining the fundamentals and advances of GIS-based MPM (e.g., Carranza, 2008; Zuo, 2020). This chapter further summarizes the main aspects of MPM from the authors' point of view.

A geological framework describing certain types of mineral deposits, a GIS for managing and manipulating the spatial datasets gathered from different sources, and mathematical frameworks, including numerical and machine-learning techniques, for generating predictive models constitute the building blocks of MPM. A geological framework helps determine a suite of predictive variables that are used for modeling. Here, predictive variables, or the so-called evidence layers, are a set of spatially modeled exploration targeting vectors that describe certain features of a mineral deposit type or processes involved in the formation,

Geospatial Analysis Applied to Mineral Exploration. DOI: https://doi.org/10.1016/B978-0-323-95608-6.00003-2

deposition, and preservation of ore deposits. A GIS is employed to visualize and manipulate these variables. Mathematical frameworks are then used for integrating these variables into predictive models of mineral prospectivity.

Initially, a general description of GIS is given in this chapter, followed by describing the geological framework commonly employed in MPM. Then, two major categories of mathematical frameworks, including conceptual and empirical methods, are discussed. The final section describes a GIS-based predictive model of porphyry copper exploration in Central Iran.

3.2 Geographical information system

As discussed in the introduction, a GIS is used to generate, visualize, and manipulate predictive variables derived from geochemical, geophysical, geological, drilling, and remote sensing datasets, helping determine areas with commonalities in geophysical, geochemical, and geological anomalies. Given the definition mentioned earlier, one should select a unique cell size for all predictive variables, and the GIS is then used to overlay the variables. Several factors contribute to selecting a suitable unit cell size for MPM. These include the type of datasets being used, the extent of the study area, and the complexity of bedrock geology in a given study area (Hengl, 2006; Carranza, 2009; Zuo, 2012).

With respect to geochemical data, sampling density (d) determines the unit cell size. Sampling density is defined by $d = A/n$, where A and n are the total area and the total number of samples, respectively. From a geostatistical point of view, selecting a coarse unit cell size for a relatively dense geochemical sampling scheme can negatively affect the recognition of geochemical anomalies. Likewise, selecting a fine cell size for a sparse set of geochemical samples can lead to biased visualization and modeling of geochemical anomalies. The denser the sampling scheme, the finer the unit cell size required. As a rule of thumb, a proper unit cell size is derived by $0.05(d^{0.5})$ (Hengl, 2006). In addition, line spacing (l) of airborne geophysical surveys plays a pivotal role in selecting a proper unique cell size, where $\sim 0.25l$ has been recognized as the proper cell size (Isles and Rankin, 2013).

Predictive variables are determined based on the geological characteristics of the particular mineral deposit type being studied. Two types of geological frameworks, descriptive deposit models (Cox and Singer, 1986) and mineral systems (Knox-Robinson and Wyborn, 1997), are employed to define predictive variables. These intrinsic aspects of GIS-based MPM are explained in the following subsections. It is worth noting that the format for predictive variables employed for MPM can be binary, multiclass, or continuous. Binary variables depict the presence or absence of a given geological setting. For example, a binary variable can model the presence or absence of geochemical anomalies. Multiclass variables are useful when it is necessary to have values for different geological settings. For example, lithological diversity can be modeled by a multiclass variable where different values are assigned to different lithologies. Continuous variables such as geophysical information are used for representing a continuous pattern.

3.2.1 Descriptive models of mineral deposits

Descriptive deposit models focus on local features of mineral deposits. The features described by these models are alteration patterns, geochemical depletion and enrichment patterns of certain elements, local geological features, local structures, and feeder zones (i.e., faults). These are usually key exploration vectors that have long been used to define a set of predictive variables (e.g., Porwal et al., 2003a). For example, it has long been recognized that porphyry copper deposits display certain alteration patterns, leading to key characteristics that can often be recognized by remote sensing and geophysical surveys (Sillitoe, 2010). Also, felsic–siliciclastic volcanic-hosted massive sulfide deposits might locally be capped by an exhalative iron formation (Peter, 2003), leading to key geophysical and geochemical features that can aid GIS-based MPM.

3.2.2 Mineral systems framework

The presence of mineral deposit models does not mean that mineral deposits of a certain style are identical. Local geological settings exert substantial control over the quality and the extent of certain features of mineral deposits. In other words, there are key differences in size, tonnage, shape, alteration patterns, and local geochemical and geophysical signatures linked to mineral deposits. Accordingly, one might fail to define effective predictive variables using descriptive deposit models because (1) these models strictly adhere to local-scale features and (2) they are feature-oriented; that is, descriptive deposit models fail to define effective predictive variables for regional- and district-scale studies. To account for the preceding problems, many researchers use a process-oriented, scale-free framework, namely, the mineral systems approach, to define predictive variables. They usually start by grouping key ore-forming processes into a suite of components, namely, (1) drivers, (2) sources, (3) pathways, and (4) traps (Mccuaig et al., 2010). A series of geochemical or tectonic processes that trigger magma or fluid flow, namely, drivers, leads to accumulating ligands or metals in certain regions, called sources. Ore-bearing fluids then migrate through certain pathways until they are trapped and precipitate their ores/metals. A series of maps that represent the mineral deposit processes is then developed and used as predictor variables in a GIS-based MPM.

3.3 Mathematical frameworks used for geographical information system-based mineral prospectivity mapping

Mathematical frameworks are used within a GIS system to (1) assign weights to predictive variables and (2) integrate various predictive variables into models of mineral prospectivity. Two major categories of mathematical frameworks are used for GIS-based MPM: conceptual and empirical frameworks (Carranza, 2008). The former or the so-called knowledge-driven methods are used for greenfield areas with few known mineralized zones. The latter methods comprise data-driven techniques and are employed for areas with a sufficient number of known mineralized zones that can be used to train and validate the algorithm. These two categories are explained in the upcoming subsections.

3.3.1 Knowledge-driven methods

In the absence of known mineralized zones, knowledge-driven methods are used for assigning weights to predictive variables. Expert opinions are translated into numerical weights directly or through simple or more complex mathematical procedures. A general workflow for knowledge-driven MPM is shown in Fig. 3.1 Traditionally, boolean logic (Harris, 1989; Harris et al., 2001), index-overlay (Rencz et al., 1994), and fuzzy logic (An et al., 1991) have been used for knowledge-driven MPM. Besides these techniques, multicriteria decision-making systems, such as *analytic hierarchy process* (Harris et al., 1995), and *the technique for order of preference by similarity to ideal solution* (TOPSIS: Abedi and Norouzi, 2016), have also been used for MPM. In addition, hybrid techniques combining two numerical techniques have been used (Zuo et al., 2009). Despite differences in these methods, they share a number of commonalities, namely, (1) the initial weighting of predictive variables depends on expert opinion and (2) the use of specific operators, such as fuzzy operators (Dombi, 1982).

One can evaluate the performance of knowledge-driven models of mineral prospectivity in the presence of mineralized zones with fitting-rate (Chung and Fabbri, 2003) curves, also referred to as success-rate (Agterberg and Bonham-Carter, 2005) curves or efficiency of classification curves, and measure how the model matches known mineralized zones. In the fitting-rate curves, horizontal and vertical axes represent the portion of the area delineated by a given threshold value and the percentage of mineralized zones within the delineated area, respectively (Fig. 3.2). The higher a curve is positioned in these plots, the better the performance of a model. In Fig. 3.2, for example, the model depicted by the black curve is more robust compared to the one depicted by the gray curve.

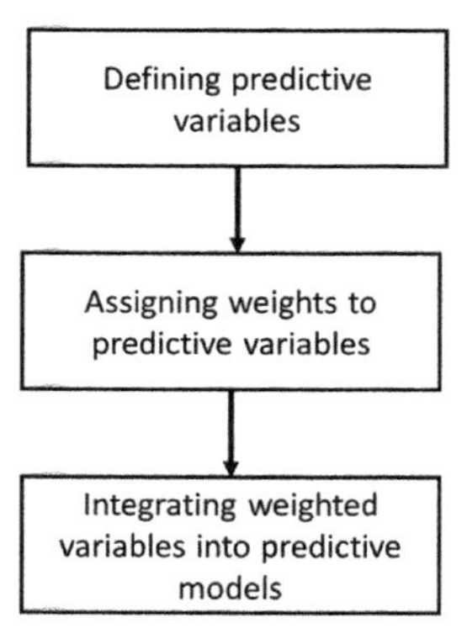

FIGURE 3.1 The general framework of knowledge-driven models of mineral prospectivity.

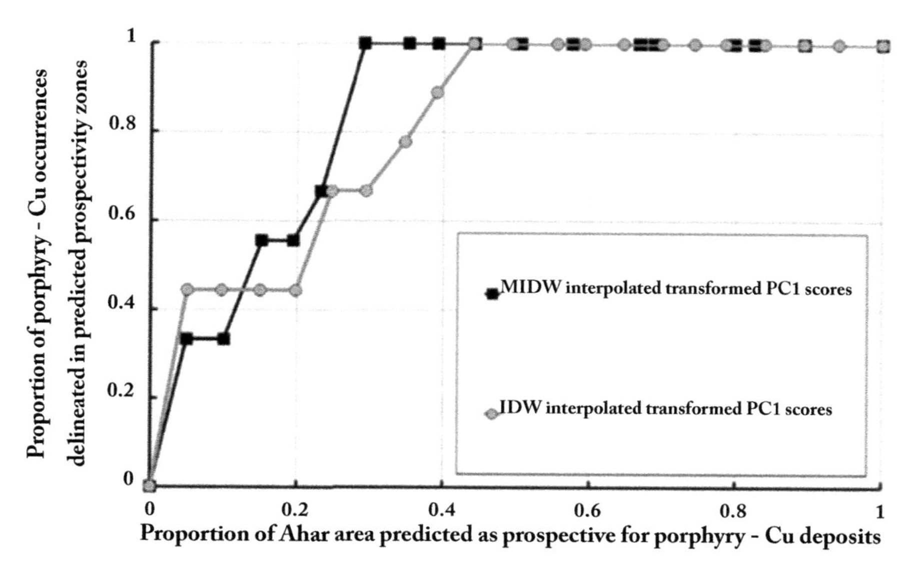

FIGURE 3.2 An example of fitting-rate curves (after Parsa et al., 2017). The model depicted by the black line outperforms the other model as its fitting-rate curve is positioned higher in the diagram.

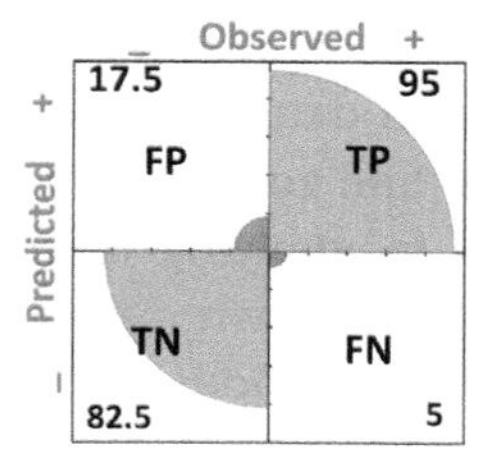

FIGURE 3.3 An example of a contingency plot showing that *TPs* and *TNs* are generally higher than *FNs* and *FPs* (after Parsa and Carranza, 2021). *FNs*, False negatives; *FPs*, false positives; *TNs*, true negatives; *TPs*, true positives.

Besides the fitting-rate curves, receiver operating characteristic (ROC) curves have also been frequently applied to assess a model's performance by measuring four quantities: true positives (TPs), true negatives (TNs), false positives (FPs), and false negatives (FNs) (Swets, 1988) (Fig. 3.3). As shown in Fig. 3.3, these quantities can be visualized by contingency plots. AROC curve plots the FP rate or "1 − specificity" versus the TP rate of "sensitivity" (Fig. 3.4). The former and the latter are derived by FP/(FP + TN) and TP/(TP + FN). The area under the ROC curve (AUC) measures the model's performance. AUC values range between 0.5 and 1, with the former and the latter representing a random pattern and a perfect match, respectively. Although knowledge-driven models of mineral prospectivity remain a feasible alternative for frontier regions and grassroots projects, an element of subjectivity is inextricably intertwined with these models, impeding their reliability (Zuo et al., 2021), which is further explained in the upcoming subsections.

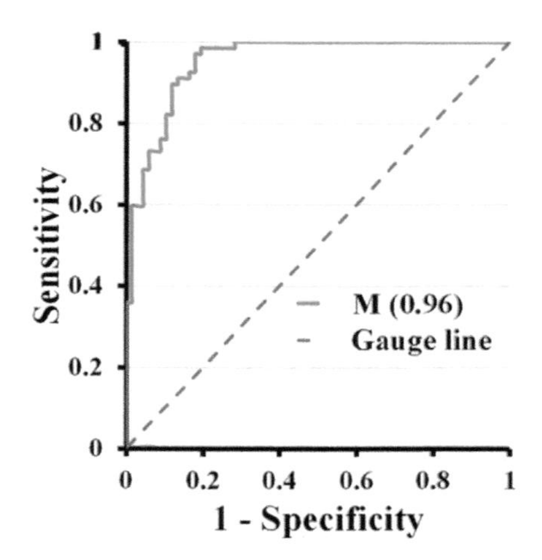

FIGURE 3.4 The ROC curve for the contingency plot of Fig. 3.3 shows an AUC value of 0.96 (after Parsa and Carranza, 2021). M shows a very good model, whereas Gauge represents a model that is no better than random. *AUC*, Area under the receiver operating characteristic curve; *ROC*, receiver operating characteristic.

3.3.2 Data-driven methods

With a sufficient number of known mineralized zones, be they mineral deposits, occurrences, or even drill cores intersecting mineralization, one can use data-driven techniques to create maps of mineral prospectivity. In data-driven methods the known mineralized zones are used to train a given algorithm to find zones of known and unknown potential mineralization. Despite the recent advances in applying these techniques, researchers are yet to reach a consensus on the number of known mineral deposits/occurrences required to develop a data-driven mineral prospectivity model (Zuo, 2020).

The weights of evidence technique (WofE) marks one of the earliest applications of MPM data-driven models (Bonham-Carter, 1989). Using the Bayes theorem, this method estimates the posterior probability of discovering mineralized zones. The locations of known mineralized zones are the only requirement of this method. Presently supervised machine-learning techniques continue (i.e., random forest, neural networks, SVM) to gain popularity in data-driven MPM. Besides known mineralized zones, these techniques also require a set of nonmineralized locations (Zuo, 2020). Although generating a coherent set of nonmineralized locations remains a challenge, there are some guidelines for generating these locations.

Foremost among the features of nonmineralized zones is their spatial distribution; these locations should be randomly distributed in space (Carranza et al., 2008; Carranza and Laborte, 2015). One of the primary justifications for this is that mineral deposits usually form distinctive spatial patterns. Volcanic-hosted massive sulfide deposits, for example, usually appear as clusters (Galley et al., 2007). Also, using a random spatial pattern helps promote the diversity of multiattribute features of nonmineralized zones, which also appears to be essential for data-driven MPM. Important to note is that in regional- to district-scale studies, mineralized and nonmineralized zones are considered points. In addition, some studies suggest an equal number of nonmineralized and mineralized zones are required to establish a balance in data-driven MPM (Porwal et al., 2003b). Also, some researchers suggest selecting nonmineralized zones far enough away from mineralized locations so that they do not share commonalities with mineralized zones is desirable (Carranza and Laborte, 2015). In addition to the previous procedure, for MPM of a given type of mineral deposit, researchers have used the location of other types of mineral deposits as negative labeled samples (e.g., Nykänen et al., 2015). According to this scenario, for example, when the goal is to target areas of high potential for epithermal Au mineralization, one can use the location of carbonate-hosted Zn–Pb deposits as negative labeled samples.

Fig. 3.5 depicts a general framework for developing data-driven models of mineral prospectivity. As shown in this figure, this process starts with selecting positive labeled samples or mineralized zones, followed by selecting negative labeled samples or nonmineralized zones using the earlier mentioned procedures. Labeled samples are then divided into training and testing sets. The former set is employed for training models, while the latter is employed for measuring the performance of models. Different methods for dividing labeled samples into training and testing sets are discussed in Chapter 7.

Bias and variance are used for assessing the performance of data-driven models. Bias evaluates how the training set fits the developed model, and variance evaluates how the

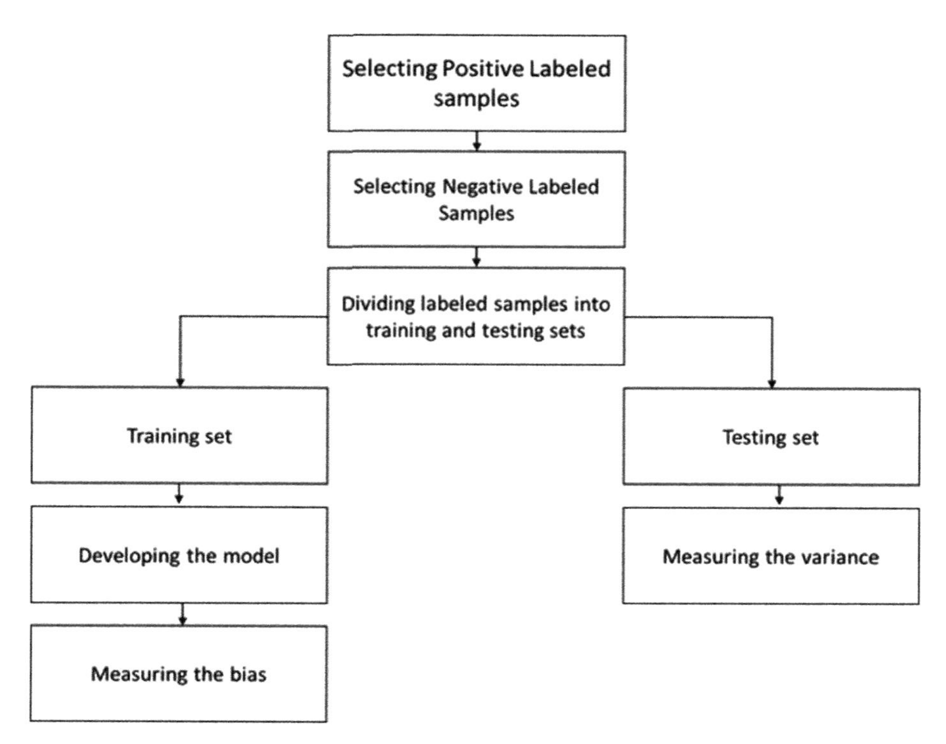

FIGURE 3.5 The general framework of data-driven models of mineral prospectivity.

test data fit the developed model (Parsa and Carranza, 2021). ROC and fitting-rate curves can measure bias and variance. If the latter is used, the fitting-rate curve is called the prediction-rate curve. Chapter 7 further describes these concepts (Parsa, 2021; Parsa and Carranza, 2021).

Using data-driven methods can help avoid subjectivity in MPM. However, since mineralization is generally a scarce phenomenon, the lack of mineralized zones might pose additional problems to data-driven MPM, especially while applying supervised machine-learning methods (Zuo, 2020). These problems are further discussed in Chapter 7. In addition, Chapter 7 discusses methods for verifying data-driven models of mineral prospectivity.

3.4 Uncertainty in geographical information system-based mineral exploration

GIS-based MPM is intrinsically subject to multiple errors and uncertainties (Zuo et al., 2021). Important to note is the difference between the concepts of uncertainty and error. The latter refers to the difference between a real value of something one tries to measure and the measured value. The former, however, refers to the status while there could be tons of measurements for something in the absence of its realistic value, which is the case for predictions (Oberkampf et al., 2002).

As illustrated in Fig. 3.6, different stages of GIS-based MPM are not without errors and uncertainties (Zuo et al., 2021). These will inevitably affect the final predictive model, leading to unreliable predictive models of mineral prospectivity. One should always try to identify

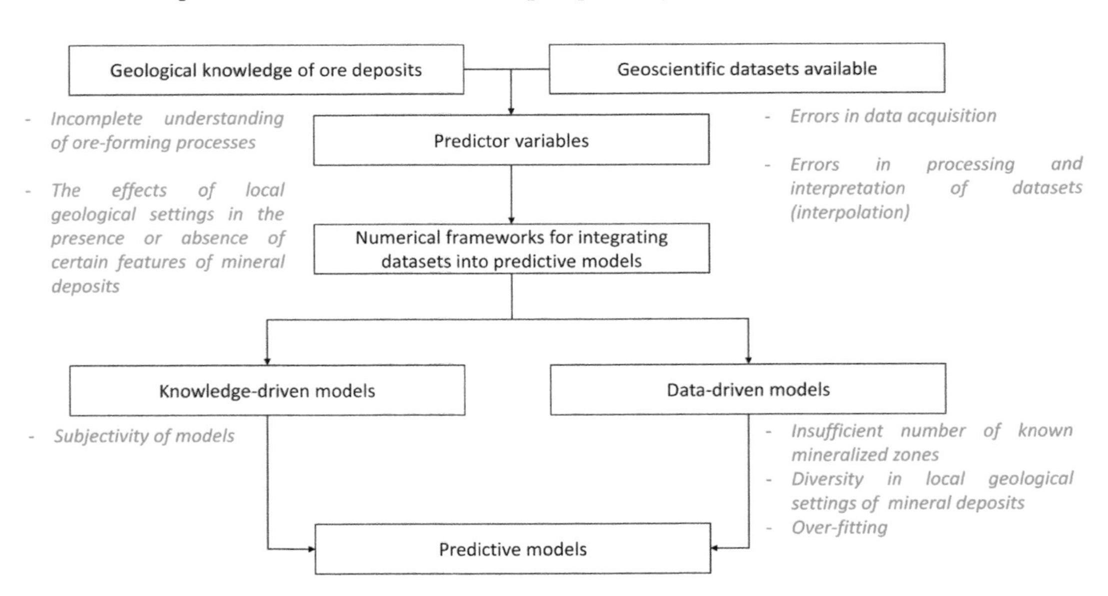

FIGURE 3.6 Uncertainties of GIS-based MPM. *GIS*, Geographical information system; *MPM*, mineral prospectivity modeling/mapping.

the main sources of errors and uncertainty to reduce their effects on MPM. Generally, a significant difference between different predictions for a given cell undermines the reliability of predictions. Such a cell, therefore, should be identified by a high value of uncertainty (Jurado et al., 2015). Uncertainty can be defined by the standard deviation of different predictions for a given cell (Jurado et al., 2015).

Different errors and uncertainties arising from various stages of MPM (Fig. 3.6) will inevitably propagate to the final predictive model, posing challenges to their reliability. Therefore it is important to understand the sources of these uncertainties and try to reduce their effect on the final prospectivity model.

To begin with, our geological assumptions for defining a suite of exploration targeting criteria, be they descriptive deposit models or mineral systems framework, might be erroneous, which is rooted in our incomplete perception of geological, geochemical, and thermodynamic processes that lead to the precipitation of ores from ore-bearing fluids (Mccuaig et al., 2010). Researchers are still pondering multiple questions linked to the formation of mineral deposits, and research is yet to unravel many aspects of the ore-forming processes. Added to this are the effects of local geological settings in the accumulation and preservation of mineralized zones, which could bring about major differences in mineral deposits of the same style (cf. Parsa and Carranza, 2021). Therefore care should be taken while using general assumptions in defining exploration targeting criteria for MPM.

Database errors are perhaps the most serious problems in GIS-based MPM, rendering predictive models unreliable (Zuo et al., 2021). Therefore care should be taken to ensure that the data collected are reliable and rigorous. In addition to the errors linked to databases, data processing is yet another source of uncertainty in GIS-based MPM. For example, geochemical datasets are usually sparsely sampled. One should, therefore, interpolate geochemical samples to create raster maps for MPM. However, there could be multiple outcomes while estimating an interpolated model, meaning that the interpolated models are also uncertain due to the lack of a sufficient number of samples (Wang and Zuo, 2019). A possible solution for reducing these effects is generating multiple scenarios using simulation (e.g., Wang et al., 2020). Sequential Gaussian simulation, for example, has been used to enhance the reliability of geochemical anomaly mapping and MPM (Wang and Zuo, 2019). It should be noted, however, that if there are an insufficient number of samples, the data should not be interpolated. A variogram can determine whether the geochemical data should be interpolated or not or, alternatively, use a small zone of influence around each geochemical sample point (Yuan et al., 2012); lake sediment and stream sediment geochemical data can use the drainage basin as a zone of influence (e.g., Parsa et al., 2016).

Eventually, data integration models are yet another source of uncertainty in GIS-based MPM. As mentioned in the previous sections, knowledge-driven models are plagued by the issue of subjectivity and bias. Human input used for prioritizing and ranking exploration targeting criteria is governed by the geologists' cognitive heuristics and mental traits. Therefore different opinions on weighting and prioritizing exploration targeting criteria result in different predictive models, meaning that there is an uncertainty linked to the subjectivity of opinions in knowledge-driven models of mineral prospectivity (Wang et al., 2020; Parsa and Pour, 2021).

Simulation-based methods have tried to simulate diverse opinions on weighting predictor variables for MPM (e.g., Parsa and Pour, 2021). These frameworks yield an array of predictive models. Average voting is then commonly used for deriving a final model of mineral prospectivity (Parsa and Pour, 2021). These methods also deliver a map of uncertainty, which could be further used for delineating low-risk exploration targets. Low-risk targets have a relatively high average value of prospectivity and low uncertainty (Parsa and Pour, 2021).

Using data-driven methods also leads to other types of uncertainties. First, different sets of random nonmineralized zones could be defined for data-driven models. Each set could lead to a different predictive model (Zuo and Wang, 2020). Besides, despite commonalities in mineral deposits of the same genetic type, local geological settings significantly affect mineral deposits' geochemical and geophysical signatures. While using the location of mineral deposits for training the models, larger mineral deposits with stronger geochemical and geophysical signatures tend to dominate the trends in predictive models. Also, the presence or absence of a given mineral deposit in the training set could significantly change a predictive model. Therefore there is uncertainty linked to the presence or absence of certain mineral deposit locations in the training set. This uncertainty has been accounted for by taking a suite of random subsamples from the training set and developing an individual predictive model with each set. Average voting can be used to select a modulated predictive model from the developed suite of models (Parsa and Carranza, 2021).

3.5 Case studies

The concepts of MPM are explained herein with a case study of knowledge-driven MPM for porphyry copper exploration in Central Iran. The study area is located in the Central Iranian volcano plutonic zone, namely, the Urumieh–Dokhtar magmatic belt. This area hosts a number of prominent porphyry copper deposits, namely, Sarcheshmeh and its neighboring porphyry copper deposits (Fig. 3.7), accounting for significant production of copper. However, the ever-rising demand for copper linked to the green transition and achieving net-zero carbon emissions require discovering further copper resources.

The concepts of mineral systems are applied to translate essential ore-forming processes, namely, source, transport, trapping, and deposition, into a suite of vectors or predictive variables for MPM using geochemical and geological datasets (Table 3.1).

A data-driven model with only five mineral deposits as training locations is insufficient, thus leading to the use of a knowledge-driven method. With knowledge-driven modeling, however, comes the issue of subjectivity as was mentioned previously. A simulation-based framework, summarized in Fig. 3.8, is applied to control the effects of subjectivity in the MPM of this study area. This framework starts with transferring all predictive variables into a [0,1] range, thus helping to standardize the variables. A Monte Carlo simulation (Janssen, 2013) adhering to a normal probability distribution function with the mean and standard deviation of 0.5 and 0.2 was formulated for generating a set of 100 weights between 0 and 1, which were later assigned to individual variables (Fig. 3.9).

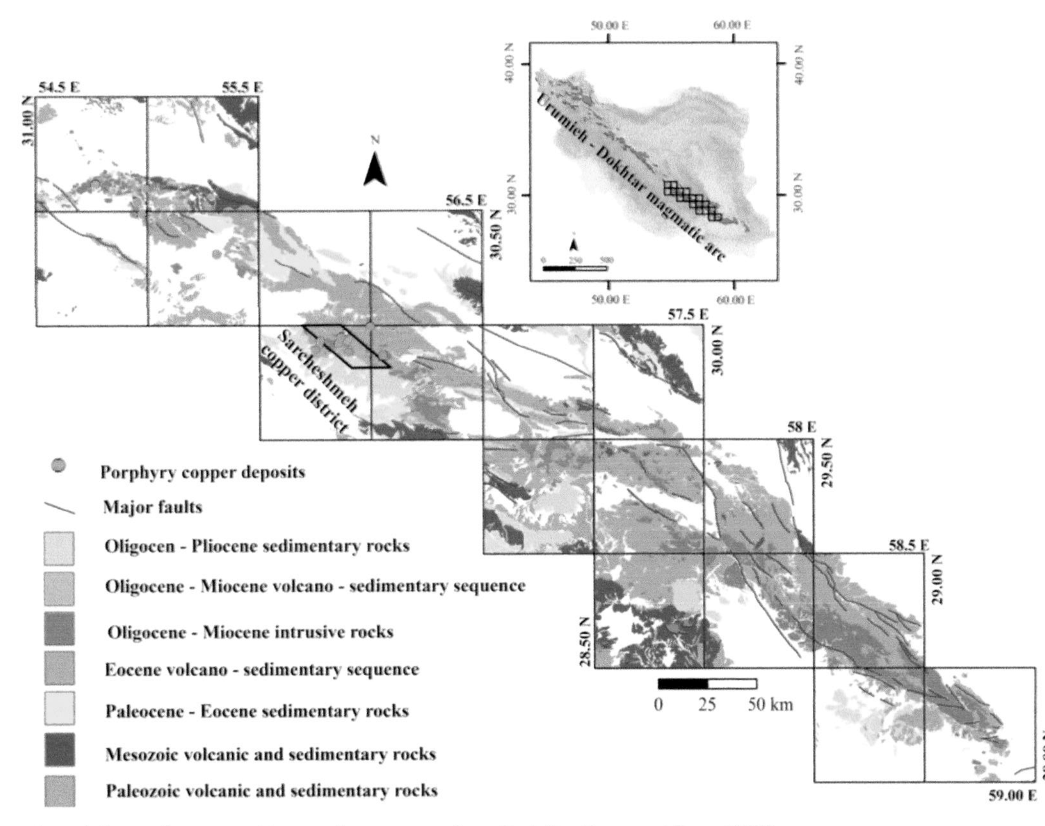

FIGURE 3.7 The location of the study area and its porphyry copper deposits (after Parsa and Pour, 2021).

Table 3.1 Predictive variables derived by mineral systems analysis.

GIS-based exploration targeting criterion	Description	The critical ore-forming process linked to the criterion	The constituent process is linked to the criterion	Data used for generating the criterion
W1	Distance to intrusive rocks	Source	Plutonic and subvolcanic lithologies serve as heat and metal sources for porphyry Cu systems (Sillitoe, 2010)	Outcrops of Oligocene–Miocene plutonic rocks were compiled from the published geological maps of the study district
W2	The density of fault intersections	Transport	Fault intersections are low-pressure zones allowing for the circulation and ascent of ore-bearing fluids (Pirajno, 2012)	Fault intersections were compiled from the digitized geological maps of the study district
W3	N-trending fault density	Transport	Faults serve as pathways for transporting ore-bearing fluids (Pirajno, 2012)	Fault traces were the constituent compiled from the digitized geological maps of the study area
W4	NNW-trending faults' density			
W5	ENE-trending faults' density			
W6	Argillic and phyllic altered zones	Trap	Circulation of hydrothermal fluids, acidic in compositions, promotes chemical reactions with country rocks, leading to the development of certain mineral assemblages in the form of argillic, phyllic, and iron oxide alterations (Sillitoe, 2010)	Remotely sensed ASTER imagery data
W7	Iron oxide-bearing zones			
W8	Multielement geochemical signature linked to mineralization	Deposition	Geochemical halos can represent mineralized zones of porphyry systems (Sillitoe, 2010)	Stream sediment geochemical data were specifically collected by the Geological Survey of Iran

GIS, Geographical information system.

The following demonstrates the set of simulated weights where $w\ j\ k$ is the kth weight assigned to the jth predictor variable ($j = 1, 2, 3, \ldots, n$), and t is the number of iterations:

$$\left(w_j^1, w_j^2, w_j^3, \ldots, w_j^t\right). \quad (3.1)$$

For the kth iteration, the weighted value of each cell, x_i^j (i is the id value assigned to each cell, and j is the number of predictor variables), is calculated using the following equation:

$$x_{ij}^k = \frac{w_j^k \times x_{ij}}{\sum_{j=1}^{n} w_j^k} (j = 1, 2, \ldots, n), (k = 1, 2, \ldots, t). \quad (3.2)$$

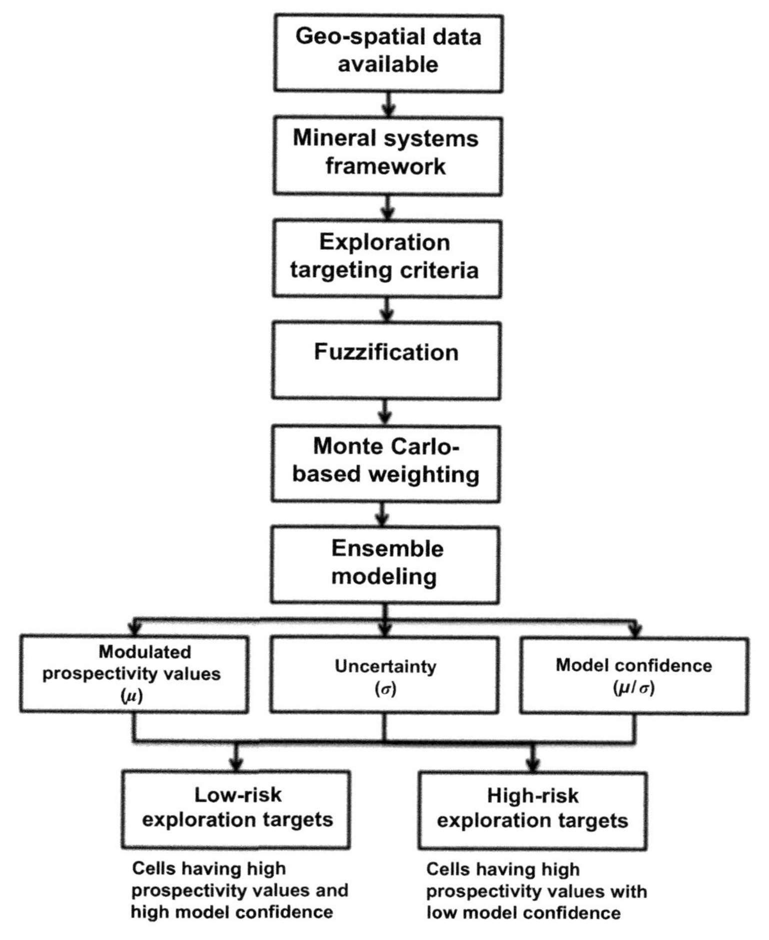

FIGURE 3.8 The framework used for modulating the effect of subjectivity in MPM. *MPM*, Mineral prospectivity modeling/mapping.

The prospectivity value of the *i*th cell in the *k*th iteration is then derived using the following:

$$E^k = \sum_{j=1}^{n} x_{ij}^k (j = 1, 2, \ldots, n), (k = 1, 2, \ldots, t). \tag{3.3}$$

While average voting is used for estimating the modulated prospectivity values, μ, the standard deviation of simulated prospectivity scores for a given cell, is deemed its uncertainty value, U; confidence is also defined by $C = \mu/U$ (Jurado et al., 2015). Maps of modulated prospectivity values, uncertainty, and confidence are demonstrated in Fig. 3.10 Plots of μ against U and μ against C are useful tools for selecting low-risk targets (Fig. 3.11).

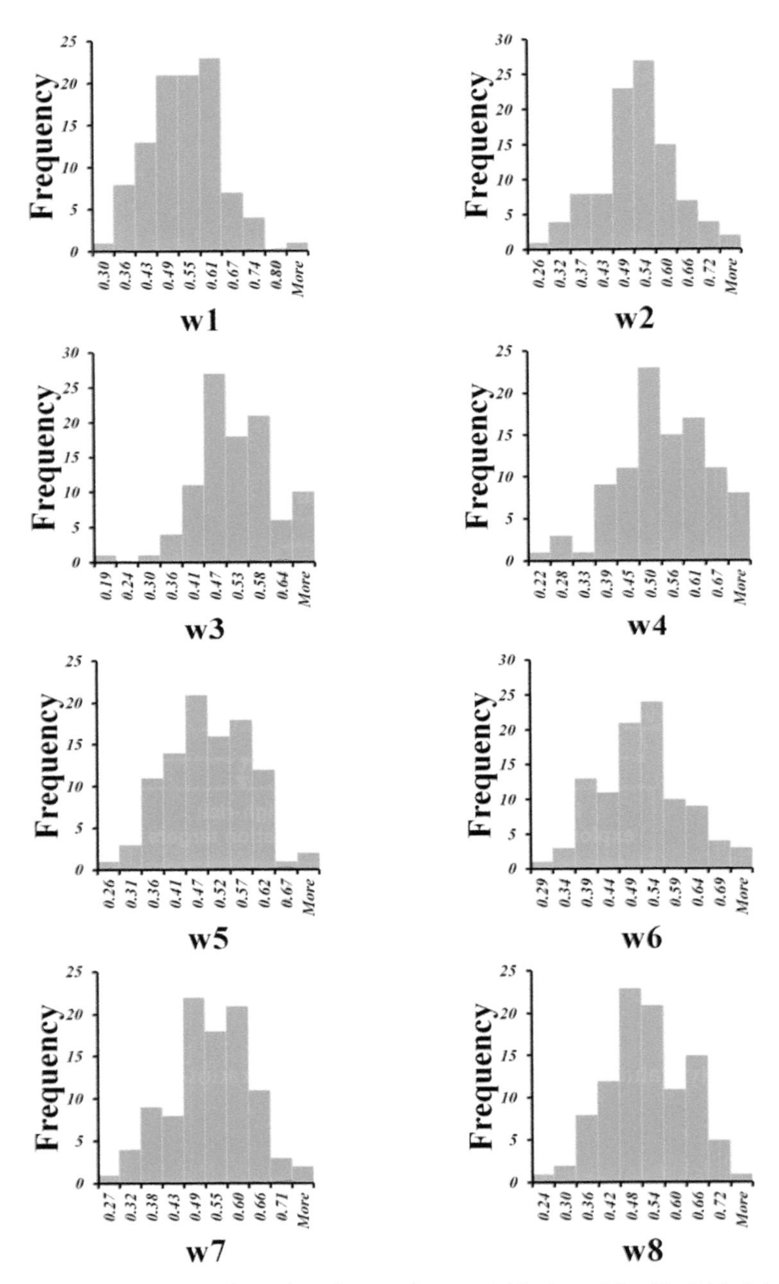

FIGURE 3.9 The suite of simulated weights assigned to predictor variables is explained in Table 3.1.

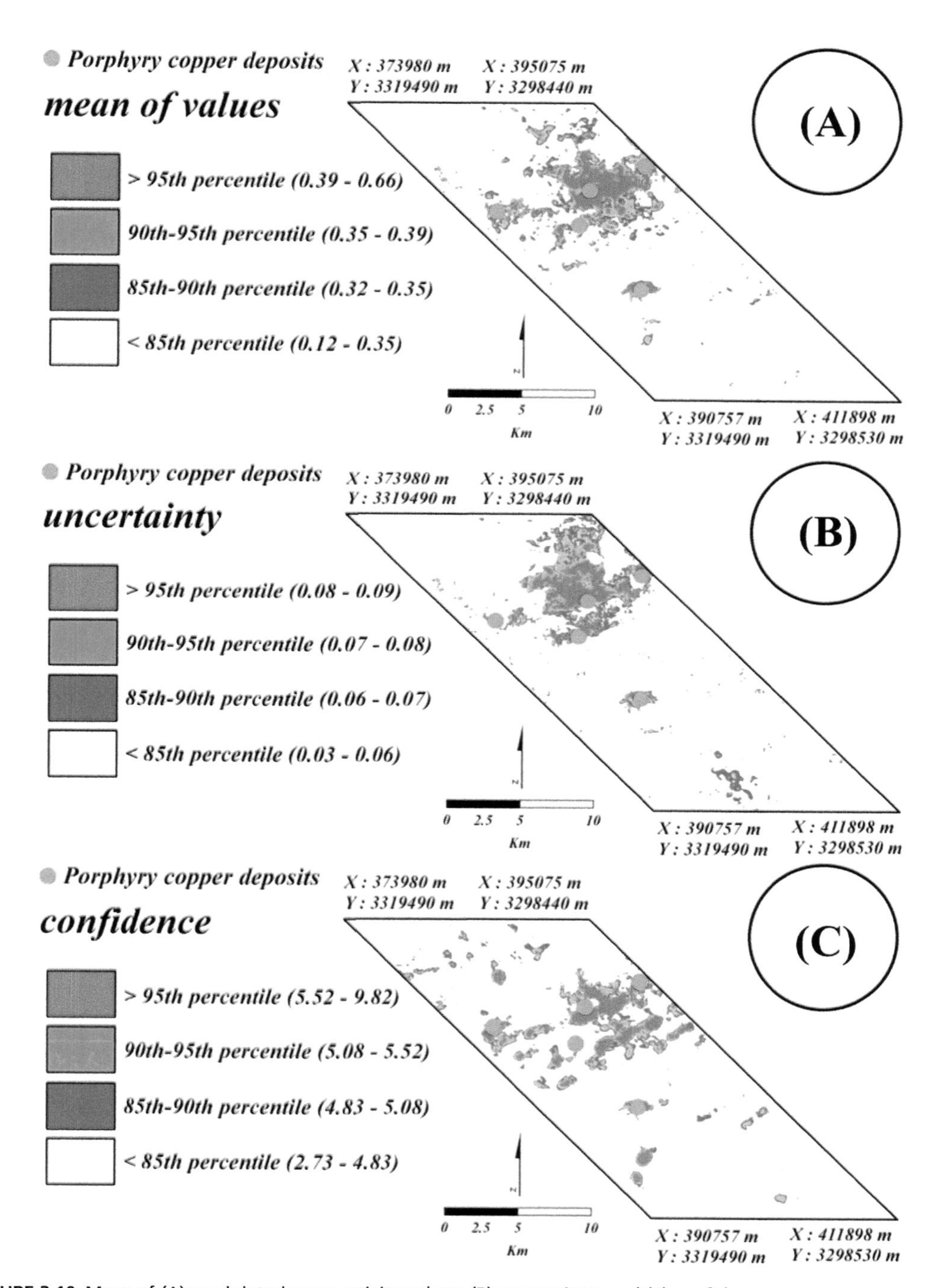

FIGURE 3.10 Maps of (A) modulated prospectivity values, (B) uncertainty, and (C) confidence.

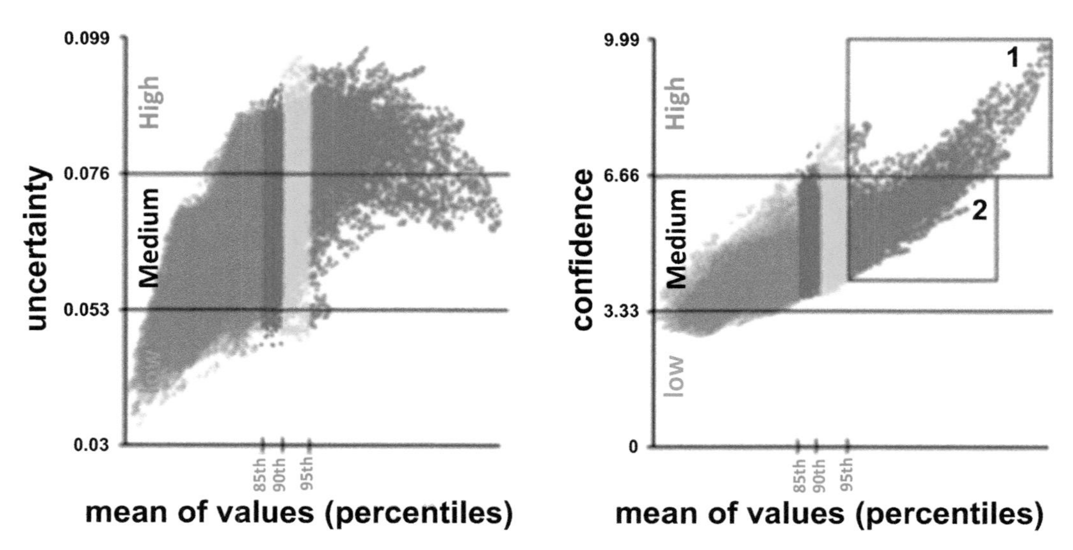

FIGURE 3.11 Plots of modulated prospectivity values versus uncertainty and confidence values. Rectangles 1 and 2 are low- and high-risk exploration targets, respectively.

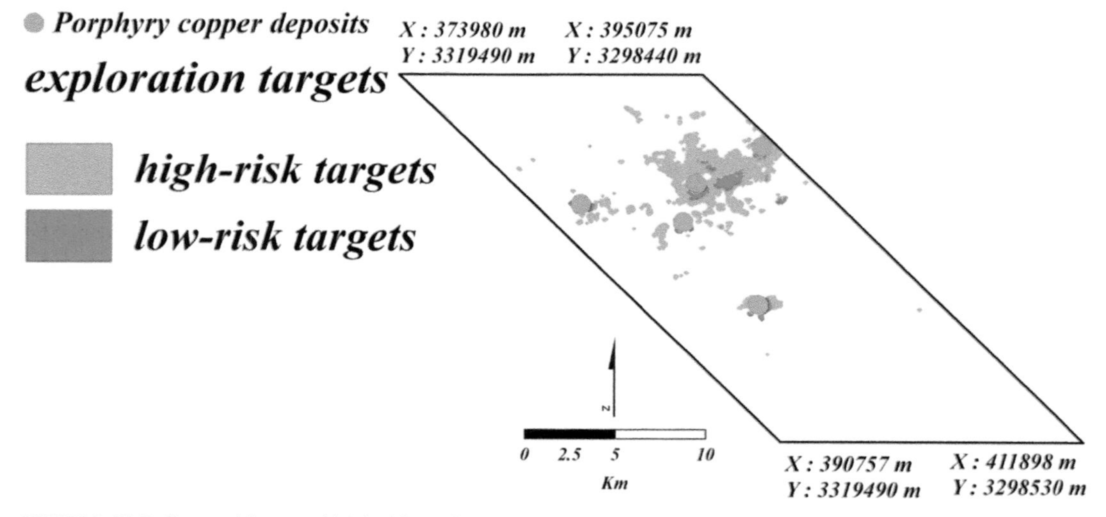

FIGURE 3.12 Delineated low- and high-risk exploration targets.

While having high modulated prospectivity values, low-risk targets are also marked by high confidence and low uncertainty values. High-risk exploration targets, on the other hand, are those marked by high modulated prospectivity values while having high uncertainty values. These targets are selected and demonstrated on the map of exploration targets (Fig. 3.12).

3.6 Summary and conclusion

This chapter has provided an introduction to mineral potential modeling. A GIS is fundamental for compiling, visualizing, and analyzing the data to create predictor variables (vectors to mineralization) for input to knowledge or data-driven algorithms for producing an MPM. The choice of an appropriate unit cell size is critical to the modeling process, and the performance of an MPM can be measured using the efficiency of classification or ROC curves. The MPM process can be characterized by different uncertainties which the geologist must understand to interpret the results correctly. Simulations can reduce the subjectivity of the weights. The deposit model is the most important aspect of the MPM process as it governs what predictor maps are generated as vectors to possible mineralization. An MPM is not a target map per se but rather a map that shows zones of higher mineral potential for a given commodity. Such a map should predict zones of known mineralization, as well as areas of possible new mineralization worthy of exploration, and follow-up.

References

Abedi, M., Norouzi, G.H., 2016. A general framework of TOPSIS method for integration of airborne geophysics, satellite imagery, geochemical and geological data. Int. J. Appl. Earth Observ. Geoinform. 46, 31–44. Available from: https://doi.org/10.1016/j.jag.2015.11.016, http://www.elsevier.com/locate/jag.

Agterberg, F.P., Bonham-Carter, G.F., 2005. Measuring the performance of mineral-potential maps. Nat. Resour. Res. 14 (1), 1–17. Available from: https://doi.org/10.1007/s11053-005-4674-0.

An, W.M., Moon, A.N., Rencz, 1991. Application of fuzzy theory for integration of geological, geophysical and remotely sensed data. Can. J. Explor. Geophys. 27, 1–11.

Bonham-Carter, G.F., 1989. Weights of evidence modelling: a new approach to mapping mineral potential. Stat. Appl. Earth Sci. 13, 171–183.

Bonham-Carter, G.F., Agterberg, F.P., 1990. Application of a Microcomputer-based Geographic Information System to Mineral-Potential Mapping. Elsevier BV, pp. 49–74. Available from: https://doi.org/10.1016/b978-0-08-040261-1.50012-x.

Carranza, E.J.M., Hale, M., Faassen, C., 2008. Selection of coherent deposit-type locations and their application in data-driven mineral prospectivity mapping. Ore Geol. Rev. 33 (3–4), 536–558. Available from: https://doi.org/10.1016/j.oregeorev.2007.07.001.

Carranza, E.J.M., Laborte, A.G., 2015. Data-driven predictive mapping of gold prospectivity, Baguio district, Philippines: application of random forests algorithm. Ore Geol. Rev. 71, 777–787. Available from: https://doi.org/10.1016/j.oregeorev.2014.08.010, http://www.sciencedirect.com/science/journal/01691368.

Carranza, E.J.M., 2009. Objective selection of suitable unit cell size in data-driven modeling of mineral prospectivity. Comput. Geosci. 35 (10), 2032–2046. Available from: https://doi.org/10.1016/j.cageo.2009.02.008.

Carranza, E.J.M., 2008. Geochemical Anomaly and Mineral Prospectivity Mapping in GIS. Elsevier.

Chung, C.J.F., Fabbri, A.G., 2003. Validation of spatial prediction models for landslide hazard mapping. Nat. Hazards 30 (3), 451–472. Available from: https://doi.org/10.1023/B:NHAZ.0000007172.62651.2b.

Cox, D.P., Singer, D.A., 1986. Bulletin: US Gov. Print. Office Mineral Depos. Model.

Dombi, J., 1982. A general class of fuzzy operators, the demorgan class of fuzzy operators and fuzziness measures induced by fuzzy operators. Fuzzy Sets Syst. 8 (2), 149–163. Available from: https://doi.org/10.1016/0165-0114(82)90005-7.

Galley, A.G., Hannington, M.D., Jonasson, I.R., 2007. Volcanogenic massive sulphide deposits. Miner. Depos. Canada: a Synth. major. deposit-types, Dist. metallogeny, evolution Geol. provinces, exploration methods. 5.

Harris, J.R., Wilkinson, L., Broome, J., Fumerton, S., 1995. Proceedings of the Canadian Geomatics Conference. National Defense Mineral Exploration using GIS-Based Favourability Analysis.

Harris, J.R., Wilkinson, L., Heather, K., Fumerton, S., Bernier, M.A., Ayer, J., et al., 2001. Application of GIS processing techniques for producing mineral prospectivity maps – a case study: mesothermal Au in the Swayze Greenstone Belt, Ontario, Canada. Nat. Resour. Res. 10 (2), 91–124. Available from: https://doi.org/10.1023/A:1011548709573, http://www.kluweronline.com/issn/1520-7439.

Harris, J.R., 1989. Data integration for gold exploration in Eastern Nova Scotia using a GIS. Proc. Remote Sens. Explor. Geol. 233–249.

Hengl, T., 2006. Finding the right pixel size. Comput. Geosci. 32 (9), 1283–1298. Available from: https://doi.org/10.1016/j.cageo.2005.11.008.

Isles, D.J., Rankin, L.R., 2013. Geological Interpretation of Aeromagnetic Data. The Australian Society of Exploration Geophysicists.

Janssen, H., 2013. Monte-Carlo based uncertainty analysis: sampling efficiency and sampling convergence. Reliab. Eng. Syst. Saf. 109, 123–132. Available from: https://doi.org/10.1016/j.ress.2012.08.003.

Jurado, K., Ludvigson, S.C., Ng, S., 2015. Measuring uncertainty. Am. Econ. Rev. 105, 1177–1216. American Economic Association, United States. Available from: http://pubs.aeaweb.org/doi/pdfplus/10.1257/aer.20131193.

Knox-Robinson, C.M., Wyborn, L.A.I., 1997. Towards a holistic exploration strategy: using geographic information systems as a tool to enhance exploration. Austr. J. Earth Sci. 44 (4), 453–463. Available from: https://doi.org/10.1080/08120099708728326.

Mccuaig, T.C., Beresford, S., Hronsky, J., 2010. Translating the mineral systems approach into an effective exploration targeting system. Ore Geol. Rev. 38 (3), 128–138. Available from: https://doi.org/10.1016/j.oregeorev.2010.05.008.

Nykänen, V., Lahti, I., Niiranen, T., Korhonen, K., 2015. Receiver operating characteristics (ROC) as validation tool for prospectivity models – a magmatic Ni–Cu case study from the Central Lapland Greenstone Belt, Northern Finland. Ore Geol. Rev. 71, 853–860. Available from: https://doi.org/10.1016/j.oregeorev.2014.09.007, http://www.sciencedirect.com/science/journal/01691368.

Oberkampf, W.L., DeLand, S.M., Rutherford, B.M., Diegert, K.V., Alvin, K.F., 2002. Error and uncertainty in modeling and simulation. Reliab. Eng. Syst. Saf. 75 (3), 333–357. Available from: https://doi.org/10.1016/S0951-8320(01)00120-X.

Parsa, M., Carranza, E.J.M., 2021. Modulating the impacts of stochastic uncertainties linked to deposit locations in data-driven predictive mapping of mineral prospectivity. Nat. Resour. Res. 30 (5), 3081–3097. Available from: https://doi.org/10.1007/s11053-021-09891-9, https://link.springer.com/journal/11053.

Parsa, M., Maghsoudi, A., Yousefi, M., Carranza, E.J.M., 2017. Multifractal interpolation and spectrum–area fractal modeling of stream sediment geochemical data: implications for mapping exploration targets. J. Afr. Earth Sci. 128, 5–15. Available from: https://doi.org/10.1016/j.jafrearsci.2016.11.021, http://www.sciencedirect.com/science/journal/1464343X.

Parsa, M., Maghsoudi, A., Yousefi, M., Sadeghi, M., 2016. Recognition of significant multi-element geochemical signatures of porphyry Cu deposits in Noghdouz area, NW Iran. J. Geochem. Explor. 165, 111–124. Available from: https://doi.org/10.1016/j.gexplo.2016.03.009, http://www.sciencedirect.com/science/journal/03756742.

Parsa, M., Pour, A.B., 2021. A simulation-based framework for modulating the effects of subjectivity in greenfield mineral prospectivity mapping with geochemical and geological data. J. Geochem. Explor. 229, 106838. Available from: https://doi.org/10.1016/j.gexplo.2021.106838.

Parsa, M., 2021. A data augmentation approach to XGboost-based mineral potential mapping: an example of carbonate-hosted Zn Pb mineral systems of Western Iran. J. Geochem. Explor. 228, 106811. Available from: https://doi.org/10.1016/j.gexplo.2021.106811.

Peter, J.M., 2003. Ancient Iron Formations: Their Genesis and Use in the Exploration for Stratiform Base Metal Sulphide Deposits, with Examples from the Bathurst Mining Camp. pp. 145–176.

Pirajno, F., 2012. Hydrothermal Mineral Deposits: Principles and Fundamental Concepts for the Exploration Geologist. Springer Science + Business Media.

Porwal, A., Carranza, E.J.M., Hale, M., 2003a. Artificial neural networks for mineral-potential mapping: a case study from Aravalli Province, Western India. Nat. Resour. Res. 12, 155–171.

Porwal, A., Carranza, E.J.M., Hale, M., 2003b. Knowledge-driven and data-driven fuzzy models for predictive mineral potential mapping. Nat. Resour. Res. 12 (1), 1–25. Available from: https://doi.org/10.1023/A:1022693220894.

Rencz, A.N., Harris, J., Watson, G.P., Murphy, B., 1994. Data integration for mineral exploration in the Antigonish Highlands, Nova Scotia: application of GIS and RS. Can. J. Remote Sens. 20 (3), 257–267.

Sillitoe, R.H., 2010. Porphyry copper systems. Econ. Geol. 105 (1), 3–41. Available from: http://economicgeology.org/cgi/reprint/105/1/3.10.2113/gsecongeo.105.1.3United Kingdom.

Swets, J.A., 1988. Measuring the accuracy of diagnostic systems. Science 240 (4857), 1285–1293. Available from: https://doi.org/10.1126/science.3287615.

Wang, Z., Yin, Z., Caers, J., Zuo, R., Monte, A., 2020. Carlo-based framework for risk-return analysis in mineral prospectivity mapping. Geosci. Front. 11 (6), 2297–2308. Available from: https://doi.org/10.1016/j.gsf.2020.02.010, https://www.sciencedirect.com/journal/geoscience-frontiers.

Wang, J., Zuo, R., 2019. Recognizing geochemical anomalies via stochastic simulation-based local singularity analysis. J. Geochem. Explor. 198, 29–40. Available from: https://doi.org/10.1016/j.gexplo.2018.12.012, http://www.sciencedirect.com/science/journal/03756742.

Yuan, F., Li, X., Jowitt, S.M., Zhang, M., Jia, C., Bai, X., et al., 2012. Anomaly identification in soil geochemistry using multifractal interpolation: a case study using the distribution of Cu and Au in soils from the Tongling mining district, Yangtze metallogenic belt, Anhui province, China. J. Geochem. Explor. 116–117, 28–39. Available from: https://doi.org/10.1016/j.gexplo.2012.03.003.

Zuo, R., Cheng, Q., Agterberg, F.P., 2009. Application of a hybrid method combining multilevel fuzzy comprehensive evaluation with asymmetric fuzzy relation analysis to mapping prospectivity. Ore Geol. Rev. 35 (1), 101–108. Available from: https://doi.org/10.1016/j.oregeorev.2008.11.004.

Zuo, R., Kreuzer, O.P., Wang, J., Xiong, Y., Zhang, Z., Wang, Z., 2021. Uncertainties in GIS-based mineral prospectivity mapping: key types, potential impacts possible solutions. Nat. Resour. Res. 30 (5), 3059–3079. Available from: https://doi.org/10.1007/s11053-021-09871-z, https://link.springer.com/journal/11053.

Zuo, R., Wang, Z., 2020. Effects of random negative training samples on mineral prospectivity mapping. Nat. Resour. Res. 29 (6), 3443–3455. Available from: https://doi.org/10.1007/s11053-020-09668-6, https://link.springer.com/journal/11053.

Zuo, R., 2020. Geodata science-based mineral prospectivity mapping: a review. Nat. Resour. Res. 29 (6), 3415–3424. Available from: https://doi.org/10.1007/s11053-020-09700-9, https://link.springer.com/journal/11053.

Zuo, R., 2012. Exploring the effects of cell size in geochemical mapping. J. Geochem. Explor. 112, 357–367. Available from: https://doi.org/10.1016/j.gexplo.2011.11.001.

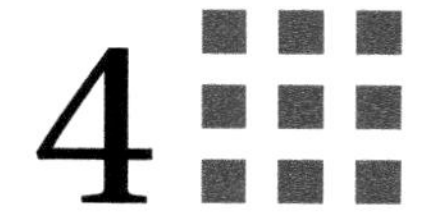

4 Processing and interpretation of geochemical data for mineral exploration

Mohammad Parsa[1], Adel Shirazi[2], Aref Shirazi[2], Amin Beiranvand Pour[3]

[1]MINERAL EXPLORATION RESEARCH CENTRE, HARQUAIL SCHOOL OF EARTH SCIENCES, LAURENTIAN UNIVERSITY, SUDBURY, ON, CANADA [2]FACULTY OF MINING ENGINEERING, AMIRKABIR UNIVERSITY OF TECHNOLOGY, TEHRAN, IRAN [3]INSTITUTE OF OCEANOGRAPHY AND ENVIRONMENT (INOS), UNIVERSITY MALAYSIA TERENGGANU (UMT), KUALA NERUS, TERENGGANU, MALAYSIA

4.1 Introduction

Geochemical surveys are bound to collect samples from diverse types of media, including bedrock, soil, till, stream sediments, and lake sediments, at various extents, ranging from continental to deposit scales. Samples are then analyzed chemically for single- or multielements using the methods that suit the purpose of geochemical surveys (Govett et al., 1975).

The purpose of a geochemical survey, bedrock geology, and geomorphological settings of the study are among the components exerting control over the scale, method of chemical analysis, and sampling media (Levinson, 1974). These surveys reveal beneficial information on the enrichment and depletion of elements that might be associated with mineral deposits. Geochemical surveys can also reveal information on the bedrock geology of the area being studied (e.g., Kirkwood et al., 2016), making them invaluable tools for the study of covered regions.

Generally, the goal of geochemical surveys is to identify some anomalies that are linked to mineral deposits (Carranza, 2008; Carranza et al., 2012). However, there are multiple challenges associated with meeting the foregoing objective. First, the emplacement and preservation of mineral deposits generate different multi- and unielement geochemical signatures (cf., Pirajno, 2012). It is, therefore, of utmost importance to recognize geochemical signatures that describe mineral deposits and ore-forming processes (Parsa et al., 2016). The dispersion pattern of uni- and multielements is a function of the sampling medium, the size fraction of samples being analyzed, the method of chemical analysis, and secondary processes involved in the migration of chemical complexes (cf., McClenaghan et al., 2013).

Geospatial Analysis Applied to Mineral Exploration. DOI: https://doi.org/10.1016/B978-0-323-95608-6.00004-4

It is worth noting that a given mineral deposit may show different geochemical signatures in different sampling media (Levinson, 1974). Thus selecting the right sampling media is a crucial aspect of geochemical surveys. In addition to the previous aspects, defining geochemical anomalies is yet another challenge that requires considering multiple factors (Carranza et al., 2012).

The previous aspects of geochemical datasets are discussed herein. This chapter considers the geospatial aspects of geochemical data and explains how these datasets are collected and how one should analyze and process these datasets to get the most of them for mineral exploration purposes.

This chapter is organized in five sections. After this section the most common geochemical sampling media are described in the second section. The third section describes the common methodologies used for assessing the chemical composition of geochemical samples. This is followed by the interpretation of geochemical datasets and a case study representing the use of a set of geochemical data for mineral exploration.

4.2 Geochemical sampling media

The main goal of a sampling process is to collect a representative sample—a sample that collectively describes the chemical composition of the sampling medium and suits the purpose of the study (McClenaghan et al., 2013). In other words, one should consider the purpose of a geochemical survey and the chemical characteristics of the sampling medium when conducting a sampling survey.

The complexity of geological settings, the extent of the study area, the budget, and the purpose of the study are among the factors that control the density of geochemical sampling. As an example, only a few samples might be enough while trying to determine the average composition of a homogenous rock formation. On the other hand, while studying an area with variable lithological units, different alteration halos, and complex geomorphological setting, a dense sampling scheme might be required to account for the variability and complexity of the area. In the latter example an enough number of samples must be collected from each geological facies to reveal their average chemical composition (Haldar, 2013).

Generally, denser sampling schemes are well suited for mineral exploration purposes. This is owing to the fact that a dense sampling scheme is useful for (1) describing the underlying geological variability of an area and (2) recognizing anomalous dispersion patterns of elements and multielements (Garrett, 1969).

Common geochemical sampling media (Govett et al., 1975), namely, water samples, rock samples, soils samples, till samples, and stream sediment samples are explained herein. It should be noted that there are other sampling media, such as vegetation, that are not explained in this section. Readers are referred to the literature on biogeochemistry for finding information on this type of samples (Kovalevsky, 1987; Dunn, 2011; de Verneil et al., 2018).

There are two general geochemical environments that are considered and studied in exploration geochemistry—primary and secondary (Grigoryan, 1974; Beus and Grigorian, 1977). Primary environments are those derived from rock samples. The term "primary" used herein refers to the premise that fresh rock samples do not describe weathering, mechanical transportation, and erosion (Boyle, 1982). Therefore these samples are suitable for assessing mineralized zones or their adjacent rock units (Haldar, 2013). In diamond drilling surveys, for example, rock samples are analyzed for delineating mineralized zones. The geochemistry of rock samples is usually referred to as lithogeochmeistry (Alexandre, 2021). It is worth noting that lithogeochmeistry is applicable in terrains with outcropping rock units, meaning that lithogeochmeical surveys are not usually applicable in glaciated zones or zones covered by regolith (McClenaghan et al., 2013).

In addition to primary environments, secondary geochemical halos, derived from the analysis of till, soil, lake sediment, and stream sediment samples, represent a secondary migration of elements in their sampling media. Dispersion pattern of elements in secondary environment is controlled by the degree of mechanical dispersion, mobility of elements, climate, topography, bedrock geology, and even anthropological factors (cf., Kelley et al., 2006). Given the previous premise, it is of utmost importance to recognize geochemical dispersion patterns that might be rooted in the presence of underlying mineralized zones. The upcoming subsections are focused on the sampling procedures used for the most common sampling media in exploration geochemistry.

4.2.1 Water samples

Collecting water samples is relatively easy as water is homogenous, and large volumes of samples are not typically required. A plastic vial is usually filled with approximately 100 cm^3 of water collected even from a certain depth or the surface of eater. Some additives, including HNO_3, are commonly introduced to the collected sample. These additives are used as a deterrent to the separation of water's solid phase (Haldar, 2013).

Most sampling campaigns also measure the pH of water samples. Lake and stream samples are suitable for regional- to district-scale geochemical surveys. Hydrogeochemistry, the study of water geochemical samples, has been frequently employed for mineral exploration (Simpson et al., 1993; Taufen and Gubins, 1997; Leybourne et al., 2003; Caron et al., 2008).

4.2.2 Rock chip samples

Rocks are the most common sampling media, especially in detailed mineral exploration phases (Govett, 2013; Alexandre, 2021). While sampling from outcrops, usually a geological hammer is used to remove the weathered parts and take a fresh sample. Usually, several midsized samples are taken from a given rock unit. Lithological diversity exerts control on the number of samples taken from a given unit (Haldar, 2013).

A significant difference between the geochemistry of rock samples and that of other sampling media is the fact that rock samples represent primary geological halos, meaning that the signatures revealed by rock samples are indicators of primary geological environments.

4.2.3 Soil geochemical samples

Soil geochemical surveys are also among common surveys in mineral exploration (e.g., Cohen et al., 2010; Bini et al., 2011). There are three soil horizons available: From the top to bottom, there are postorganic horizon A, transitional horizon B, and also the bottom horizon C, containing most of fragmented rock particles (Haldar, 2013). Horizons B and C appeal the most to exploration geochemists given their enrichment in rock fragments.

The soil profile should be entirely exposed to determine different horizons. Factors controlling the depth and the presence of different soil horizons include climate, topography, and age of soil. It goes without saying that the depth of soil horizons and their presence vary in different areas. Arid and semiarid areas, for example, might lack horizon A (Haldar, 2013).

Considering the heterogenous nature of soil and that there are variable grain sizes in a soil profile, soil samples could be fairly heavy. After selecting the appropriate horizon for sampling, a sample of around 2–5 kg is collected and placed in a plastic bag. One should be careful about keeping the initial conditions of samples as any changes may significantly affect the results of geochemical analysis. Soil surveys are conducted in various scales, ranging from regional- to local-scale studies.

4.2.4 Till geochemical samples

Till, a cycled sediment deposited directly by glacier ice, is deemed an effective sampling medium for mineral exploration in glaciated terrains as it can reflect the initial composition of its underlying bedrock (McClenaghan et al., 2013). Till is a combination of crushed bedrock mixed with older sediments and is subject to a mechanical transportation mechanism (McMartin and McClenaghan, 2001). This transportation occurs in an orientation related to the movement of glacier ice and can vary from few meters to several kilometers. Therefore the interpretation of till geochemical data requires knowledge of the Quaternary geology of the area and the movement of ice glaciers (Shilts, 1984). Therefore drift sampling of tills is essentially useful when applied in combination with boulder tracing and striation mapping (McClenaghan et al., 2013).

Usually, the finest fraction of soil's C horizon is subject to sampling and chemical analysis. Although the sampling techniques vary in permafrost and forested areas, a fine fraction of weakly oxidized till (<2 mm) is usually selected in both cases as this fraction size can reveal concentrations of ore-related elements. This fraction is usually derived by abrasion and physical crushing of primary minerals, distributing metal-bearing components into different grain sizes. The accumulation of metal-bearing components of till in clay-sized fraction of soil is highly likely, which is mostly linked to the large surface area of clay minerals, their cation exchange capacity, and that they can accommodate various ranges of ionic radii (McClenaghan et al., 2013).

In the case of a thin draft covers, samples can be collected by trenching or hand excavation at a depth of less than 5 m. However, while dealing with a thick draft cover, portable drills, reverse circulation rotary drills, or rotosonic drills are employed for collecting till samples at depth. Compositional variability of bedrocks, the targeted elements, and the budget are among the factors affecting the choice of analytical techniques and size fraction of samples (McClenaghan and Paulen, 2018).

Geochemical anomalies derived by till samples usually represent a thin plume extending in the direction of glacier movement from the underlying ore zone toward the surface of glaciers; that is, ore-bearing components are more frequent near their source, and their population usually decreases while moving away from their source. Dispersion pattern of elements in till samples is usually a function of the degree of mechanical transportation, the mobility of elements, and the degree of weathering and contamination processes (McClenaghan et al., 2013; McClenaghan and Paulen, 2018).

Postglacial weathering and its subsequent soil formation affect till geochemistry, especially in oxidization zones above the permafrost table or groundwater. These processes help decompose minerals liable to clastic dispersion, such as carbonates and sulfides, into the oxidization zone. This leads to the migration of elements that are transported by the percolating rainwater. As a result, elements in the oxidized zones are (1) immediately precipitated in the soil, (2) trapped in organic matter, secondary oxides/hydroxides, or clay-sized phyllosilicates, or (3) completely leached out from the soil profile. Conditions are different under the permafrost table where ore minerals can remain intact. Therefore, a solid understanding of vertical variations in soil profiles is essential prior to interpreting till geochemical patterns (McClenaghan et al., 2013; McClenaghan and Paulen, 2018).

Overall, because of the effects of biogeochemical enrichment in the surface organic layer, postdepositional mobilization of elements, differences in elements' mobility, and their mechanical dispersion patterns, the interpretation of till geochemical anomalies is a complex process. Prior to sampling, one should determine the thickness, distribution, glacial deposit types, and glacial history of sampling sites. Till surveys are also conducted in various scales, ranging from regional- to local-scale studies (McClenaghan et al., 2013).

4.2.5 Stream sediment geochemical samples

Stream sediments contain particles derived from the upstream weathered rocks and are dispersed by water (Fletcher, 1997). Stream sediment geochemistry is a popular tool for mineral exploration, especially for reconnaissance surveys (Painter et al., 1994), as stream sediments represent a relatively large area of their upstream catchment basin (Spadoni, 2006). Several parameters, including climate, lithological diversity, water chemistry, topography, and anthropological factors, contribute to the chemical composition of stream sediments (Parsa et al., 2016). Dispersion pattern of elements in stream sediments is also a function of element's mobility in surficial environments (Levinson, 1974).

It goes without saying that the interpretation of stream sediment geochemical samples is different from that of rock samples. First, stream sediment samples represent secondary geochemical halos that are under the effect of multiple parameters. Second, it has been recommended to draw the upstream sample catchments and assign the values of samples to the catchments for their spatial interpretation (Spadoni, 2006).

Designing the sampling scheme and choosing the right sampling density are materials to the breadth and depth of study, scale, budget, and access to sampling sites. It is worth nothing that selecting the right fraction of samples is important for chemical analysis. Generally, many elements tend to accumulate in finer fractions of stream sediments. These include but are not limited to Mn, Cu, Zn, and Fe (Levinson, 1974). Given this fact, the collection of <176-μm portion of sediments is a common practice in stream sediment surveys (Haldar, 2013). An example of using stream sediment geochemistry for mineral exploration is presented in the fifth section of this chapter.

4.3 Instrumental techniques applied to geochemical data

Geochemical samples are usually sent to a laboratory for chemical analysis. Except for using water samples, regardless of sampling media, samples are usually subject to dissolution. Element particles are already dissolved in water, meaning that, in most cases, water samples can be subject to direct chemical analysis.

A jaw or disk crusher is usually used to crush rock samples. These are later sieved to a given size that meets the requirement of study and the chemical analysis method. Important to note is the fact that the crushing process, itself, is a source of contamination.

Soil samples are usually subject to drying, and some external particles, such as wood, are removed from samples. Generally, sieving is not required for samples, and immediately after drying and removing external particles, soil samples can be directly used for dissolution. The same logic can be applied to stream sediment and till samples. Sample dissolution is meant to convert the insoluble component of samples into soluble chemical species.

Acid digestion is a frequently used method for sample preparation in exploration geochemistry. Usually, HCl, HNO_3, $HClO_4$, H_2SO_4 and HF, or a combination of them, are used for acid digestion. Aqua regia, a combination of HCl and HNO_3, is among the popular solutions used for the analysis of trace elements. Further information on sample preparation methods can be found in Balaram and Subramanyam (2022). It is not the purpose of this chapter to delve into different analytical techniques, but a brief summary of analytical techniques is given in the next subsections followed by general guidelines on the quality control of geochemical data.

4.3.1 Instrumental techniques

There are numerous instrumental techniques that can be applied to the chemical analysis of geochemical samples. Some of the commonly used instrumental techniques in exploration geochemistry, namely, X-ray fluorescence (XRF) spectrometry, atomic absorption spectrophotometry (AAS), inductively coupled plasma mass spectrometry (ICP-MS), and inductively coupled plasma optical emission spectrometry (ICP-OES), are briefly explained herein. It should be noted that each instrumental technique comes with its pros, and it is for geochemists to decide on the type of method they want to use.

XRF is well suited for the analysis of solid materials, meaning that it generally does not require sample digestion. Most of geochemical elements can be identified by XRF at concentrations below parts per million (ppm). This technique is useful for mineral exploration, especially when a quick analysis of samples is required. Portable XRF techniques, for example, can result in rapid, in situ preparation of geochemical datasets. The limits of detection for XRF are not as good as other techniques, including ICP-OES and ICP-MS. ICP-OES is well suited for the analysis of solutions. Very low limits of detection and the multielement analysis of samples are the pros of using ICP-OES. Likewise, ICP-MS also works on solutions and results in the analysis of multielements. It is marked by lower limits of detection compared to ICP-OES. ICP-MS can also detect the isotopic composition of some elements. ICP-OES and ICP-MS are among the commonly used techniques in regional geochemical surveys. This is owing to the fact that these techniques can determine the concentration of multielements with an acceptable detection limit. AAS is a cost-effective option compared to ICP-OES and ICP-MS and is suitable for the analysis of single elements. General guidelines and conclusions on the use of instrumental techniques in geochemistry can be found in Willis (1986).

4.3.2 Quality control

The reliability of geochemical data is very important for mineral exploration surveys. Accuracy and precision of chemical analysis are two factors that are generally assessed in the quality control stage. The latter is assessed by using duplicate samples, two identical samples submitted to the laboratory. Usually, one duplicate sample is considered for every 10 samples. A well-known method for assessing the precision of chemical analysis using duplicate samples is described by Thompson and Howarth (1976).

The former, however, is assessed by placing standard samples in the batch of samples. These are samples with predetermined concentrations of elements. Standard samples, or certified reference materials, typically constitute 1% of samples one submits to the laboratory. It means that 11% of samples placed in a batch of samples submitted for chemical analysis are for the mere purpose of quality control. Standard samples can be either purchased from reputable sources or produced internally during the exploration surveys.

4.4 Interpretation of geochemical data

Interpretation of geochemical data is not a trivial process owing to intrinsic complexities embedded in these datasets. This section is centered on the interpretation of geochemical data as a part of geospatial analysis applied to mineral exploration. Several factors should be taken into consideration while analyzing geochemical data. The complex interrelationship between element concentrations in different sampling media (Templ et al., 2008; Zuo et al., 2021), the compositional nature of geochemical data (Aitchison, 1982; Buccianti and Grunsky, 2014), and the presence of extreme (Reimann et al., 2005) and censored values (Sanford et al., 1993; Grunsky, 2010; Carranza, 2011) are among the common problems associated with geochemical data.

As mentioned earlier in Section 4.1 and will be discussed in the following subsections, ore-forming processes lead to the enrichment of a suite of elements in the mineralized zones, followed by the depletion of other elements (Pirajno, 2012). Therefore, one should recognize multielement associations that describe mineralization (e.g., Parsa et al., 2016). This is a fundamental step one should take while interpreting multielement geochemical data.

It is worth mentioning that geochemical data contain censored values. These are element concentrations that are lower than (or higher than) the detection limit of the instrumental technique employed (Sanford et al., 1993; Grunsky, 2010; Carranza, 2011). For dealing with censored values, one should employ a substitution procedure to replace values that are below the detection limit of the instrument used. There are substitution procedures proposed in the literature to which readers are referred (Sanford et al., 1993; Carranza, 2011).

Besides the previous aspect, many traditional techniques employed for the interpretation of geochemical data require the dataset to follow a normal distribution (Shuguang et al., 2015). While employing such techniques, the presence of extreme values (i.e., outliers) presents a further challenge to processing geochemical data as these values deter the data from following a normal distribution. Therefore, in the case of employing those techniques, treating extreme values also becomes important.

Traditionally, unielement concentrations are analyzed by exploratory data analysis and box plots (Tukey, 1977) to determine geochemical anomalies. The problem with this procedure is that these methods only assess the statistical distribution of geochemical samples and completely neglect the spatial distribution of samples (Cheng et al., 1994). Also, these methods are merely focused on single-element compositions. However, ore-forming processes usually result in the depletion and enrichment of a suite of elements rather than single elements.

For multielement analysis of geochemical datasets, one should also consider the using multivariate and machine learning techniques. Care must be taken while applying these methods to geochemical data as these datasets are compositional in nature (Aitchison, 1982). The concept of compositional analysis of geochemical data is explained in the upcoming subsection followed by general guidelines on delineating geochemical anomalies.

4.4.1 Compositional data analysis

Geochemical data contain only positive values that are reported in proportions, such as ppm, ppb, and percentage. Ideally, for a given sample, the sum of all the analytes, including elements analyzed and the loss of ignition (LOI), should reach a constant value. In theory, this value must equal 1 or 100%. This means that an increase in the concentration of a group of elements should result in a decrease in the concentration of other elements. This complies with the nature of geochemical anomalies; ore-forming processes result in the enrichment and subsequent depletion of mineralized zones from certain elements/metals.

From a statistical point of view, the earlier issue means that we are dealing a number of nonindependent variables. According to Aitchison (1982), this is usually referred to as the closure problem of compositional geochemical data. The compositional nature of geochemical data renders their statistical indices, including the covariance and correlation between elements, biased. To address the previous challenge, one should treat geochemical data with a series of transformation, namely, log-ratio transformation. Applying these transformations is a prerequisite to many geochemical analysis.

Log-ratio transformations, initially introduced by Aitchison (1982), have been used as a method for projecting compositional datasets into the real number space, addressing the problem of spurious correlations and, thus, allowing for the application of standard statistical techniques. These include the additive (alr: Aitchison, 1982), the centered (clr: Aitchison, 1982), and the isometric (ilr: Egozcue et al., 2003) log-ratio transformations. Full details of these techniques and the pros and cons of using them have been described in numerous publications (e.g., Grunsky, 2010; Buccianti and Grunsky, 2014), to which the readers are referred.

As applied in numerous research papers (e.g., Carranza, 2011), log-ratio transformations help enhance our understanding of geochemical data through enhanced interpretation (Grunsky and Caritat, 2020) and visualization (e.g., Somma et al., 2021) of the interrelationship between element concentrations.

Herein, a common compositional-based approach for processing geochemical data is explained. Factor analysis (FA) employs a correlation or covariance matrix of measured variables (Treiblmaier and Filzmoser, 2010) to assist in unraveling the underlying patterns of geochemical data and providing links between elements.

The outcome of applying FA to a set of geochemical data is a suite of secondary factors summarized in two matrices, namely, loadings and scores. Information on the interrelationship between elements and factors is summarized in the former, and information on the relationship between samples and factors is given by scores (Tripathi, 1979).

FA should be coupled to log-ratio transformation techniques for interpreting geochemical data (Aitchison, 1982; Egozcue et al., 2003). In addition, the presence of outliers in geochemical data adversely affects the estimation of correlation and covariance matrix, and thus, the results of FA carry exploration bias (Filzmoser et al., 2009). Therefore, this problem should be modulated by the robustification of factor estimation using robust estimators, for example, the minimum covariance determinant (MCD), instead of using a correlation or covariance matrix (Pison et al., 2003). Robust factor analysis (RFA) of compositional data considers the previous aspects and is suitable for processing and interpreting geochemical data. An example of using RFA is explained in the case study presented in this chapter.

Among log-ratio transformations, the clr transformation results in symmetric results and an one-to-one explanation of transformed variables (Aitchison, 1982). Thus the clr transformation is the most proper transformation for multivariate geochemical data analyses (Filzmoser et al., 2009). However, due to the singularity of transformed variables, the application of MCD and the other robust estimators and thus robustification of FA is not possible in the clr space (Filzmoser et al., 2009). According to Filzmoser et al. (2009), one way out of this is to apply MCD on ilr-transformed data for estimating the covariance matrix. Then, for interpreting the results, the estimated matrix should be backtransformed to the clr-space where interpretation of factors is possible via compositional biplots and loading matrices.

4.4.2 Techniques used for delineating geochemical anomalies

To delineate geochemical anomalies, one should consider geochemical data's statistical and spatial characteristics (Cheng et al., 1994). The latter, however, is neglected in traditional methods of anomaly recognition, namely, setting subjective threshold values (Hawkes and Webb, 1962), exploratory data analysis (Tukey, 1977), and probability plots (Sinclair, 1974).

The concept of fractal theory (Mandelbrot, 1982) has been adopted in exploration and environmental geochemistry to address the issue mentioned before. The idea of using this concept in applied geochemistry lies in the premise that geochemical data can reveal self-similarity, helping delineate anomalous patterns (Cheng et al., 1994). This idea sparked the development of many fractal/multifractal methods for interpreting geochemical data, namely, the concentration-area (Cheng et al., 1994), singularity mapping (Cheng, 2007), and other variants of fractal modeling. Fractal tools consider both spatial and statistical characteristics of geochemical data (Zuo et al., 2021), making them invaluable tools for delineating geochemical data, as practiced in many studies (e.g., Parsa et al., 2017 and references therein).

A common methodology applied to delineating geochemical anomalies is summarized herein. This method, the concentration–area (C–A) fractal model (Cheng et al., 1994), is based on the fact that in an interpolated map of a geochemical signature, each contour, representing a given concentration (C), is marked by an occupied area (A). This method is explained by Cheng et al. (1994) and summarized in many publications (e.g., Parsa et al., 2017). As per Cheng et al. (1994), there is a power-law relationship between these two quantities. This relationship becomes a linear equation that can be applied to delineating geochemical anomalies. The slopes of straight lines, which could be fitted to the log–log plots of concentration versus the area, denote the fractal dimensions of different geochemical populations. The intersections of straight lines determine the threshold, which may delineate geochemical populations (Cheng et al., 1994). This is further explained by drawing an example from the literature in the upcoming subsection.

4.5 Case study

The case study presented here explains how a suite of stream sediment samples can help guide exploration programs using the methods explained in this chapter (Fig. 4.1). This case study belongs to the eastern part of Iran with a few minor occurrences of Ni mineralization (Parsa et al., 2017). A simplified outcrop map of this area can be observed in Fig. 4.1.

A total of 188 stream sediment samples have been collected. The samples were sieved by a 180-μm screen and the fractions *b*180 μm were selected for chemical analysis. The sieved fractions were digested in HNO_3 + HCl and analyzed by ICP-OES for multielement analysis.

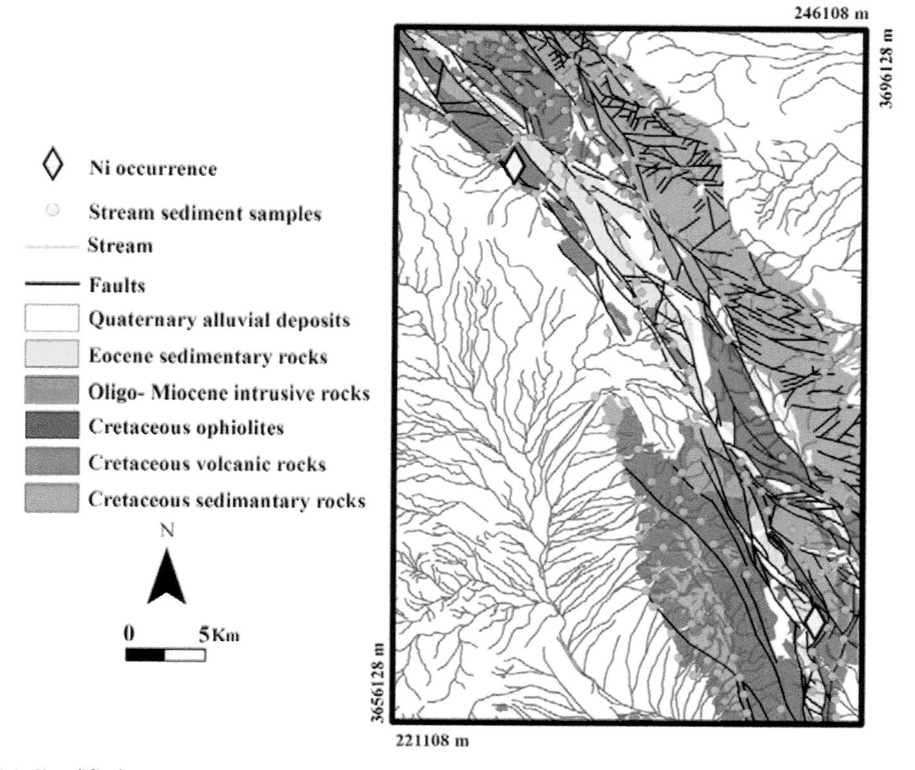

FIGURE 4.1 Simplified outcrop map of the area being investigated and the location of geochemical samples.

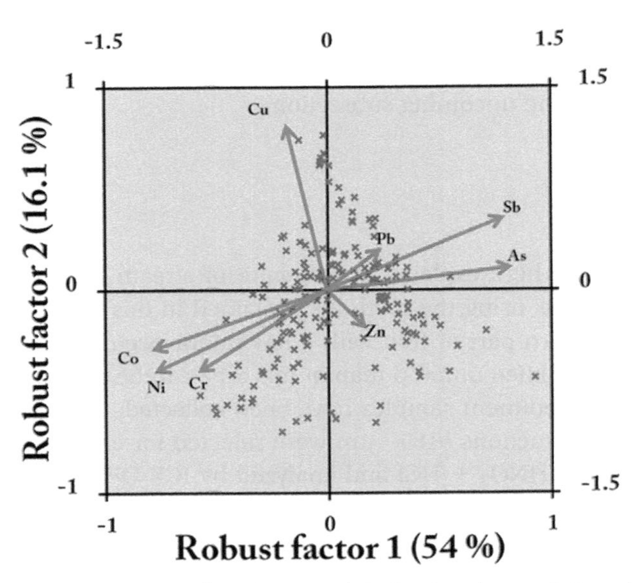

FIGURE 4.2 Biplots of elements derived by robust factor analysis showing the association of Cr, Co, and Ni.

The detection limits were 0.2 ppm for Cu, 0.2 ppm for Pb, 0.2 ppm for Zn, 0.2 ppm for Co, 2 ppm for Ni, 0.5 ppm for As, 0.1 ppm for Sb, and 2 ppm for Cr.

RFA was applied to the dataset according to the following procedure. Initially, ilr-transformation was applied on the raw data of seven analyzed elements for the estimation of the covariance matrix (Filzmoser et al., 2009). Then, as the ilr-transformation sacrificed one variable (Egozcue et al., 2003), the resultant covariance matrix was backtransformed to the clr-space for deriving interpretable robust factors (Filzmoser et al., 2009). The number of factors was set to be three, because these three factors not only explained 85.6% of the total variability but also yielded in a precise discrimination of element associations. Fig. 4.2 represents the biplots derived from this analysis. As per this figure, Ni, Cr, and Co could be deemed mineralization-related element associations. The readers are referred to Parsa et al. (2017) for details on the interpretation of these results. According to the histogram of these elements Fig. 4.3, these are marked by positively skewed distributions, showing the existence of geochemical enrichment in the area. Spatial distribution of these three elements is also depicted in Fig. 4.4. These maps are derived by the application of inverse distance weighting method for the interpolation of element concentrations.

The *C–A* method was further applied to interpret these maps. As per Fig. 4.5, the *C–A* log–log plot was resulted in the delineation of four separate geochemical populations for Ni. These were labeled as low-background, background, anomaly, and high anomaly. Concentrations above 570 ppm have been marked as highly anomalous zones for Ni in the area. Likewise, *C–A* plots separated four geochemical populations for Co in the area, with values above 45 ppm as highly anomalous values. However, only three populations were

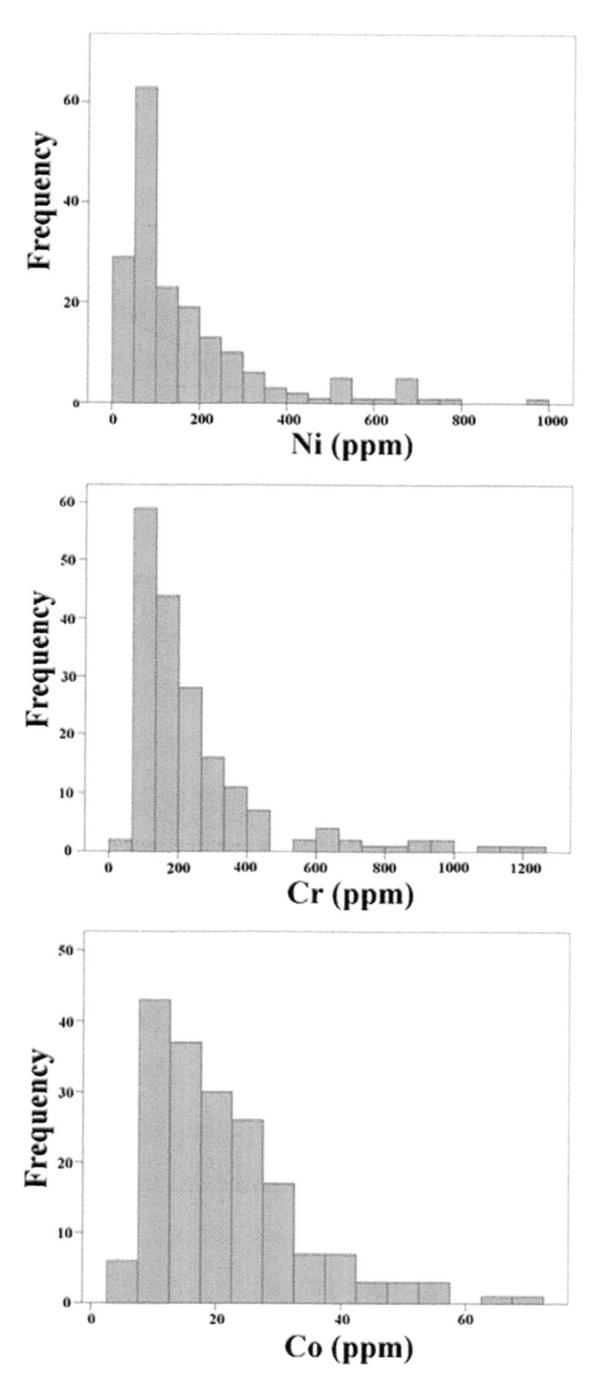

FIGURE 4.3 Histograms of Ni, Cr, and Co, showing positively skewed distributions for these elements.

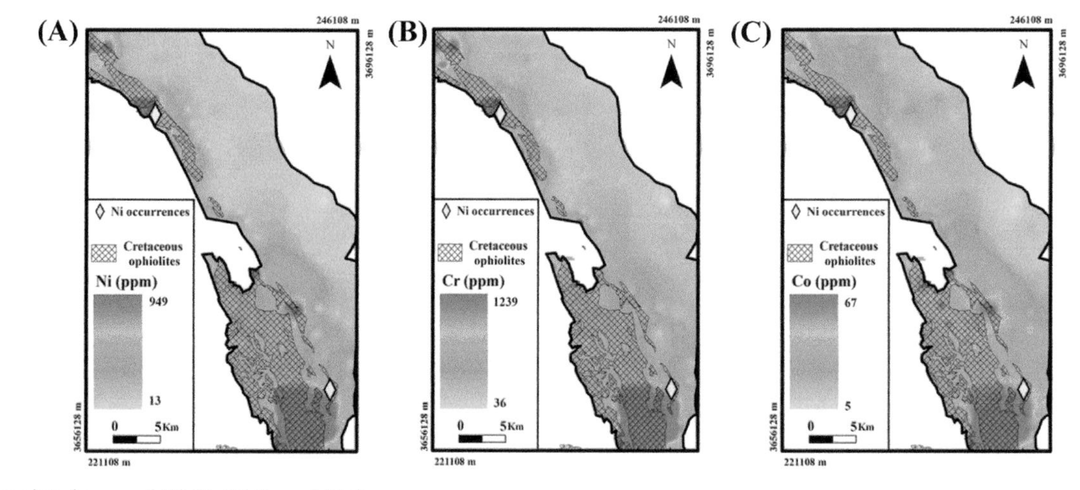

FIGURE 4.4 Interpolated maps of (A) Ni, (B) Cr, and (C) Co.

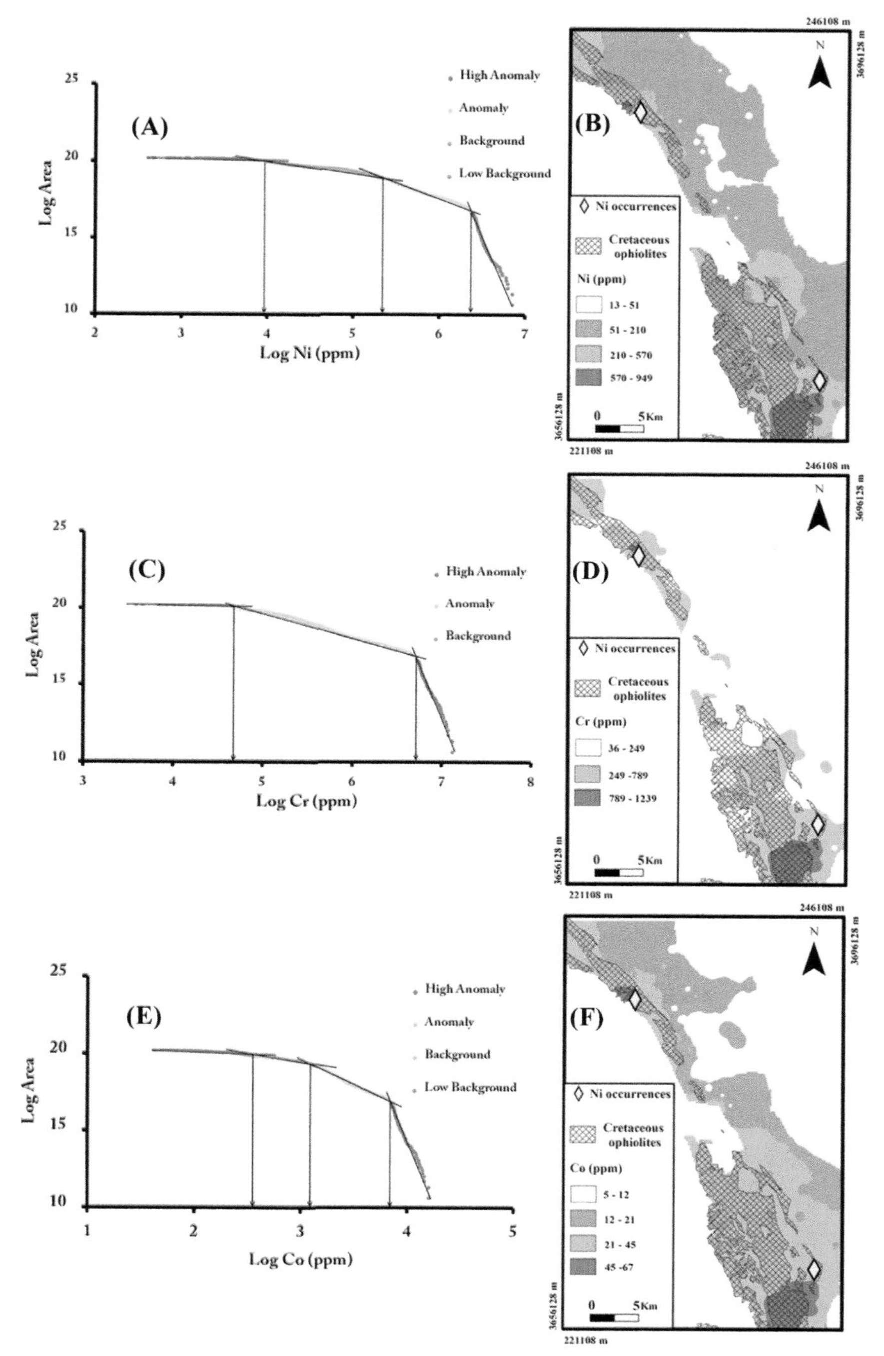

FIGURE 4.5 The results of *C–A* fractal method in delineating geochemical populations of (A, B) Ni, (C, D) Cr and (E, F) Co.

recognized for Cr. These were marked as background, anomaly, and high anomaly, with values above 789 ppm pertaining to the highly anomalous zones. Field work was carried out for ground-truthing of geochemical anomalies, revealing the existence of mineralized zones in a highly anomalous area.

References

Aitchison, J., 1982. The statistical analysis of compositional data. J. R. Stat. Soc.: Ser. B (Methodol.) 44 (2), 139–160.

Alexandre, P., 2021. Lithogeochemistry. Practical Geochemistry. Springer, Cham, pp. 35–60.

Balaram, V., Subramanyam, K.S.V., 2022. Sample preparation for geochemical analysis: strategies and significance. Adv. Sample Prep. 1, 100010.

Beus, A.A., Grigorian, S.V., 1977. Geochemical Exploration Methods for Mineral Deposits. Geoscience Publishing

Bini, C., Sartori, G., Wahsha, M., Fontana, S., 2011. Background levels of trace elements and soil geochemistry at regional level in NE Italy. J. Geochem. Explor. 109 (1–3), 125–133.

Boyle, R.W., 1982. Geochemical methods for the discovery of blind mineral deposits. Part 2. CIM Bull.; (Canada), 75(845).

Buccianti, A., Grunsky, E., 2014. Compositional data analysis in geochemistry: are we sure to see what really occurs during natural processes? J. Geochem. Explor. 141, 1–5.

Caron, M.E., Grasby, S.E., Ryan, M.C., 2008. Spring water trace element geochemistry: a tool for resource assessment and reconnaissance mineral exploration. Appl. Geochem. 23 (12), 3561–3578.

Carranza, E.J.M., 2008. Geochemical Anomaly and Mineral Prospectivity Mapping in GIS. Elsevier.

Carranza, E.J.M., 2011. Analysis and mapping of geochemical anomalies using logratio-transformed stream sediment data with censored values. J. Geochem. Explor. 110 (2), 167–185.

Carranza, E.J.M., Zuo, R., Cheng, Q., 2012. Fractal/multifractal modelling of geochemical exploration data. J. Geochem. Explor. 122, 1–3.

Cheng, Q., 2007. Mapping singularities with stream sediment geochemical data for prediction of undiscovered mineral deposits in Gejiu, Yunnan Province, China. Ore Geol. Rev. 32 (1–2), 314–324.

Cheng, Q., Agterberg, F.P., Ballantyne, S.B., 1994. The separation of geochemical anomalies from background by fractal methods. J. Geochem. Explor. 51 (2), 109–130.

Cohen, D.R., Kelley, D.L., Anand, R., Coker, W.B., 2010. Major advances in exploration geochemistry, 1998–2007. Geochem.: Explor. Environ. Anal. 10 (1), 3–16.

de Verneil, A., Rousselet, L., Doglioli, A.M., Petrenko, A.A., Maes, C., Bouruet-Aubertot, P., et al., 2018. OUTPACE long duration stations: physical variability, context of biogeochemical sampling, and evaluation of sampling strategy. Biogeosciences 15 (7), 2125–2147.

Dunn, C.E., 2011. Biogeochemistry in Mineral Exploration. Elsevier.

Egozcue, J.J., Pawlowsky-Glahn, V., Mateu-Figueras, G., Barcelo-Vidal, C., 2003. Isometric logratio transformations for compositional data analysis. Math. Geol. 35 (3), 279–300.

Filzmoser, P., Hron, K., Reimann, C., Garrett, R., 2009. Robust factor analysis for compositional data. Comput. Geosci. 35 (9), 1854–1861.

Fletcher, W.K., 1997. Stream sediment geochemistry in today's exploration world. In: Proceedings of Exploration, vol. 97, pp. 249–260.

Garrett, R.G., 1969. The determination of sampling and analytical errors in exploration geochemistry. Econ. Geol. 64 (5), 568–569.

Govett, G.J.S. (Ed.), 2013. Rock Geochemistry in Mineral Exploration. Elsevier.

Govett, G.J.S., Goodfellow, W.D., Chapman, R.P., Chork, C.Y., 1975. Exploration geochemistry—distribution of elements and recognition of anomalies. J. Int. Assoc. Math. Geol. 7 (5), 415–446.

Grigoryan, S.V., 1974. Primary geochemical halos in prospecting and exploration of hydrothermal deposits. Int. Geol. Rev. 16 (1), 12–25.

Grunsky, E.C., Caritat, P.D., 2020. State-of-the-art analysis of geochemical data for mineral exploration. Geochem.: Explor. Environ. Anal. 20 (2), 217–232.

Grunsky, E.C., 2010. The interpretation of geochemical survey data. Geochem.: Explor. Environ. Anal. 10 (1), 27–74.

Haldar, S.K., 2013. Mineral exploration. Miner. Explor. 1, 193–222.

Hawkes, H.E., Webb, J.S., 1962. Geochemistry in Mineral Exploration.

Kelley, D.L., Kelley, K.D., Coker, W.B., Caughlin, B., Doherty, M.E., 2006. Beyond the obvious limits of ore deposits: the use of mineralogical, geochemical, and biological features for the remote detection of mineralization. Econ. Geol. 101 (4), 729–752.

Kirkwood, C., Everett, P., Ferreira, A., Lister, B., 2016. Stream sediment geochemistry as a tool for enhancing geological understanding: an overview of new data from south west England. J. Geochem. Explor. 163, 28–40.

Kovalevsky, A.L. (Ed.), 1987. Biogeochemical Exploration for Mineral Deposits. VSP.

Levinson, A.A., 1974. Introduction to Exploration Geochemistry.

Leybourne, M.I., Boyle, D.R., Goodfellow, W.D., 2003. Interpretation of Stream Water and Stream Sediment Geochemistry in the Bathurst Mining Camp. Application to Mineral Exploration, New Brunswick, Canada.

Mandelbrot, B.B., 1982. The Fractal Geometry of Nature, vol. 1. WH freeman, New York.

McClenaghan, M.B., Paulen, R.C., 2018. Application of till mineralogy and geochemistry to mineral exploration. Past Glacial Environments. Elsevier, pp. 689–751.

McClenaghan, M.B., Plouffe, A., McMartin, I., Campbell, J.E., Spirito, W.A., Paulen, R.C., et al., 2013. Till sampling and geochemical analytical protocols used by the Geological Survey of Canada. Geochem.: Explor. Environ. Anal. 13 (4), 285–301.

McMartin, I., McClenaghan, M.B., 2001. Till geochemistry and sampling techniques in glaciated shield terrain: a review. Geol. Soc. Lond. Spec. Publ. 185 (1), 19–43.

Painter, S., Cameron, E.M., Allan, R., Rouse, J., 1994. Reconnaissance geochemistry and its environmental relevance. J. Geochem. Explor. 51 (3), 213–246.

Parsa, M., Maghsoudi, A., Yousefi, M., Sadeghi, M., 2016. Recognition of significant multi-element geochemical signatures of porphyry Cu deposits in Noghdouz area, NW Iran. J. Geochem. Explor. 165, 111–124.

Parsa, M., Maghsoudi, A., Yousefi, M., Sadeghi, M., 2017. Multifractal analysis of stream sediment geochemical data: Implications for hydrothermal nickel prospection in an arid terrain, eastern Iran. J. Geochem. Explor. 181, 305–317.

Pirajno, F., 2012. Hydrothermal Mineral Deposits: Principles and Fundamental Concepts for the Exploration Geologist. Springer Science & Business Media.

Pison, G., Rousseeuw, P.J., Filzmoser, P., Croux, C., 2003. Robust factor analysis. J. Multivar. Anal. 84 (1), 145–172.

Reimann, C., Filzmoser, P., Garrett, R.G., 2005. Background and threshold: critical comparison of methods of determination. Sci. Total. Environ. 346 (1–3), 1–16.

Sanford, R.F., Pierson, C.T., Crovelli, R.A., 1993. An objective replacement method for censored geochemical data. Math. Geol. 25 (1), 59–80.

Shilts, W.W., 1984. Till geochemistry in Finland and Canada. J. Geochem. Explor. 21 (1–3), 95–117.

Shuguang, Z., Kefa, Z., Yao, C., Jinlin, W., Jianli, D., 2015. Exploratory data analysis and singularity mapping in geochemical anomaly identification in Karamay, Xinjiang, China. J. Geochem. Explor. 154, 171–179.

Simpson, P.R., Edmunds, W.M., Breward, N., Cook, J.M., Flight, D., Hall, G.E.M., et al., 1993. Geochemical mapping of stream water for environmental studies and mineral exploration in the UK. J. Geochem. Explor. 49 (1–2), 63–88.

Sinclair, A.J., 1974. Selection of threshold values in geochemical data using probability graphs. J. Geochem. Explor. 3 (2), 129–149.

Somma, R., Ebrahimi, P., Troise, C., De Natale, G., Guarino, A., Cicchella, D., et al., 2021. The first application of compositional data analysis (CoDA) in a multivariate perspective for detection of pollution source in sea sediments: the Pozzuoli Bay (Italy) case study. Chemosphere 274, 129955.

Spadoni, M., 2006. Geochemical mapping using a geomorphologic approach based on catchments. J. Geochem. Explor. 90 (3), 183–196.

Taufen, P.M., Gubins, A.G., 1997. Ground waters and surface waters in exploration geochemical surveys. Proceedings of Exploration 97, 271–284.

Templ, M., Filzmoser, P., Reimann, C., 2008. Cluster analysis applied to regional geochemical data: problems and possibilities. Appl. Geochem. 23 (8), 2198–2213.

Thompson, M., Howarth, R.J., 1976. Duplicate analysis in geochemical practice. Part I. Theoretical approach and estimation of analytical reproducibility. Analyst 101 (1206), 690–698.

Treiblmaier, H., Filzmoser, P., 2010. Exploratory factor analysis revisited: how robust methods support the detection of hidden multivariate data structures in IS research. Inf. Manage. 47 (4), 197–207.

Tripathi, V.S., 1979. Factor analysis in geochemical exploration. J. Geochem. Explor. 11 (3), 263–275.

Tukey, J.W., 1977. Exploratory Data Analysis. Addison-Wesley, Reading, 688 pp.

Willis, J.P., 1986. Instrumental analytical techniques in geochemistry: requirements and applications. Fresenius' Z. für analytische Chem. 324 (8), 855–864.

Zuo, R., Wang, J., Xiong, Y., Wang, Z., 2021. The processing methods of geochemical exploration data: past, present, and future. Appl. Geochem. 132, 105072.

5

Geophysical data for mineral exploration

Ahmed M. Eldosouky[1], Luan Thanh Pham[2], Reda A.Y. El-Qassas[3], Thong Duy Kieu[4], Hassan Mohamed[3], Cuong Van Anh Le[5]

[1]GEOLOGY DEPARTMENT, FACULTY OF SCIENCE, SUEZ UNIVERSITY, SUEZ, EGYPT [2]FACULTY OF PHYSICS, UNIVERSITY OF SCIENCE, VIETNAM NATIONAL UNIVERSITY, HANOI, VIETNAM [3]EXPLORATION DIVISION, NUCLEAR MATERIALS AUTHORITY, MAADI, CAIRO, EGYPT [4]HANOI UNIVERSITY OF MINING AND GEOLOGY, HANOI, VIETNAM [5]UNIVERSITY OF SCIENCE, VIETNAM NATIONAL UNIVERSITY HO CHI MINH CITY, HO CHI MINH, VIETNAM

5.1 Introduction to geophysical exploration

Geophysical exploration is the applied branch of geophysics for economic purposes. This depends on the use of physical methods, such as gravitational, electromagnetic, seismic, magnetic, and electrical at the surface of the Earth to study the physical properties of the subsurface. It is used to infer or detect the existence and position of economic geological deposits, such as fossil fuels, other hydrocarbons, and geothermal and groundwater reservoirs. Exploration geophysics can be utilized to delineate the target style of mineralization, via estimating its physical characteristics directly (Elkhateeb et al., 2021; Melouah et al., 2021; Eldosouky et al., 2021a, 2022a; Ekwok et al., 2022).

Mineral exploration concentrates on the activities that are interested in searching for a new mineral and evaluating economic ones (Haldar, 2018). The main target of mineral exploration is to map the economic deposits with minimum time and cost. The geophysical survey represents an effective tool to detect subsurface objects and preferred sites for drilling (Haldar, 2018). The geophysical survey uses physical techniques such as electric, electromagnetic, seismic, magnetic, and gravity to measure the physical characteristics of the Earth and search for anomalous zones. It is also used to detect the locations of economical deposits, such as ore minerals, hydrocarbons, geothermal and groundwater reservoirs, and archeological remains. In addition, it is used to delineate the structures and rock units' distribution.

Geophysical exploration can be a cost-effective tool by supplying information about mineralization, structures that contain mineralized zones, and archeological targets. In general, there are promising areas for mineralization (uranium, gold, manganese, iron, copper, zinc, lead, cobalt, nickel, silver, and others) (Abbas et al., 2005; Abdallatif et al., 2010; Odah et al., 2013; Assran et al., 2019; Ahmed et al., 2020).

Geospatial Analysis Applied to Mineral Exploration. DOI: https://doi.org/10.1016/B978-0-323-95608-6.00005-6

5.2 Geophysical methods

5.2.1 Gravity methods

Gravity exploration is based on the measurement of variations in the gravitational field caused by density structures within the subsurface (Hinze et al., 2012). The gravity methods are used frequently in oil, gas, and mineral exploration to map subsurface geological structures and to directly determine ore reserves for some massive sulfide (Ms) orebodies (Nabighian et al., 2005). Many different techniques have been introduced to interpret gravity data. These techniques can be classified into four main groups. The first one is the edge detection techniques. The second group consists of the inversion techniques. The third group is the semiquantitative techniques.

The group of edge detection techniques leads to two subgroups of gravity interpretation techniques. The first one is known as the unbalanced filters and the second one comprises the balanced filters. The unbalanced filters are based on the derivatives of gravity data (Evjen, 1936; Cordell, 1979; Roest et al., 1992), while the balanced filters use the ratio of the derivatives of gravity data (Miller and Singh, 1994; Wijns et al., 2005; Cooper and Cowan, 2006, 2008; Pham et al., 2019, 2020a,b, 2021; Pham, 2021; Oksum et al., 2021). The second group of techniques aimed at estimating depth-to-interface estimates is based on the inversion of gravity data. These methods are based on the Bott (1960) stacked prism model or fast Fourier transform (FFT)-based algorithm of Parker (1973) (Oldenburg, 1974; Granser, 1986, 1987; Santos et al., 2015; Chakravarthi et al., 2016; Silva and Santos, 2017; Pham et al., 2018, 2022; Oksum, 2021). The inversion techniques require a mean depth, density contrast, or low-pass filtering. The group of semiquantitative techniques includes the Euler deconvolution, Werner deconvolution, and tilt-depth methods that became widely used tools for interpreting gravity data. The mathematical basis of these methods was presented by Werner (1949), Thompson (1982), and Oruç (2011). The major advantage of the semiquantitative methods is that they can determine the source depths without specifying prior information about the mean depth, density contrast, or low-pass filtering (Pham et al., 2022).

5.2.2 Magnetic methods

The magnetic method is one of the oldest and most widely used geophysical techniques for the exploration of the Earth's subsurface. It is a relatively inexpensive and easy tool to apply to subsurface exploration problems with subsurface magnetic property variations from the uppermost layer of soil to the base of the Earth's crust. The magnetic method maps this variation in the Earth's normal magnetic field (Eldosouky et al., 2017; Eldosouky et al., 2020a).

Magnetic anomalies can be produced based on the lateral contrast in rock composition (Lithology) and in rock structure (Millegan and Bird, 1998; Lyatsky, 2010). Magnetic field data come from a few rock types, such as volcanic, intrusive, and basement rocks (Okiwelu and Ude, 2012).

Magnetic susceptibility is a dimensionless quantity in SI. Further, magnetic susceptibility depends on the composition, size, and domain properties of magnetic grains. Generally, larger

Table 5.1 Magnetic susceptibility values for common mineral and rock types (Sharma, 1997).

Mineral or rock type	Magnetic susceptibility ($k \times 10^{-6}$ SI)
Granite (with magnetite)	20–40,000
Slates	0–1200
Basalt	500–80,000
Oceanic basalt	300–36,000
Limestone (with magnetite)	10–25,000
Gneiss	0–3000
Sandstone	35–950
Hematite (ore)	420–10,000
Magnetite (ore)	$7 \times 10^4 - 14 \times 10^6$
Magnetite (crystal)	150×10^6

grains have a greater susceptibility than smaller magnetic grains (Reynolds et al., 1990). The more susceptible a material, the more highly magnetized it becomes when placed in an inducing field. Table 5.1 lists the magnetic susceptibility values for common minerals and rocks.

The magnetic method requires an in-depth understanding of its basic principles, and careful data collection, reduction, and interpretation. Traditionally, useful information was extracted from geophysical field results by examining line profiles of raw or filtered survey data and maps. Enhanced maps are useful for estimating quantities of buried materials and locations, for example, maps of magnetic data often show the geologic structure or identify an anomalous region that might be associated with the desired target (Nabighian et al., 2005; Eldosouky et al., 2021b).

Interpretations may be limited to qualitative approaches that simply map the spatial location of anomalous subsurface conditions, but under favorable circumstances, the technological status of the method will permit more quantitative interpretations involving a specification of the nature of the anomalous sources (Nabighian et al., 2005).

5.2.2.1 Instrumentation

The magnetic survey data are acquired with the ground and airborne magnetic surveys at just about every conceivable scale and for a wide range of purposes. In exploration, the magnetic survey has been employed in the search for minerals. Detailed and regional magnetic surveys continue to be a primary mineral exploration tool, used for the direct detection of mineralization such as gold–iron oxide–copper deposits, Mss, and heavy mineral sands; for locating favorable host rocks or environments such as kimberlites, porphyritic intrusions, carbonatites, hydrothermal alteration, faulting, and for mapping geological structure in the study area (Nabighian et al., 2005).

5.2.3 Magnetotelluric methods

Magnetotellurics (MTs) uses a natural electromagnetic field to image electrical conductivity beneath the Earth (Tikhonov, 1950; Cagniard, 1953). Recently, developments in theories, instrumentation, data processing, and interpretation techniques have yielded MT to be a competitive geophysical method, suitable to image a broad range of geological targets.

The MT technique involves measuring fluctuations in the natural electric, E, and magnetic, B, the field in orthogonal directions and at the surface of the Earth. When electromagnetic fields diffuse into a ground medium, according to the behavior of electromagnetic waves, the penetration of the electromagnetic wave depends on the oscillation frequency. Hence, the frequency of the electromagnetic fields being measured determines the study depth.

The orthogonal components of horizontal electric and magnetic fields are related via a complex impedance tensor (Simpson and Bahr, 2005):

$$\begin{pmatrix} E_x \\ E_y \end{pmatrix} = \begin{pmatrix} Z_{xx} & Z_{xy} \\ Z_{yx} & Z_{yy} \end{pmatrix} \begin{pmatrix} B_x/\mu_0 \\ B_y/\mu_0 \end{pmatrix}. \tag{5.1}$$

The apparent resistivity and phase are defined as follows:

$$\rho_{ij}(\omega) = \frac{1}{\mu_0 \omega} \left| Z_{ij}(\omega) \right|^2, \tag{5.2}$$

$$\phi_{ij}(\omega) = a\tan\left(\frac{\mathrm{Im}\left(Z_{ij}(\omega)\right)}{\mathrm{Re}\left(Z_{ij}(\omega)\right)} \right), \tag{5.3}$$

where Re and Im are real and imaginary components of the Z, respectively.

5.2.4 Seismic methods

Although the seismic method is mainly employed for hydrocarbon exploration, there have been many recent applications to mineral prospecting. Reflection seismic techniques are utilized to deliver high-resolution pictures of lithological boundaries and subsurface structures, as well as details about the physical properties of rocks, such as their seismic velocity and density. To date, several hundred [two dimensional (2D)] lines and tens of [three dimensional (3D)] surveys have been obtained and documented in significant metallogenic sections worldwide. The researcher has a variety of geophysical strategies to map the subsurface. His preference depends on multiple factors, including the depth and physical properties of the target body, the properties of the overburden and host rock, and the logistic necessities and expenditure of surveys. Magnetic, electromagnetic, and gravity methods have been successfully employed to explore metallic ore bodies. Regardless, reflection seismic procedures provide a foremost trade-off between depth and resolution. This technology, mainly 3D seismic, is speedy evolving as an established strategy for mine planning and deep exploration (Malehmir et al., 2014).

Moreover, the current trend in exploring and exploiting mineralization at greater depths favors the usage of seismic techniques, not only for targeting deep-seated deposits and deep mine planning but also for attaining a more acceptable interpretation of the prevailing architecture of mineralized areas. This information is particularly essential when developing strategies to investigate mining camps where most surface deposits have been discovered and exploited. Ms bodies have more increased seismic densities and velocities than their host rocks. If the bodies are large enough, they can provide rise to strong reflections (Eaton et al., 2010; Malehmir et al., 2014).

5.2.5 Radiometric method

Gamma-rays are emitted during the decay of some naturally occurring elements on the Earth. The original discovery of radioactivity was made by Becquerel in 1896 after Röentgen had announced in 1895 the discovery of X-ray. Curie in 1898, investigating minerals of uranium, extracted two new elements, polonium, and radium that were much more active than uranium. The conventional approach to the acquisition and processing of gamma-ray spectrometric data is to monitor three or four relatively broad spectral windows [International Atomic Energy Agency (IAEA), 2003] (Table 5.2 and Fig. 5.1).

Table 5.3 indicates the concentration of the three radioelements (potassium, uranium, and thorium) in common rock types. Within igneous rocks, potassium, uranium, and thorium are increased in overall abundance with increasing silica content up to and including the pegmatite phase [International Atomic Energy Agency (IAEA), 1979].

The gamma-ray spectrometric survey used to map the radioelements (potassium, thorium, and uranium) rather than the total radiometry of rocks is also widely used in environmental, geological, and soil mapping (El-Qassas and Abu-Donia, 2019). It is also used to estimate and assess the terrestrial radiation dose to the population and to identify areas of potential natural radiation hazard [International Atomic Energy Agency (IAEA), 2003]. Moreover, the airborne gamma-ray survey is used to determine the heat production maps (Richardson and Killeen, 1980; Granar, 1982; Salem et al., 2005).

5.2.6 Ground-penetrating radar

Ground-penetrating radar (GPR) is a real-time nondestructive testing technique that uses high-frequency electromagnetic waves and can provide data with very high resolution for a short amount of time. This technique operates by transmitting electromagnetic waves into the probed material and receiving echo signals back from subsurface layer interfaces that could be boundaries between materials with different dielectrics. The corresponding arrival times and the amplitudes of the obtained echo signals can then be used to detect the location and nature of the interfaces. Since the GPR is able to detect buried objects, it is recommended as a helpful tool for mapping structures, shallow geological features, and subsurface

Table 5.2 Standard gamma-ray energy windows are recommended for natural radioelement mapping [International Atomic Energy Agency (IAEA), 1991].

Window	Nuclide	Energy range (MeV)
Total count	–	0.400–2.810
Potassium	^{40}K (1.460 MeV)	1.370–1.570
Uranium	^{214}Bi (1.765 MeV)	1.660–1.860
Thorium	^{208}Tl (2.614 MeV)	2.410–2.810

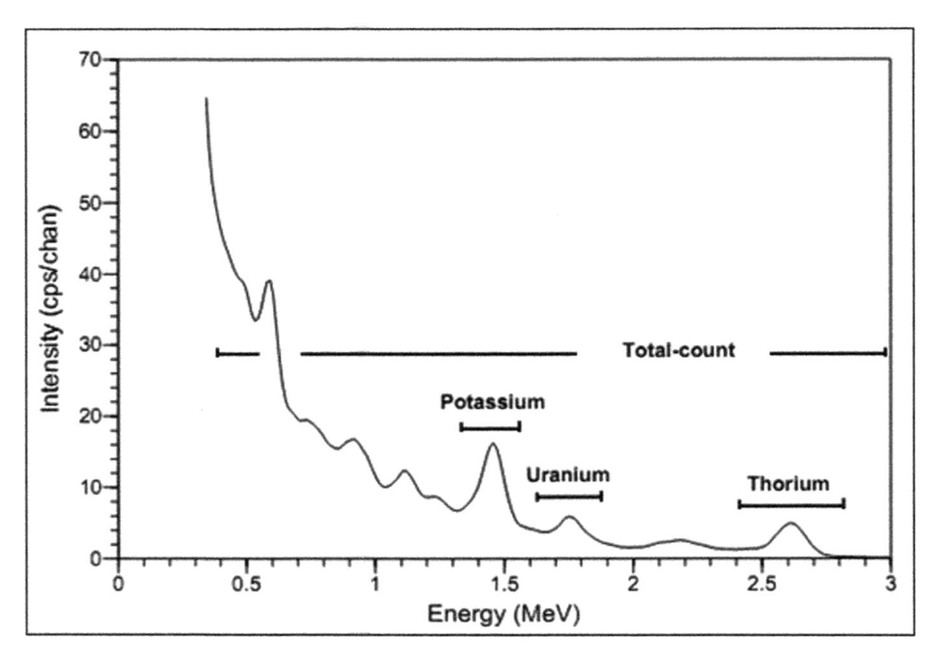

FIGURE 5.1 Typical airborne gamma-ray spectrum showing the positions of the conventional energy windows [International Atomic Energy Agency (IAEA), 2003].

Table 5.3 Radioelement concentrations in crustal rocks (Clark et al., 1966).

Rock type	K (%) average	U (ppm) average	U (ppm) range	Th (ppm) average	Th (ppm) range	Th:U average	Th:U range
Crustal average	2.1	3		12		4	
Mafic igneous	0.5	1	0.2–3	3	0.5–10	3	3–5
Inter. igneous	1–2.5	2.3	0.5–7	9	2–20	4	2–6
Acidic igneous	4	4.5	1–12	18	5–20	4	2–10
Arenaceous sediments	1.4	1	0.5–2	3	2–6	3	
Argillaceous sediments	2.7	4	1–13	16	2–47	4	1–12
Limestone	0.3	2	1–10	2	–	1	–
Black shale	2.7	8	3–250	16	–	2	Wide
Laterites	Low	10	3–40	50	8–132	5	Wide
Metamorphism	Depends on the parent rock type						

Source: Adapted from Clark, S.P., Peterman, Z.E., Heier, K.S., 1966. Abundances of uranium, thorium and potassium, handbook of physical constant. Geol. Soc. Am. Memoir. 47; and International Atomic Energy Agency (IAEA), 1979. Gamma-Ray Surveys in Uranium Exploration. International Atomic Energy Agency (IAEA).

targets (Sato, 2015; Dewi et al., 2017). In recent years, the GPR technique has gained acceptance as an important tool in the exploration of natural resources, the detection of underground pipes, and studies in archeology (Francke, 2010; Porsani et al., 2019; Mierczak and Karczewski, 2021).

5.2.7 Self-potential and induced polarization

5.2.7.1 Self-potential method

Self-potential (SP) technique is one of the oldest geoelectrical methods that is considered the cheapest of the surface geophysical surveys. This method was used in 1920 as a secondary tool in base metal exploration to detect the presence of massive ore bodies. Recently, it is used in groundwater and geothermal investigations to outline near-surface structures (shear zones, faults, and contacts). The method involves the measurement of natural potential differences as a guide to the existence of mineralization. The SP anomalies are produced from the flow of ions, heat, and fluids in the surficial layers of the Earth.

Several hypotheses have been developed to explain the mechanism of SP in mineral zones, but a reasonably complete illustration of mineralization potential is that proposed by Sato and Mooney (1960). This mechanism is illustrated in Fig. 5.2, showing the flow of electrons and ions that leave the upper surface negatively charged and the lower positively (Telford et al., 1990).

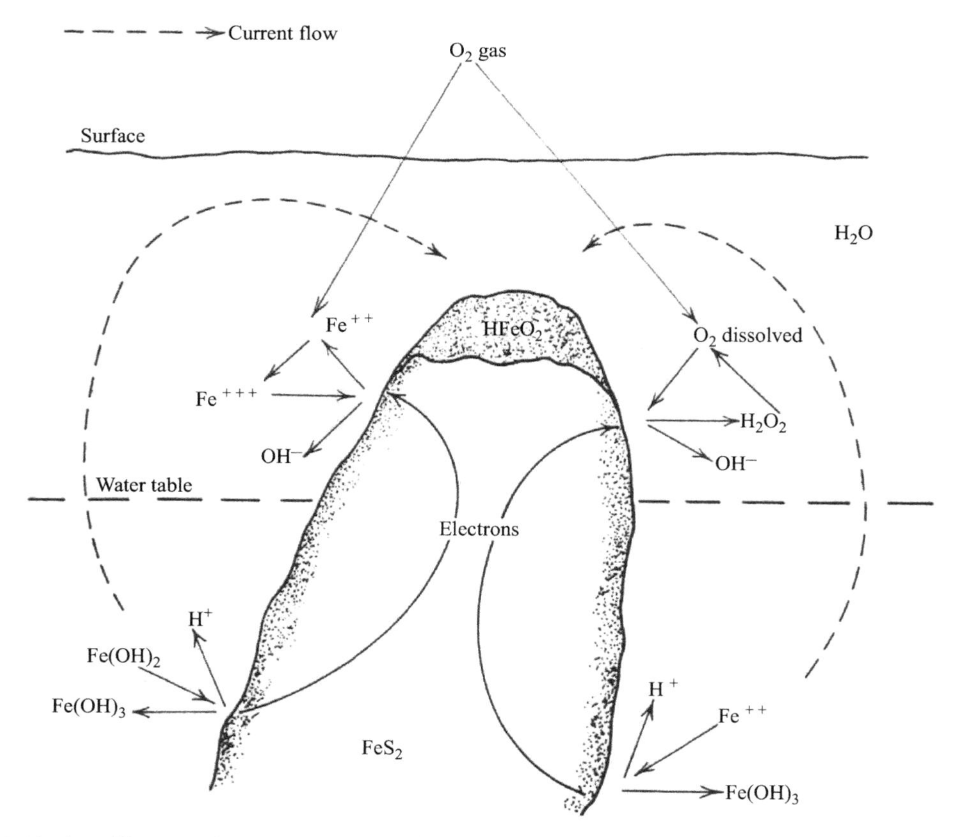

FIGURE 5.2 The self-potential mechanism in pyrite (after Sato and Mooney, 1960).

5.2.7.2 Induced polarization method

Induced polarization (IP) is one of the most widely used techniques in ore deposit exploration. It is applied due to its sensitivity to the electronic and ionic conductors. It was developed for detecting small concentrations of disseminated mineralization in base metal exploration (Sumner, 1976; Pelton et al., 1978; Parasnis, 1986; Assran et al., 2007). It is also used in groundwater, geotechnical, and environmental exploration (Anderson et al., 1990; Drascovits Hobot et al., 1990; and others). The 2D resistivity/IP prospecting can give information about both vertical and lateral variations of the Earth's properties and can be used in the delineation of the structure and depth of buried features.

IP surveys can be made in time- and frequency-domain modes. According to Seigel (1979), it is generally appreciated that the time-domain technique has the advantage of higher potential sensitivity. It also provides absolute measurements since the measurements are made after the cut-off of the current pulse. Thus it is possible to improve the signal-to-noise of such measurements by increasing the transmitter power, thereby increasing the sensitivity of measurements for low IP response. In addition, the distortions introduced by the electromagnetic coupling on the measured IP signal are very weak in the time-domain technique.

5.3 Methodologies, processing, and interpretation

5.3.1 Gravity

The enhanced horizontal gradient amplitude (EHGA) was introduced by Pham et al. (2020a,b) for detecting the edges of potential field sources. The EHGA method provides the maximum amplitudes on the horizontal boundaries of the sources. The method is given by

$$\mathrm{EHGA} = \Re\left(a\sin\left(k\left(\frac{\mathrm{HGA}_z}{\sqrt{\mathrm{HGA}_x{}^2 + \mathrm{HGA}_y{}^2 + \mathrm{HGA}_z{}^2}} - 1\right) + 1\right)\right), \tag{5.4}$$

where HGA is the total horizontal gradient amplitude of the potential field F, and given by

$$\mathrm{HGA} = \sqrt{\left(\frac{\partial F}{\partial x}\right)^2 + \left(\frac{\partial F}{\partial y}\right)^2}, \tag{5.5}$$

and k is a positive number decided by the interpreter. In this study, we used $k = 5$ (Pham et al., 2020a,b).

To determine the depth of density sources, we used the Euler deconvolution method that was introduced by Thompson (1982) and Reid et al. (1990). They calculated the 3D form of Euler's homogeneity equation that can be defined as follows:

$$(x - x_0)\frac{\partial F}{\partial x} + (y - y_0)\frac{\partial F}{\partial y} + (z - z_0)\frac{\partial F}{\partial z} = N(B - F), \tag{5.6}$$

where x_0, y_0, and z_0 is the location of the potential field source, F is the field measured at x, y, and z, B is the regional value, and N is the structural index that characterizes the source geometry.

5.3.2 Magnetic

A variety of semiautomatic and automatic methods, based on the use of gradients of the potential field, have been developed as efficient tools for the determination of geometric parameters, such as locations of boundaries and depth of the causative sources (O'Brien, 1972; Nabighian, 1972, 1974; Cordell, 1979; Thompson, 1982; Barongo, 1985; Blakely and Simpson, 1986; Hansen and Simmonds, 1993; Reid et al., 1990; Roest et al., 1992; Salem and Ravat, 2003).

Processing techniques provide subsurface imaging capabilities that are used in geological interpretations. However, no single processing technique can be applied in all areas as the geologic field can vary widely. Therefore applying multiple techniques will yield the best results. Several numerical interpretation approaches have been developed. The techniques of interpretation can be divided into three categories (Fig. 5.3). Each category has the same goal, to illuminate the spatial distribution of magnetic sources, but they approach the goal with quite different logical processes. Fig. 5.3 shows the measured anomaly that is represented by A, calculated anomaly by A_o, and transformed measured anomaly by A'. Parameters $p1$, $p2$, and $p3$ are attributes of the source, such as depth, thickness, density, or magnetization.

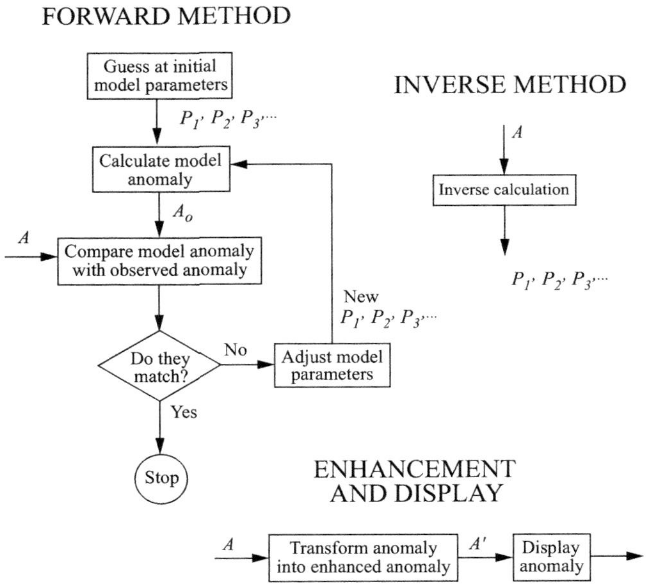

FORWARD METHOD

Guess at initial model parameters → $P_1, P_2, P_3, \ldots$ → Calculate model anomaly → A_o → Compare model anomaly with observed anomaly (A) → Do they match? — No → Adjust model parameters → New $P_1, P_2, P_3, \ldots$ → Calculate model anomaly; Yes → Stop

INVERSE METHOD

A → Inverse calculation → $P_1, P_2, P_3, \ldots$

ENHANCEMENT AND DISPLAY

A → Transform anomaly into enhanced anomaly → A' → Display anomaly →

FIGURE 5.3 Three categories of techniques to interpret potential field data (Blakely, 1995).

Forward method: An initial model for the source body is constructed based on geologic and geophysical intuition. The model's anomaly is calculated and compared with the observed anomaly, and model parameters are adjusted to improve the fit between the two anomalies. These three-step processes, body adjustment, anomaly calculation, and anomaly comparison, are repeated until calculated and observed anomalies are deemed sufficiently alike. Inverse method: One or more body parameters are calculated automatically and directly from the observed anomaly. Data enhancement and display: No model parameters are calculated, but the anomaly is processed in some way to enhance certain characteristics of the source, thereby facilitating the overall interpretation (see, e.g., Blakely, 1995, p. 182).

Traditionally, the magnetic method is used to estimate the depth of the basement structure. Additionally, the magnetic method is the oldest of the geophysical methods for locating both hidden ores and structures associated with mineral deposits, oil, and gas (Telford et al., 1990). In mineral exploration, the magnetic method is most often used to prospect for magnetic minerals directly, but also effective in the search for useful minerals that are not magnetic themselves but are associated with other minerals having magnetic effects detectable at the surface (Dobrin, 1960). Pham et al. (2020a,b) suggested the use of the improved logistic (IL) filter that is defined as

$$IL = \frac{1}{1 + \exp\left[-p(R_{HGA} - 1) + 1\right]}, \tag{5.7}$$

where R_{HGA} is the ratio of the vertical derivative/horizontal derivative of the HGA

$$R_{HGA} = \frac{(\partial \mathrm{HGA}/\partial z)}{\sqrt{(\partial \mathrm{HGA}/\partial x)^2 + (\partial \mathrm{HGA}/\partial y)^2}}, \tag{5.8}$$

and p is a positive (real) value that is decided by the researcher (between 2 and 5).

5.3.3 Inversion of magnetotelluric data

There are many models that can adequately fit the observation data regarding the noise level. To narrow the solution domain of geophysical inversion, the inverted model is chosen as the smoothest model (Constable et al., 1987; Kelbert et al., 2014). The objective function of the inversion is as follows:

$$\varnothing = \varnothing_d + \beta\varnothing_m, \tag{5.9}$$

where $\varnothing_d$ and $\varnothing_m$ are the misfit and smoothness terms, and b is the regularization parameter balancing between the misfit and the model structures.

5.3.4 Seismic attributes contributing to geology interpretation

Strong reflection seismic amplitude can be used to reveal geology boundaries. Moreover, seismic patterns can also be useful in detecting different geologies. However, seismic interpretation needs skilled scientists. In this research, we have provided some seismic texture seismic attributes for assisting seismic interpretation. The attributes can help to reveal some blurred but meaningful seismic features that one person finds difficult to detect by the naked eye.

The definition of texture attributes is well illustrated in the grey level co-occurrence matrix (GLCM) Texture Tutorial document by Hall-Beyer (2017). Characteristics of an image can refer to the level of roughness or smoothness. A seismic image that is constructed by different amplitudes can take the advantage of the texture attributes' application. Two steps are designed for computing a seismic texture. First, a GLCM term, known as a 2D matrix, computes the probability of occurrence of any two specific nearby values (Hall-Beyer, 2017). Second, a statistical equation is applied to the GLCM to compute a texture.

Three seismic texture attributes can be produced as the following equations:

$$\text{GLCMMean}\mu_i = \sum_{i,j=0}^{N-1} iP_{i,j}, \tag{5.10}$$

$$\text{GLCMVariation}\sigma_i{}^2 = \sum_{i,j=0}^{N-1} P_{i,j}\left(i-\mu_i\right)^2, \tag{5.11}$$

$$\text{GLCMCorrelation} = \sum_{i,j=0}^{N-1} \frac{(i-\mu_i)(j-\mu_j)}{\sigma_i{}^2\sigma_j{}^2}. \tag{5.12}$$

In our research, GLCMMean and GLCMCorrelation for seismic are used for enhancing seismic interpretation. The Variation shows a level of chaos/dispersion of the values within the mean (Hall-Beyer, 2017; dGB Earth Sciences, 2015). Meanwhile, GLCMCorrelation refers to the linear dependency of the seismic inputs with the nearby ones (Hall-Beyer, 2017; dGB Earth Sciences, 2015). The Correlation can be linked useful with another texture attribute (Hall-Beyer, 2017; dGB Earth Sciences, 2015).

5.3.5 Radiometric data

There are many types of gamma-ray spectrometric surveys. The survey can be achieved by aircraft, vehicles, ground, and borehole spectrometers [International Atomic Energy Agency (IAEA), 2003]. The data are presented as contour maps, and they have been subjected to a qualitative and quantitative interpretation.

A major objective of the interpretation of airborne gamma-ray spectrometer data is to define the probable boundaries of potential uraniferous provinces in which the rocks and soils are preferentially enriched in uranium (Saunders and Potts, 1976). The qualitative interpretation of gamma-ray spectrometric data depends mainly upon the correlation between the radioactivity measurements and the surface distribution of the various rock units. The quantitative interpretation deals with the processes that control the distribution of radioelements in rocks and soils [International Atomic Energy Agency (IAEA), 2003]. Besides, the statistical analysis was applied to the three radioelements (K, U, and Th) for each geologic rock unit. The statistical analysis includes minimum, maximum, arithmetic mean (X), standard deviation (S.D.), and coefficient of variability (CV%) that check the normality of each rock unit. Coefficient of variability [$CV\% = (S.D./X) \times 100$] is applied, if CV% of a certain rock unit is <100; the unit tends to exhibit normal distribution (Sarma and Koch, 1980).

The structural lineaments analysis can be deduced from the total count (TC) and three radioelements' maps to determine the significant trends that are the affecting factors on the distribution of the radiometric anomalies.

5.3.6 Self-potential

The field survey using SP is carried out by measuring the difference in electric potential between two electrodes that relate to the Earth's surface at survey stations in the interested area. One station is selected as a base station, and all measured potentials were referenced to that point. The base station is located at a calm SP potential point as possible, and one of the two electrodes is put in this station. The other electrode is moving on the grid stations.

The processing of the SP data is very simple, and it can be done simultaneously in the field. The value of the base for each profile is subtracted from the reading of each station on the same profile. The net reading could be plotted against the distance to produce the SP-processed profiles. These profiles first were drawn on millimeter papers to determine some parameters graphically with relatively more accuracy.

The results of a SP survey are presented as a set of profiles, plotting SP values against the distance. These profiles are usually used to construct a contour map showing equipotential lines. The structure to be detected is generally 3D; therefore it is better to interpret a contour map rather than profile data (Thanassoulas and Xanthopoulos, 1991). The interpretation of these data is mainly qualitative, where the interpretation procedure selected will depend on the desired goals of the investigation, the quality of the field data, and the availability of additional geophysical and hydrological data (Sharma, 1997).

The quantitative interpretation of SP is usually used to transform the conductive body into a physical model of a simple geometric shape. The parameters of the source may be evaluated by using the method of characteristic curves. In this method, mathematical relations among the parameters of the body and a few readily identifiable points on the anomaly curve are used for interpretation. There are two methods of quantitative understanding (Ram Babu and Atchuta Rao, 1988; Murty and Haricharan, 1985). They presented curves for interpreting the SP anomalies using a few characterizing points of the nomogram.

5.3.7 Induced polarization

The IP/resistivity measurements are first viewed in a traditional 2D pseudosections presentation. These measurements are influenced, to a large degree, by the rock materials nearest the surface (or, more precisely, nearest the measuring electrodes), and the interpretation of the pseudosections of IP data is mostly uncertain. This is because stronger responses that are located near the surface could mask a weaker one that is located at depth. Also, the pseudosections do not create particularly good images of the subsurface (Pelton et al., 1978; Barker, 1981;

Smith et al., 1984; Oldenburg and Li, 1994; Loke and Barker, 1996b; Ulrich and Slater, 2004). So, the apparent resistivity and chargeability measurements were inverted into models using the RES2DINV program (Loke, 2004). This program automatically determines 2D models for both the resistivity and chargeability data obtained from the IP/resistivity survey (Griffiths and Barker, 1993). This computer program uses the smoothness-constrained least-square inversion technique (Sasaki, 1992) to convert measured apparent IP/resistivity values to true resistivity and chargeability values and plot them in cross section (2D model). The program creates a resistivity and chargeability cross section, calculates the apparent resistivity and chargeability for that cross section, and compares the calculated apparent resistivity and chargeability to the measured apparent resistivity and chargeability. The iteration continues until a combined smoothness-constrained objective function is minimized. The surface topography can have a significant effect on the resistivity measurements (Tsourlos et al., 1999); for this reason, the surface topographic data of the IP profiles were included in the resistivity and chargeability models.

Recently, 3D models of resistivity and chargeability data were generated from IP/resistivity surveys or combined with 2D models of resistivity and chargeability (Chambers et al., 2006; Vieira et al., 2016; Halim et al., 2017). The 3D models are useful in delineating the lateral and vertical variations of resistivity and chargeability to get the best understanding of the ore deposit existences and contaminated sites, etc. (Ustra et al., 2012; Vieira et al., 2016).

5.4 Case studies

5.4.1 Voisey's Bay deposit (gravity data)

The Voisey's Bay nickel–copper–cobalt deposit located on the northeast coast of Labrador (Canada) is known as one of the most significant mineral discoveries in the past half-century (Farquharson et al., 2008). The main ore body is a Ms lens roughly ellipsoidal in shape with horizontal sizes of 350 m × 650 m (Farquharson et al., 2008). Fig. 5.4A shows Bouguer ground gravity over the Voisey's Bay deposit digitized from King (2007). In this study, we applied the EHGA filter to determine the horizontal boundaries of the deposit and used the Euler deconvolution method to estimate the depths of these horizontal boundaries. Fig. 5.4B shows the result obtained from applying the EHGA filter. We can see that the peaks of the EHGA demonstrated the existence of the main ore body with an approximately ellipsoidal shape as reported by Farquharson et al. (2008) and Lelièvre et al. (2012). Fig. 5.4C shows the depth estimates determined by applying the Euler deconvolution method. Fig. 5.4D shows the histogram of the depth estimates in Fig. 5.4C. It is clearly seen that most of the horizontal boundaries are in the depth range of 20–40 m. This depth result agrees well with the results given by other authors (Farquharson et al., 2008; Maag and Li, 2018).

5.4.2 Gabal Semna region, Eastern Desert of Egypt (magnetic data)

Gabal (G) Semna region, Eastern Desert (ED) of Egypt, has become a major exploration target for deep minerals in some promising zones, due to a program of the Egyptian

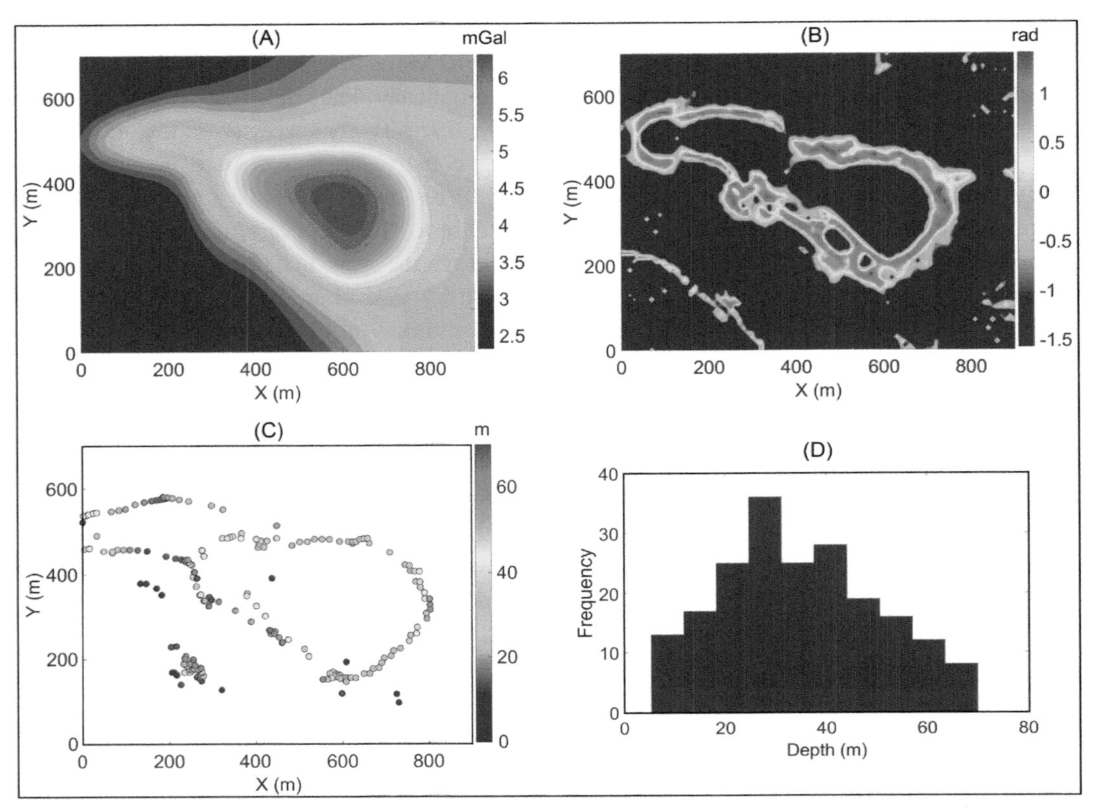

FIGURE 5.4 (A) Bouguer ground gravity over the Voisey's Bay deposit, (B) EHGA of gravity data in (A), (C) depth estimates determined from the Euler deconvolution method, and (D) histogram of the depth estimates in (C). *EHGA*, Enhanced horizontal gradient amplitude.

government that began to provide high-resolution airborne geophysical data (Fig. 5.5). Aeromagnetic survey data were used to estimate the structures and their relation to ore deposits. The 3D inversion was applied to estimate the depth and geometry of the magnetized ore bodies. The 3D magnetic inversion results revealed that high-amplitude magnetic anomalies correspond to metavolcanics and mafic rocks. High-amplitude (positive) magnetic anomalies are correlated with the large susceptibilities' metavolcanics.

Aeromagnetic data were used to map surface geology and alteration zones that can host mineralization, and the 3D inversion was applied to outline the subsurface distribution of these promising areas. Many filtering techniques have been applied to map edges, structures, boundaries, and other geologic features basis on the contrast of magnetic susceptibility and remanence (Holden et al., 2008; Elkhateeb and Eldosouky, 2016; Sehsah et al., 2019; Eldosouky et al., 2020b, 2022b, 2022c; Saada et al., 2021a,b; Eldosouky and Mohamed, 2021; Sehsah and Eldosouky, 2022; El-Sehamy et al., 2022). For deep mineral exploration, this approach can be used in the Egyptian ED and identical areas around the world. Several corrections have been

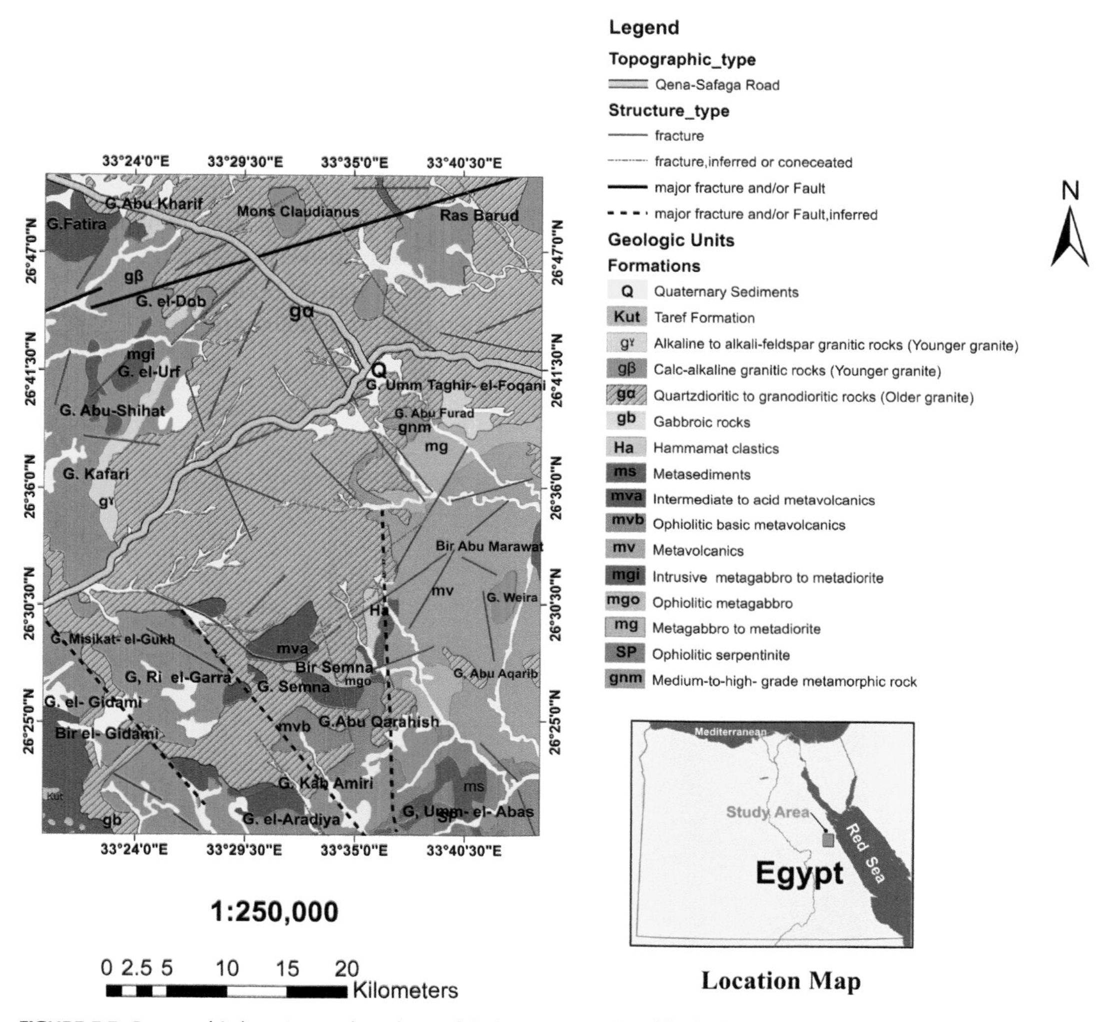

FIGURE 5.5 Geographic location and geology of G. Semna area. *Modified after Conoco Coral Corporation and the Egyptian General Petroleum Corporation EGPC (1987a).*

applied to the aeromagnetic data such as aircraft altitude, diurnal correction, and removing the regional gradient of the Earth's magnetic field (Aero-Service, 1984).

Significant progress has been made in mineral exploration, especially with advances in inversion and 3D modeling. It has become easier to integrate geophysics and geology into consistent and more reliable models (Robert et al., 2007). In mineral exploration, 3D interpretation of magnetic data has become common practice. The main objective of the 3D interpretation is to estimate the magnetic susceptibility distribution and remanence properties of the rocks that host minerals-bearing veins with a particular emphasis on the relationships among magnetic physical properties, alteration, and mineralization (Mohamed et al., 2018; Oldenburg and Pratt, 2007). Previous studies show the diverse applications of 3D inversion in ore prospecting, crustal imaging, and environmental studies (e.g., Abdelrahman and Essa, 2005; Balkaya et al., 2017; Mohamed et al., 2018).

Total magnetic field data are shown in Fig. 5.6. Total magnetic field data have been reduced to the magnetic pole (RTP) (Baranov, 1957) map (Fig. 5.7). The magnetic anomaly of the RTP map varies between −209 nT and >300 nT (Fig. 5.7). The high magnetic anomalies correspond to surface geology and are related to intrusive metagabbro, serpentinite, metagabbro, and metavolcanics (Fig. 5.7). Also, buried magnetic bodies are the primary source for these positive anomalies. The negative magnetic anomalies are correlated with

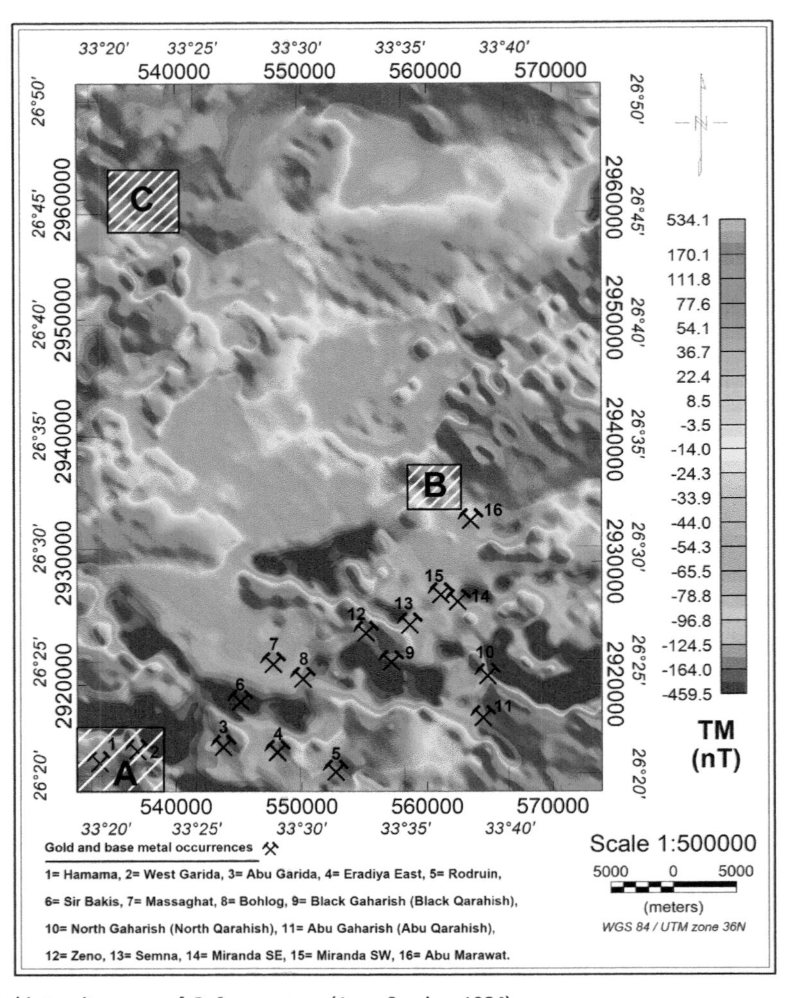

FIGURE 5.6 Total intensity maps of G. Semna area (Aero-Service, 1984).

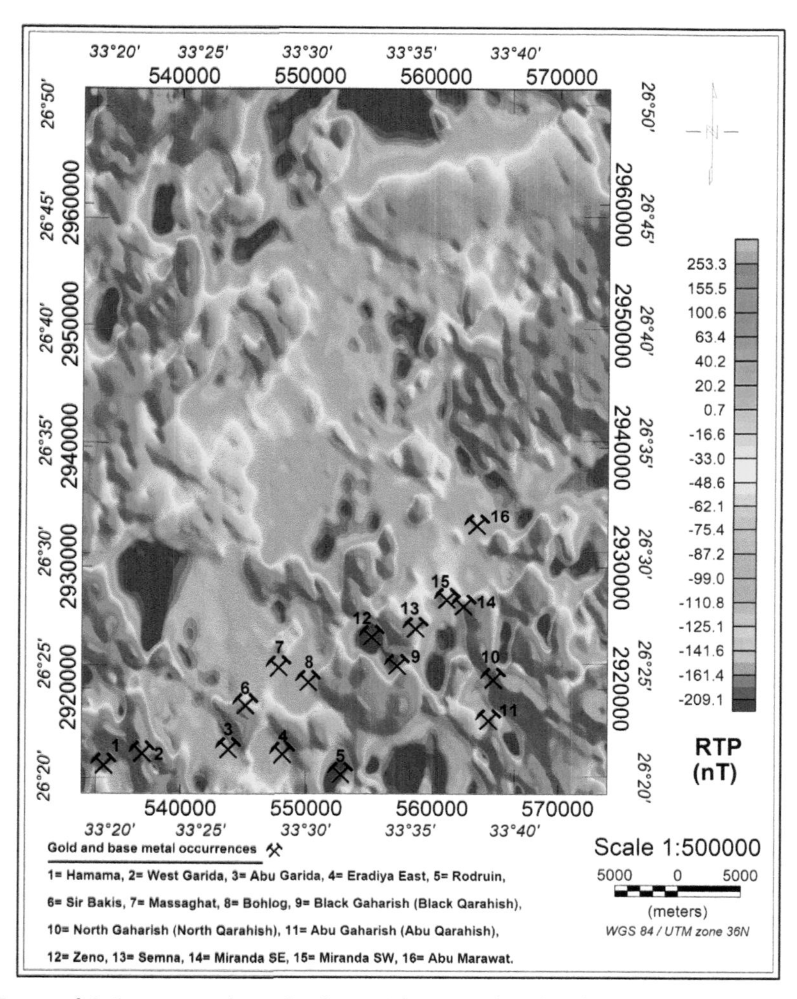

FIGURE 5.7 RTP map of G. Semna area (Aero-Service, 1984). *RTP*, Reduced to the magnetic pole.

some parts of Hammamat clastics and acidic metavolcanics, granitic rocks, and metasediments (Fig. 5.7).

The separation of regional and residual magnetic anomalies is a crucial step in mineral exploration. RTP data are separated into their residual and regional components (Fig. 5.8A and B) (Spector and Grant, 1970). The FFT was applied to the RTP magnetic data to explore the frequency content of these data. The analysis shows that the deep-seated magnetic component frequency has a long wavelength cutoff of 200 km and a short-wavelength cutoff of 17.2 km, while the near-surface magnetic component frequency has a long-wavelength cutoff of 17.2 km and a short-wavelength cutoff of 8.3 km. It is obvious that the locations of most gold mineralization deposits are related to positive magnetic anomalies and their contacts

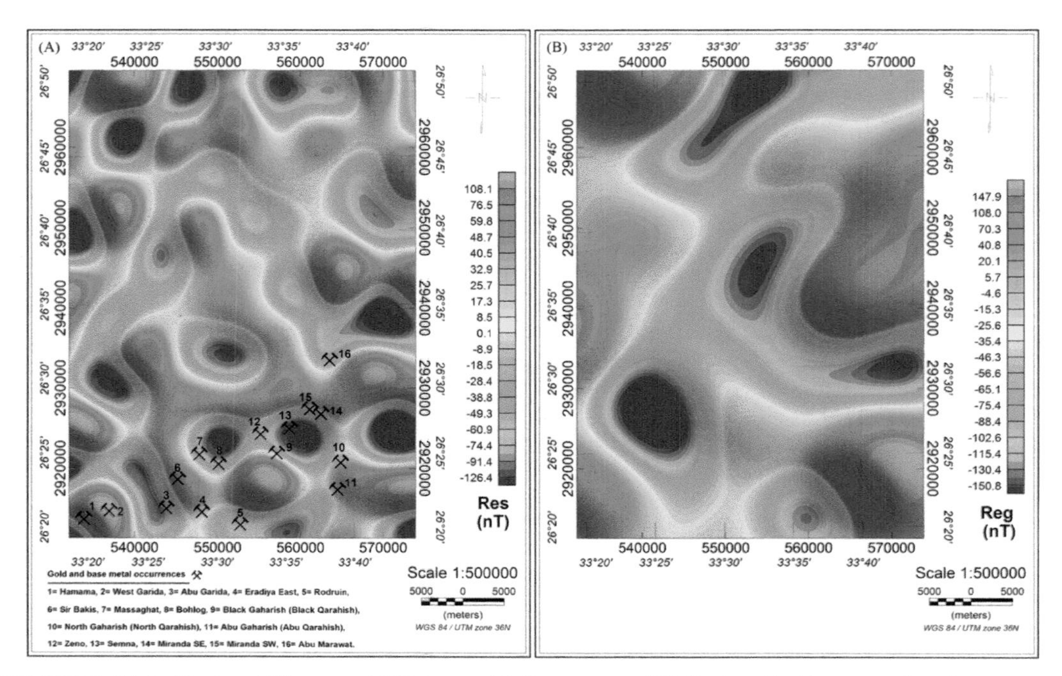

FIGURE 5.8 (A) Residual and (B) regional RTP maps of G. Semna area. *RTP*, Reduced to the magnetic pole.

with negative magnetic anomalies of the RTP and residual maps (Figs. 5.7 and 5.8). The investigation of the residual magnetic anomaly (Fig. 5.8A) shows that the southern portion of the area consists of alternative magnetic anomalies (positive and negative) that are characterized by high amplitudes and frequencies and have an E–W trend. The trend of the anomalies changes from E–W to WNW–ESE direction in the southwestern and central parts of the study area (Fig. 5.8A). Also, the frequencies and amplitudes of these anomalies change with trends changing. They were dissected by NW–SE, NE–SW, and WNW–ESE faults. Magnetic anomalies (positive and negative) belts with varying amplitudes and frequencies, trending in the NNE–SSW directions, can be observed in the northwestern portion of the G. Semna area (Fig. 5.8). Several faults intersect these belts with major trends in E–W, NNE–SSW, NWN–SES, and N–S directions. The low-pass (regional) aeromagnetic map (Fig. 5.8B) appears to be related to four magnetic anomalies (positive) with high magnitudes and frequencies situated at the eastern, southeastern, western, and northwestern parts of the study area. These anomalies trend in northeastern and northwestern directions, respectively. Moreover, three moderate positive anomalies are separated from high positive magnetic anomalies by faults trending NNW–SSE, NNE–SSW, and NW–SE (Fig. 5.8B). This reveals that the three moderate positive subanomalies are related to deeper magnetic sources than the four high positive anomalies that belong to shallow magnetic sources.

Five negative anomalies are trending in NNE–SSW, NW–SE, and ENE–WSW directions (Fig. 5.8B) and spread out over the study area, with different amplitudes and frequencies.

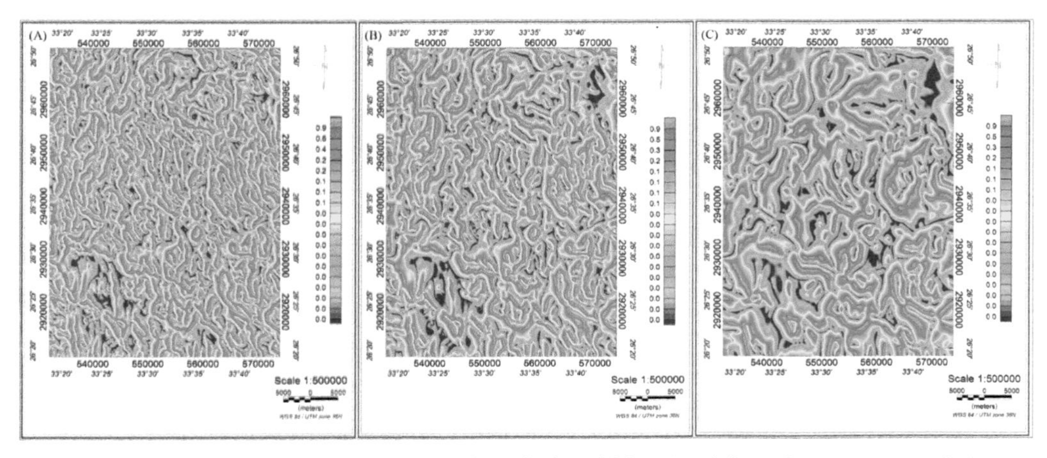

FIGURE 5.9 Improved logistic (IL) filter of (A) RTP, (B) residual, and (C) regional data of G. Semna area. *IL*, Improved logistic; *RTP*, reduced to the magnetic pole.

These anomalies are trending in NNE–SSW, NW–SE, and ENE–WSW directions (Fig. 5.8B). An IL filter is applied to the RTP, residual, and regional data (Figs. 5.7 and 5.8A and B respectively) to reveal the faults, edges, and contacts of the G. Semna area. The study area was affected by main structural trends: NW–SE, NE–SW, ENE–WSW, and N–S directions as revealed by the results of the IL filter for aeromagnetic maps (Fig. 5.9A–C). Contacts of deep and shallow features are delineated clearly with the IL filter maps in the RTP, residual, and regional data.

5.4.2.1 3D magnetic inversion

In this study, 3D inversion is applied to help target mineral exploration deposits. Inversion refers to constructing a 3D magnetic-susceptibility model that is consistent with the geology and related to the measured magnetic response. TMI field data (Fig. 5.6) provide a comprehensive 3D magnetic susceptibility model in the study area. Three different areas are selected to construct a 3D magnetic susceptibility model. The locations of two areas where known gold deposits have been studied during previous exploration, area A (Hamama and West Garida) area B (Abu Marawat), are shown in Fig. 5.6. The 3D inversion is applied to area C (Fig. 5.6) where no known gold deposit has been identified. The inversion results are shown in Fig. 5.10 for the study area (C). Measured and calculated magnetic field data overlaid the 3D magnetic susceptibility model (Table 5.4; Fig. 5.10). The subsurface of the study area was represented with cuboid cells in the x, y, and z (magnetic field components) directions (see Table 5.2).

The relation between magnetic susceptibility (Ms) and rocks or minerals is the critical link between geological and geophysical interpretation (Mohamed et al., 2018). Table 5.1 shows the Ms values for some rocks and minerals in the world that also occur in the studied areas. For this method to be successful, there must be a physical property contrast between

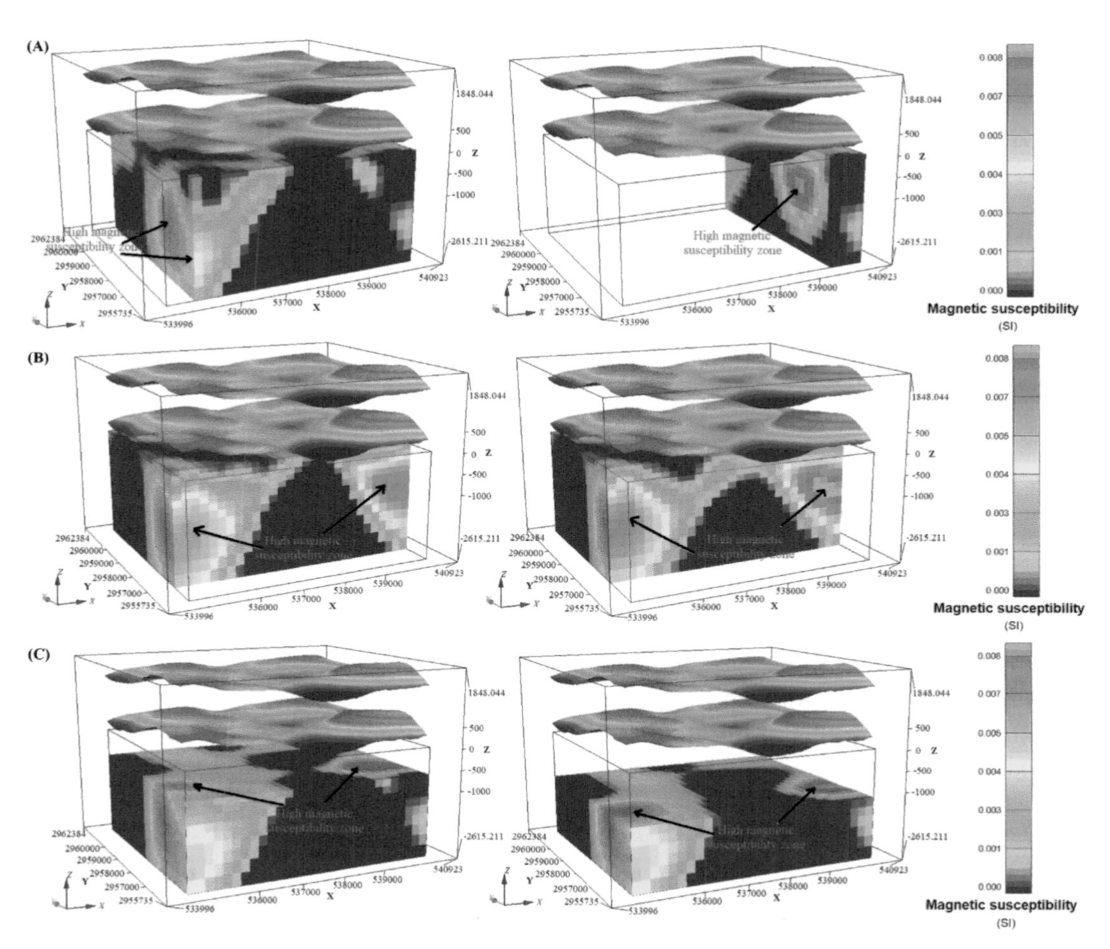

FIGURE 5.10 Cross sections (A–C) through the susceptibility model. Slices at $X = (543{,}500, 544{,}500)$, $Y = (2{,}957{,}500, 2{,}958{,}000)$, and $Z = (1000, 2000)$ for area C.

Table 5.4 Mesh parameters of the base and padding of the magnetic data (area C).

	Base	Horizontal padding	Vertical padding
Dimensions (cells)	14	5	5
Cell expansion ratios	1.08	1.5	1.5

the geologic materials. Fig. 5.10A–C shows the 3D magnetic susceptibility model with a series of slices at different positions along with eastern, northern, and depth directions in figures to show the location of the high-Ms zones. The inversion indicated the presence of continuous zones of high magnetic material at different depths. Metavolcanic rocks are the main magnetic rocks in the study area, so it seemed likely that the high magnetic region found at depth has the same lithology as that of known deposits. Wadi deposits and older granitoids are correlated with low magnetic field data. The inversion results have been used in this study, and new geologic and economic reserves have been found.

5.4.3 Cloncurry–Georgetown–Charters Towers, North Queensland, Australia (seismic and magnetotelluric)

The 2D geophysical datasets as seismic and MT data are measured in the Cloncurry–Georgetown–Charters Towers, North Queensland, Australia. The vast area, North Queensland (Korsch et al., 2012), located on the cratonic margin of Australia, has experienced a complex crustal history connecting with the successive development of several Proterozoic-to-Paleozoic orogenic systems. A deep survey seismic reflection and low-frequency MT band measurements were conducted in 2007 for a better understanding of the regional geological factors linking to the known rich mineral resources. The seismic survey was conducted by a cooperation among the Australian Government's Onshore Energy Security Program, the Queensland Government's Smart Mining and Smart Exploration initiatives, and AuScope (an unincorporated company funded by the Australian Government under the National Collaborative Research Infrastructure Strategy, NCRIS). For MT data collection and processing, Quantec Geoscience Ltd. was in charge of MT data collection and processing.

The 2D profile we use for analysis is named 07GA-IG1 (Fig. 5.11). Useful information on seismic data measurement settings can be extracted from the paper by Korsch et al. (2012). Migrated 20-s two travel time seismic section (07GA-IG1) shown in Fig. 5.12 is the input for

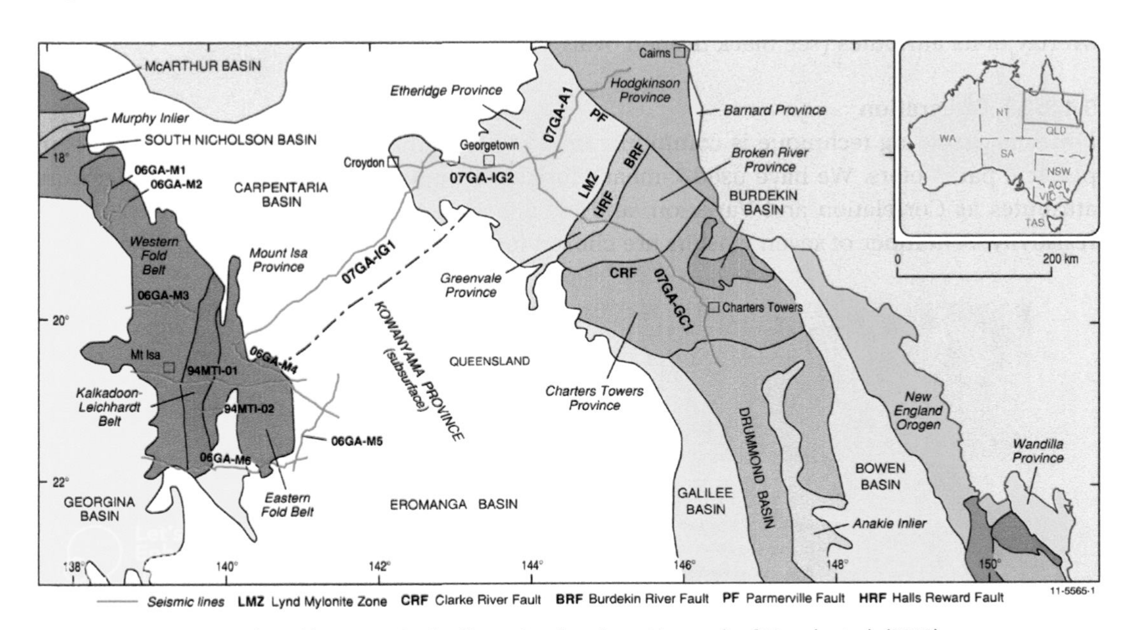

FIGURE 5.11 Location of profile 07GA-IG. The figure is taken from the work of Korsch et al. (2012).

computing its seismic texture attributes (i.e., Correlation and Variation). The equivalent maximum depth is 60 km using a medium velocity of 6000 m/s. For MT data inversion the frequency band ranges from 0.0053 to 238 Hz.

5.4.3.1 Results

5.4.3.1.1 Magnetotelluric analysis

We have run unconstrained smooth MT inversion with the prior resistivity model as 100-Ω m value for all blocks. The inverted resistivity model (Fig. 5.12) can share many similar features with the seismic interpretation result. Some importation remarks can be made as follows:

For the detection of geological faults: Shapes of resistivity models can match the trend of geological faults (see the white arrows in Gidyea Suture zone and Rowe Fossil).

High-resistivity zones can show a high correlation with geology features (i.e., Millungera Basin).

Geological faults do not always refer to low resistive zones (see cyan arrows).

5.4.3.1.2 Seismic analysis

Two seismic texture attributes, Correlation and Variation, are calculated from the processed migrated data (Fig. 5.13). It does look like the Variation can keep more details like the seismic input than the Correlation can (Fig. 5.14). Importantly, we also present an overlay image of Correlation and Variation in gray scale to see how it looks with the migrated section and the published seismic interpretation (Korsch et al., 2012) (Fig. 5.15). Note that color representation is one kind of attribute technique. Many clear strong seismic events can be seen in the three images (see green dashed ovals in Fig. 5.15). Interestingly, some weak signals that are hardly seen in the seismic original section can easily be targeted by visible events in the overlay of its attributes (see black dashed ovals in Fig. 5.15).

5.4.3.1.3 Integration

k-Means clustering technique is commonly used in evaluating and connecting different geophysical parameters. We have used *k*-means for clustering two cases: (1) two seismic texture attributes as Correlation and Variation, and (2) the two seismic attributes and inverted MT resistivity. A number of seven clusters are chosen for the two cases.

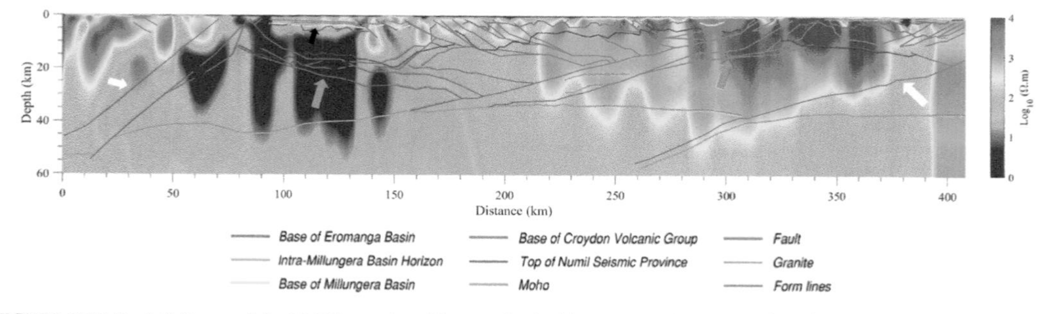

FIGURE 5.12 Resistivity model of MT inversion. The geological interpretations are taken from the work of Korsch et al. (2012). *MT*, Magnetotelluric.

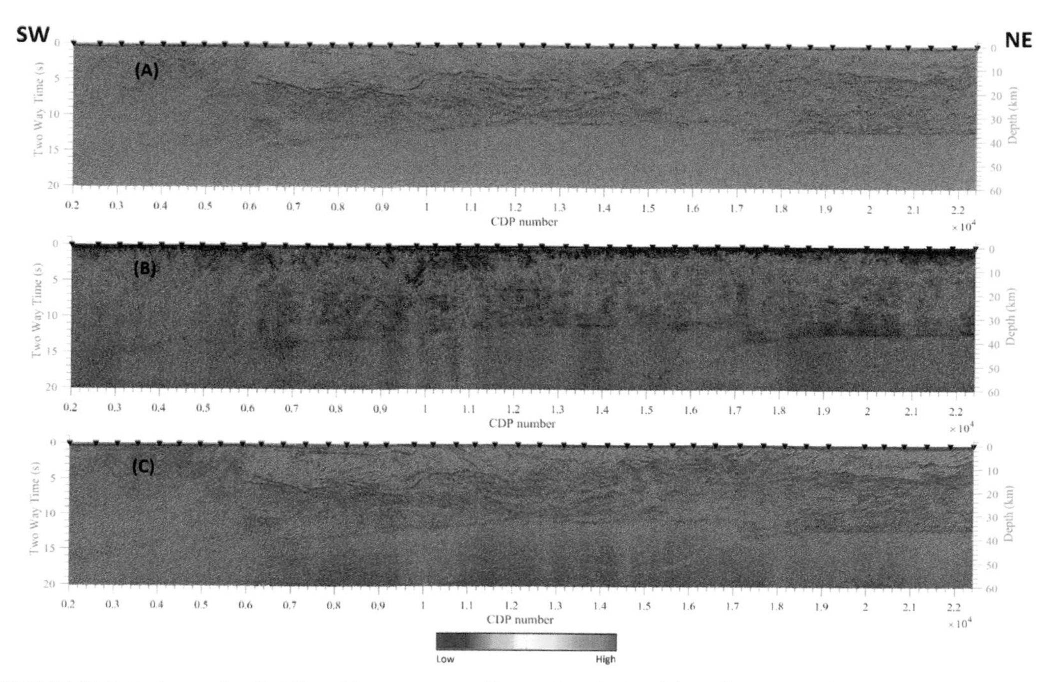

FIGURE 5.13 Seismic amplitude (A) and its texture attributes Correlation (B), and Variation (C). The triangles on the top of the profile mark the location of MT. *MT*, Magnetotelluric.

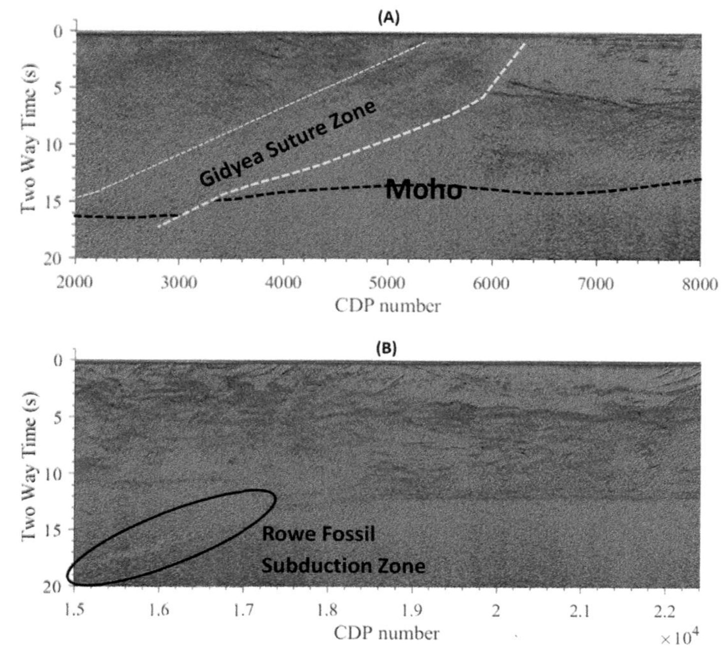

FIGURE 5.14 The seismic variation attributes illustrate clearly geological structures, Gidyea Suture Zone and Moho (A) and Rowe Fossil Subduction Zone (B).

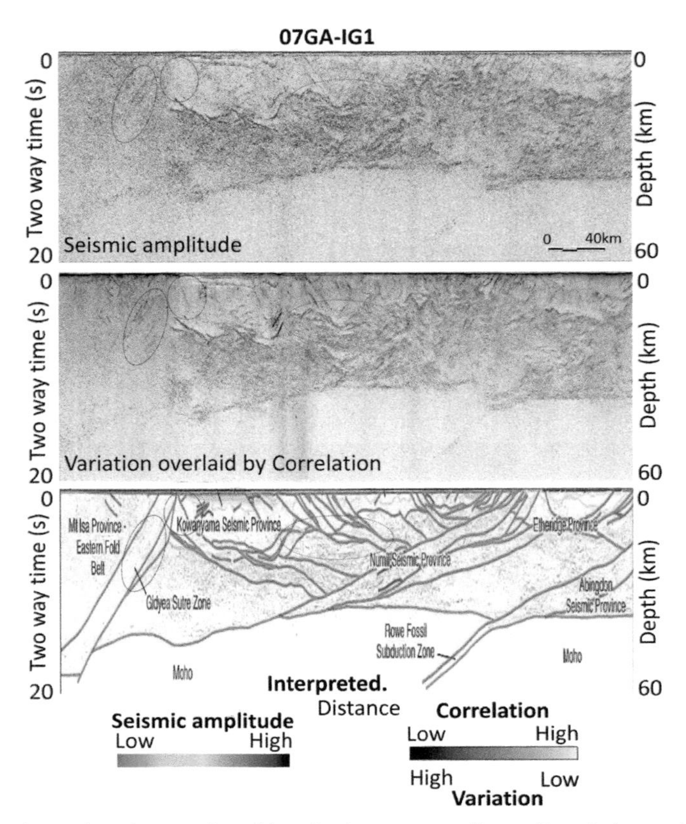

FIGURE 5.15 Seismic migrated section, overlay of its seismic texture attributes (Correlation and Variation), and geology interpretation extracted from the work of Korsch et al. (2012).

For the first case with only the seismic ones, Moho zones are well defined with the existence of clusters 5 and 6 as yellow and orange colors, respectively (Fig. 5.16). Meanwhile, its upper part (from the surface to the Moho interface) is dominated by clusters 1–3.

For the second case with the combination between seismic texture attributes and MT resistivity, more detailed features are preserved than just interpreting individual seismic or MT data (Fig. 5.17). Fantastically, we can see an orange wedge bounded by Moho and Rowe Fossil subduction zone (see the white arrow). Again, Moho can be represented by blue zones (clusters 1 and 2) for the new clustering setup. However, there are still distortions made by clusters that are conflict with seismic faults (i.e., Gidyea Suture Zone) (see the black arrow).

5.4.3.2 Conclusion

A combination of seismic and MT data can provide a useful tool for understanding the Earth. Highly detailed delineation of different geological objects is supported by deep reflection seismic, while the MT method can give the distribution of electrical resistivity parameter on large scale. Geological features such as basins, faults, and Moho zones can be detected by

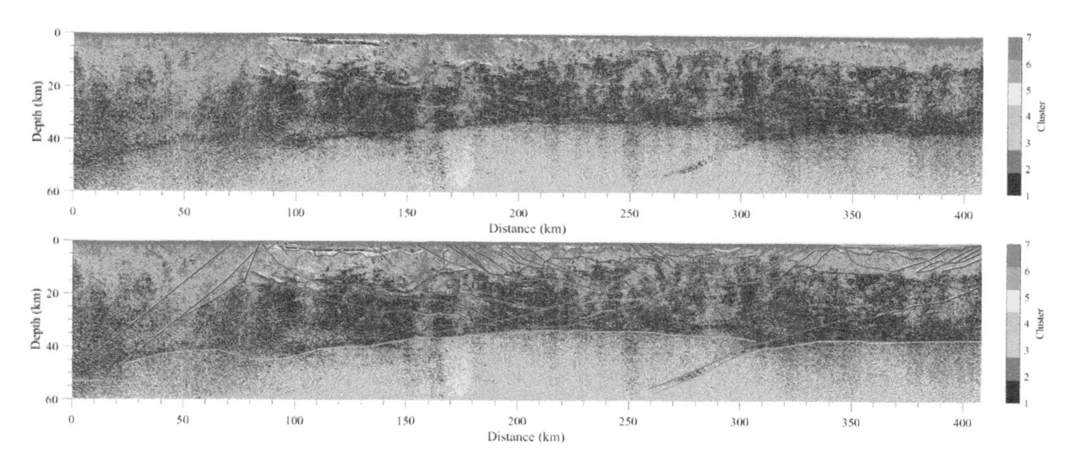

FIGURE 5.16 *k*-Means clustering result of the seismic attributes only.

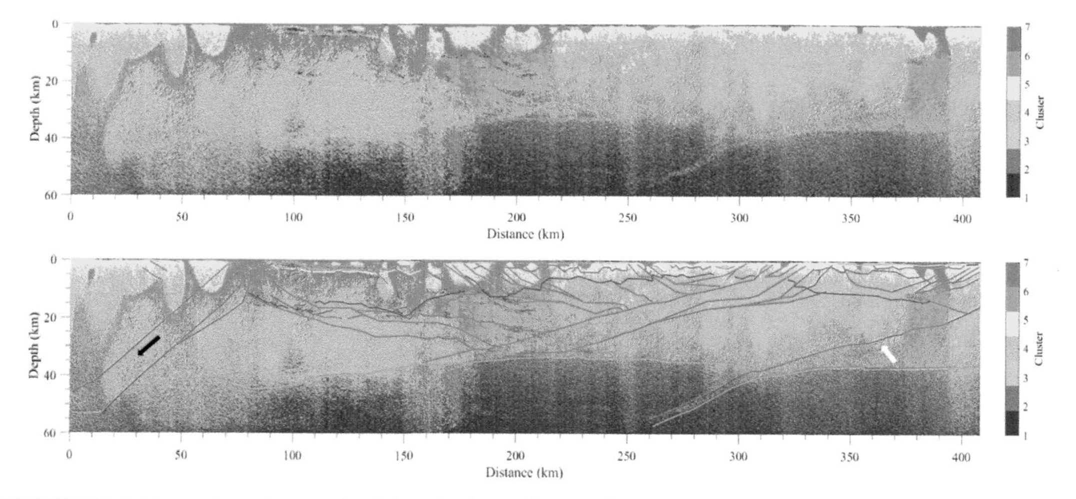

FIGURE 5.17 *k*-Means clustering result of the seismic attribute and resistivity of MT model. *MT*, Magnetotelluric.

both methods. In a nicer way, their integration can even provide more useful information with the application of the *k*-means technique. For seismic attributes analysis, representations of each attribute, their overlay, or clustered results can be good indicators for supporting geology interpretation.

5.4.4 Gebel El-Bakriyah—Wadi El Batur area, Central Eastern Desert, Egypt (radiometry)

The study area is located in the Egyptian Central Eastern Desert (Fig. 5.18). The geologic setting of this area depends on the geologic map of the Gebel Hamata area at a scale of

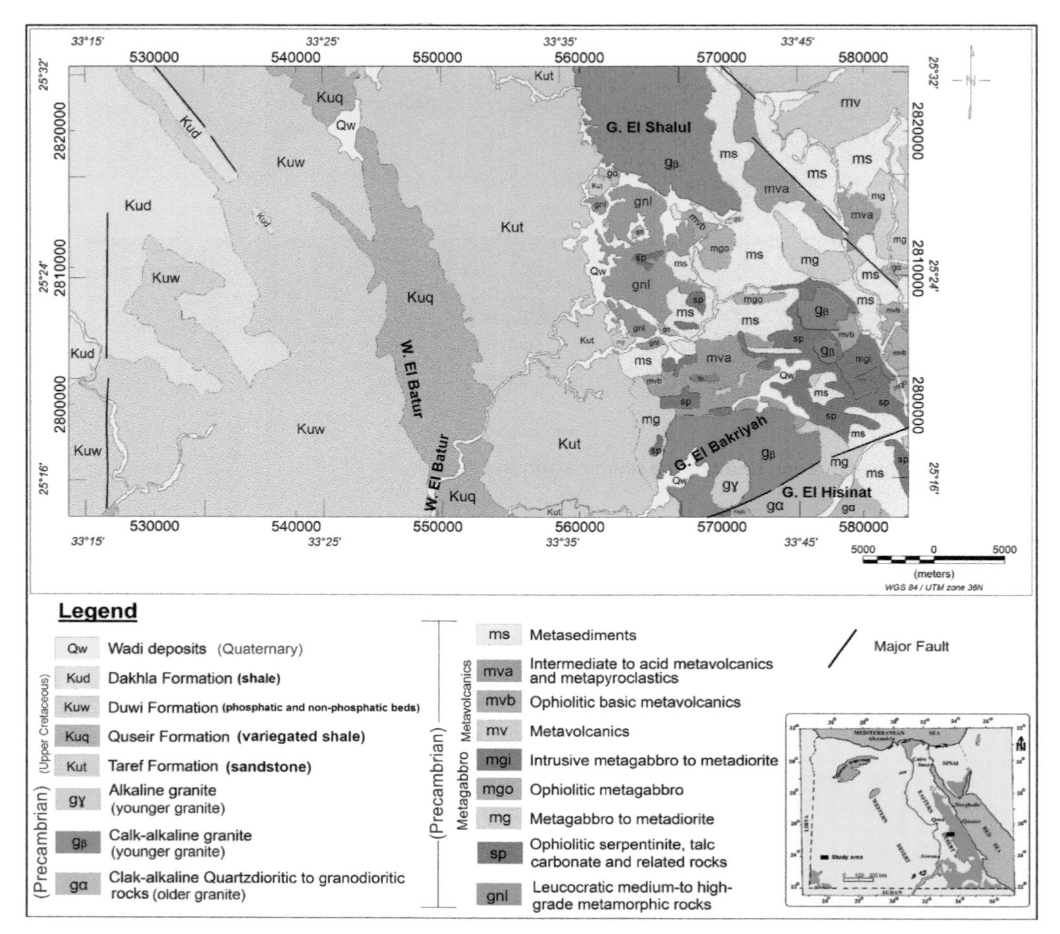

FIGURE 5.18 Location and simplified geologic map of Gebel El Bakriyah—Wadi El Batur area, Central Eastern Desert, Egypt. *Reproduced after Conoco Coral Corporation and the Egyptian General Petroleum Corporation EGPC (1987b).*

1:500,000 (Conoco Coral Corporation and the Egyptian General Petroleum Corporation EGPC, 1987b; El-Qassas, 2018). It contains a vast variety of metamorphic, igneous, and sedimentary rocks having ages ranging from Precambrian to Quaternary (Fig. 5.18). The previous geological and radiometric studies indicated that phosphates and granitic rocks contain considerable concentrations of uranium mineralization (Sadek, 1972; Salman, 1974; El-Amin, 1975; Abdel Hadi, 1978; El Shazly et al., 1981; Ammar et al., 1988; Gharieb, 1995; Hanafy, 2002; Abd El Nabi, 2009, etc.).

The study area was surveyed using airborne gamma-ray and magnetic by Aero-Service (1984). This survey represents a part of the mineral, petroleum, and groundwater assessment program (MPGAP). The main target of this case study was to construct the surface radioactive heat production (RHP) map from the airborne gamma-ray data of the Gebel El Bakriyah—Wadi El Batur area, Central Eastern Desert, Egypt (El-Qassas, 2018).

The interpretation of airborne gamma-ray spectrometric data of the study area deals with TC and the radioelements content of potassium (K), equivalent uranium (eU), and equivalent thorium (eTh), besides the ratios of eU/eTh, eU/K, and eTh/K (Figs. 5.19 and 5.20). The area possesses radiation ranging between 1.6 and 33.5 Ur as Tc, 0.07%–3.7% for K, 0.1–30.1 ppm for eU, and 0.8–28.9 ppm for eTh (El-Qassas, 2018). The TC map (Fig. 5.19A) showed that the lowest level (1.6–6.8 Ur) is situated at the eastern, central, and northwestern parts of the map and related to serpentinite, some parts of metagabbro, metavolcanics, metasediments; Taref, Quseir, and Dakhla Formations. While the highest level of TC (Fig. 5.19A) has values varying from 13 to 33.5 Ur, recorded over the younger granites of the Gebel El Bakriyah and Gebel El-Shalul and some parts of Duwi Formation that covers the southeastern, northeastern, and western parts of the map (El-Qassas, 2018).

K% map (Fig. 5.19B) displayed the lowest level (<0.5%) over the Taref Formation, some parts of the Dakhla Formation, serpentinites, and metavolcanics, and over the Duwi Formation. The highest level (>1.3%) covers the eastern part and correlated geologically with older granites, younger granites, leucocratic metamorphic rocks, and some parts of

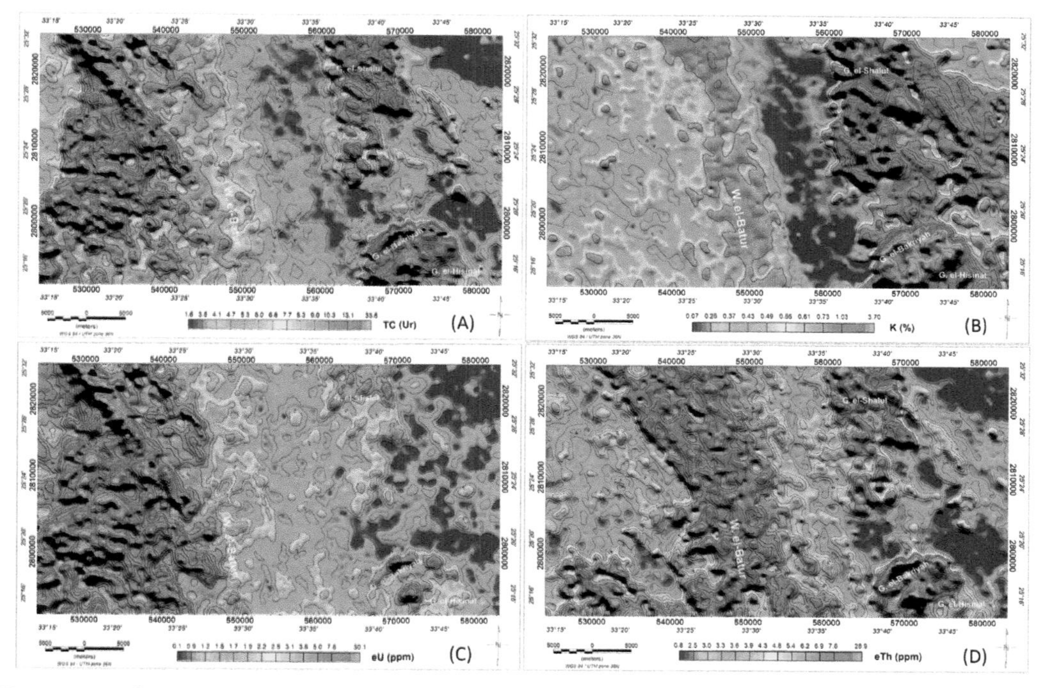

FIGURE 5.19 Airborne radiometric contour maps: (A) total-count (TC in Ur), (B) potassium (K in %), (C) equivalent uranium (eU in ppm), and (D) equivalent thorium (eTh in ppm), Gebel El Bakriyah—Wadi El Batur area, Central Eastern Desert, Egypt (Aero-Service, 1984).

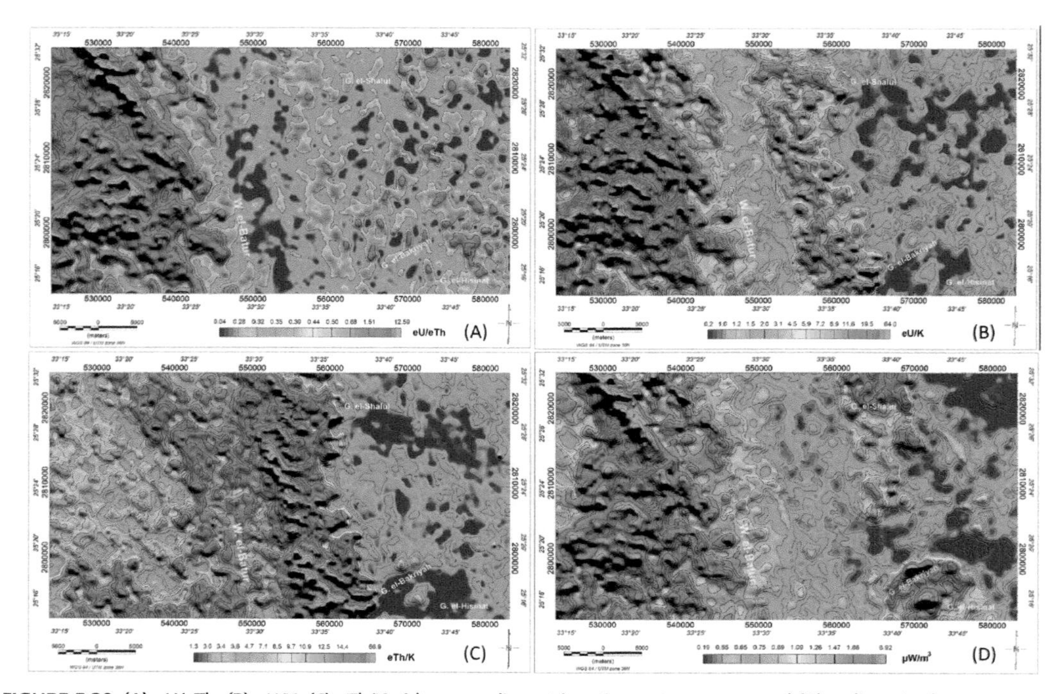

FIGURE 5.20 (A) eU/eTh, (B) eU/K, (C) eTh/K airborne radiometric ratio contour maps, and (D) radioactive heat production (RHP in $\mu W/m^3$) map, Gebel El Bakriyah—Wadi El Batur area, Central Eastern Desert, Egypt (El-Qassas, 2018).

metasediments and metavolcanics (El-Qassas, 2018). The existences of alkali-feldspars and hydrothermal alteration lead to the increase in K% values.

eU values (6–30.1 ppm) occurred over the younger granites of the Gebel El Bakriyah, Dakhla, and Duwi formations. These values represent the highest uranium content in the study area (Fig. 5.19C). The lowest eU values (>2.3 ppm) are restricted to serpentinite, metavolcanics, metagabbro, some parts of metasediments, and Taref Formation (El-Qassas, 2018).

Fig. 5.19D reveals that the lowest values of eTh content range from 0.8 to 3 ppm and the highest values (7–28.9 ppm) are related to younger granites, leucocratic metamorphic rocks, Quseir Formation, and some parts of Duwi Formation at the southeastern, northern, central, and southwestern parts (El-Qassas, 2018).

Fig. 5.20A–C describes the ratios of eU/eTh, eU/K, and eTh/K. The eU/eTh ratio map (Fig. 5.20A) is important for uranium exploration because it maps the relatively uranium-enriched areas. The highest eU/eTh values (>4) may have been produced by highly enriched formations in uranium, which have higher mobilization than thorium. These values are mainly associated with some parts of the Duwi and Dakhla formation in the western part (El-Qassas, 2018). The eU/K ratio map (Fig. 5.20B) confirms the highest eU/eTh values at the western part and agrees with eTh/K ratio (Fig. 5.20C) in determining the lowest values at the eastern part. In addition, the eTh/K map (Fig. 5.2C) displayed the contact between the sedimentary and basement rocks, which has NNW to N–S direction (El-Qassas, 2018).

Table 5.5 Airborne spectral gamma-ray data, average density, and estimated radioactive heat production values for each rock unit of Gebel El Bakriyah—Wadi El Batur area, Central Eastern Desert, Egypt (El-Qassas, 2018).

Rock units	TC (Ur)		K (%)		eU (ppm)		eTh (ppm)		Average density (ρ) (kg/m^3)	Estimated radioactive heat production (A) (μW/m^3)		
Min	Max	Min	Max	Min	Max	Min	Max	Min	Max	Average		
Dakhla Formation	4.5	28.14	0.20	0.70	2.10	27.0	2.20	6.20	2400	0.75	6.38	2.05
Duwi Formation	6.27	33.50	0.20	0.80	2.10	30.1	2.00	12.0	2350	0.91	6.92	2.09
Quseir Formation	4.00	11.25	0.30	0.85	0.70	6.00	3.30	12.0	2400	0.56	1.87	1.06
Taref Formation	2.73	9.32	0.07	1.30	0.70	3.80	2.20	10.1	2350	0.36	1.30	0.74
Granite (younger)	3.54	30.55	0.30	3.70	0.50	11.8	2.00	28.9	2670	0.32	4.83	1.50
Granite (older)	4.50	12.86	0.80	3.00	0.70	3.90	2.80	9.50	2670	0.49	1.92	1.07
Metasediments	1.66	10.29	0.20	2.20	0.10	3.40	1.20	5.70	2720	0.19	1.45	0.61
Metavolcanics	1.77	11.10	0.30	1.85	0.20	2.50	1.20	5.50	2790	0.21	1.24	0.56
Metagabbro	1.93	8.04	0.10	1.40	0.20	2.80	1.40	7.10	3030	0.29	1.31	0.68
Serpentinite	1.6	6.27	0.20	1.25	0.30	2.20	0.80	4.40	2780	0.21	0.90	0.47
Leucocratic metamorphic	4.18	14.80	0.40	2.70	0.50	3.50	3.70	9.80	2740	0.61	1.64	1.07

The radioactive heat production (A) can be calculated from the following formula (Bücker and Rybach, 1996):

$$A\ (\mu\text{W/m}^3) = 10^{-5}\rho(9.52\ C_\text{U} + 2.56\ C_\text{Th} + 3.48\ C_\text{K}), \tag{5.14}$$

where ρ is rock density (kg/m^3), and C_U (ppm), C_Th (ppm), and C_K (%) are the radioactive element contents of uranium, thorium, and potassium, respectively.

The average density for each rock unit in the area is obtained from Telford et al. (1990) and Sharma (1997) as shown in Table 5.5. Fig. 5.20D exhibits the RHP values ranging from 0.19 to 6.92 μW/m^3 (Fig. 5.20D and Table 5.5). Some parts of the younger granites at Gebel El Bakriyah, as well as Dakhla and Duwi Formation, possess the highest RHP values (2–6.92 μW/m^3), located in the southeastern and western parts of the area (Figs. 5.18 and 5.20D). Dakhla and Duwi Formation showed relatively high values of RHP (6.38 and 6.92 μW/m^3) as shown in Table 5.5. The relatively high values exist due to the higher content of uranium in shale and phosphate. Also, Fig. 5.20D reveals that the lowest values of the RHP (0.19–1 μW/m^3) are associated with serpentinite, metasediments, metagabbro, metavolcanics, Taref Formation, and some parts of Dakhla and Quseir formations. The lowest values of RHP are due to the low uranium contents in these rocks (El-Qassas, 2018).

5.4.5 Wadi El-Ghawaby, Central Eastern Desert, Egypt (induced polarization)

The study area is located about 30 km southwest of Qussier city that lies on the Red Sea coast, Central Eastern Desert, Egypt. The importance of this area and its adjacent is due to bostonite (trachyte) rocks, which represents one of the important igneous rocks that host uranium and/or thorium minerals (Assaf, 1966; Ibrahim, 1968; El Kassas, 1969). Uranium mineralizations are found commonly associated with some sulfide mineralization mainly of iron, copper, zinc, and lead (Hussein and Kassas, 1972). The highest values of eU (90 ppm) were recorded over some parts of the bostonite and its contact with slate and siltstone rocks (El-Qassas, 2014). During the field measurements, it was noticed that the values of uranium content increase with depth as recorded in the previously excavated trenches and by making small holes around the recorded high radiometric values (El-Qassas, 2014).

The main target is to identify the radiometric and conductive anomalous zones. Therefore the time-domain-IP with dipole–dipole surveys were carried out to follow the lateral and vertical extensions of the exposed mineralization and to explore any subsurface mineralization (El-Qassas, 2014). The spacing between the transmitter (Tx) and receiver (Rx) dipoles was from one to six ($n=1-6$), which allows considerable depth penetration. The depth of exploration of each measurement is controlled by the value of (n) and dipole length, while the location point occurs at the intersection of 45° lines drawn from the midpoints of each dipole (Fig. 5.21). The plotting procedure is continued from level 1 to level 6 ($n=1-6$) to attain deeper penetration in the subsurface.

An example: profile A–A^{-} was set up close to the excavated trenches in the bostonite with 600-m length (Fig. 5.22). It was carried out using dipole lengths of 20 and 40 m (Figs. 5.23–5.26) to follow the near-surface radioactive anomalies, which are recorded at the excavated trenches, for 70-m depth (El-Qassas, 2014). The 2D pseudosections of profile A–A^{-} show the filtered lines and the pseudosections of metal factor, resistivity, and chargeability (Figs. 5.23 and 5.25).

Inspection of the chargeability model reveals a gradual increase of the chargeability values downward (Fig. 5.24). Two portions can be distinguished in the chargeability model according to the chargeability values (Fig. 5.24B). The first portion extends from the near-

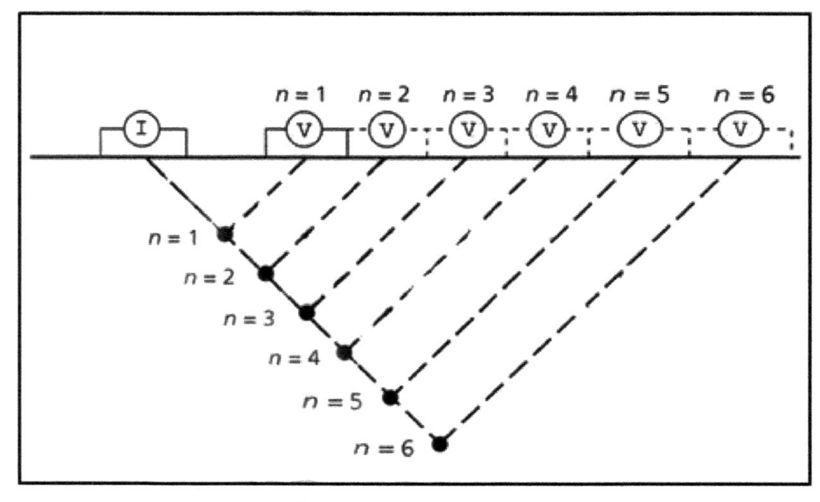

FIGURE 5.21 Diagram showing the method of construction of dipole–dipole pseudo section.

FIGURE 5.22 A photograph showing the location of A–A⁻ the induced polarization profile (after El-Qassas, 2014).

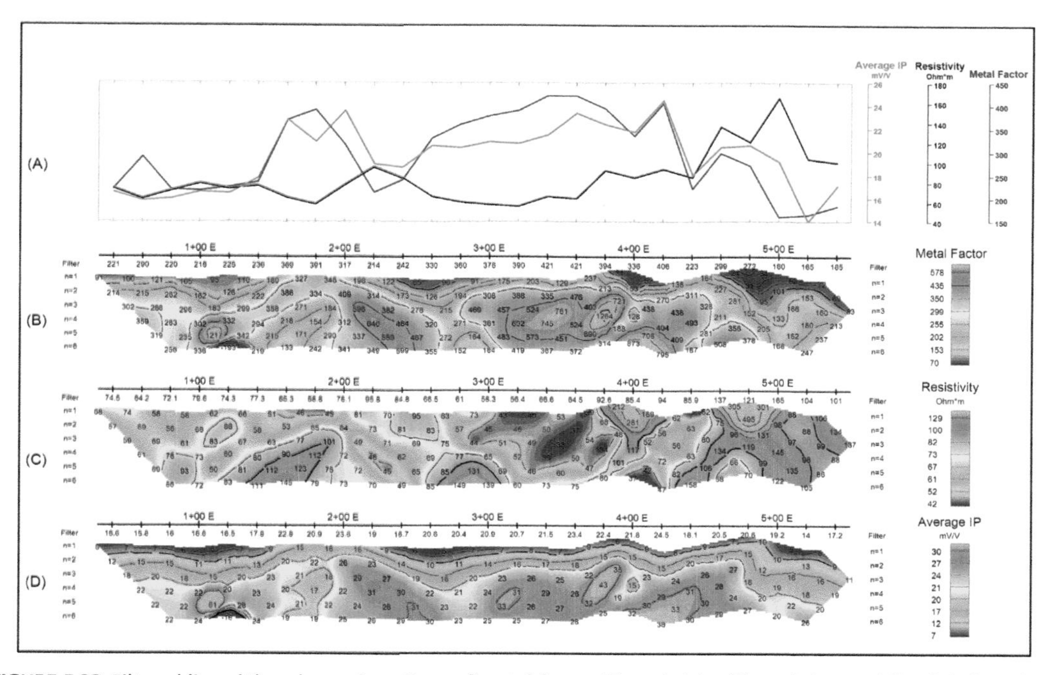

FIGURE 5.23 Filtered lines (A) and pseudosections of metal factor (B), resistivity (C), and chargeability (D) along line A–A⁻ with a dipole spacing of 20 m, Wadi El-Ghawaby area, Central Eastern Desert, Egypt.

surface to a depth of about 20 m with chargeability values varying from less than 1 to 11 mV/V. The lowest chargeability values (<1 mV/V) of this section are centered at stations 80 and 360 (El-Qassas, 2014). Meanwhile, the second portion extends from 20- to >35-m depth with chargeability values varying from 12 to >33 mV/V. It is considered a broad high chargeability zone containing six chargeability anomalies (> 33 mV/V), which may be related to higher concentrations of metallic content. They are centered at stations 100, 200, 260, 330, 410, and 470. These chargeability anomalies have lateral extensions (widths) ranging from 20 to 60 m, vertical extensions (thickness) varying from 10 to 20 m, and depths ranging from 15 to >25 m. All the six chargeability anomalies are corresponding to relatively low-to-moderate resistivity values (35–200 Ω m) (Fig. 5.24A). This may suggest that these anomalies are associated with the contact zone between the bostonite and Hammamat sediments (El-Qassas, 2014).

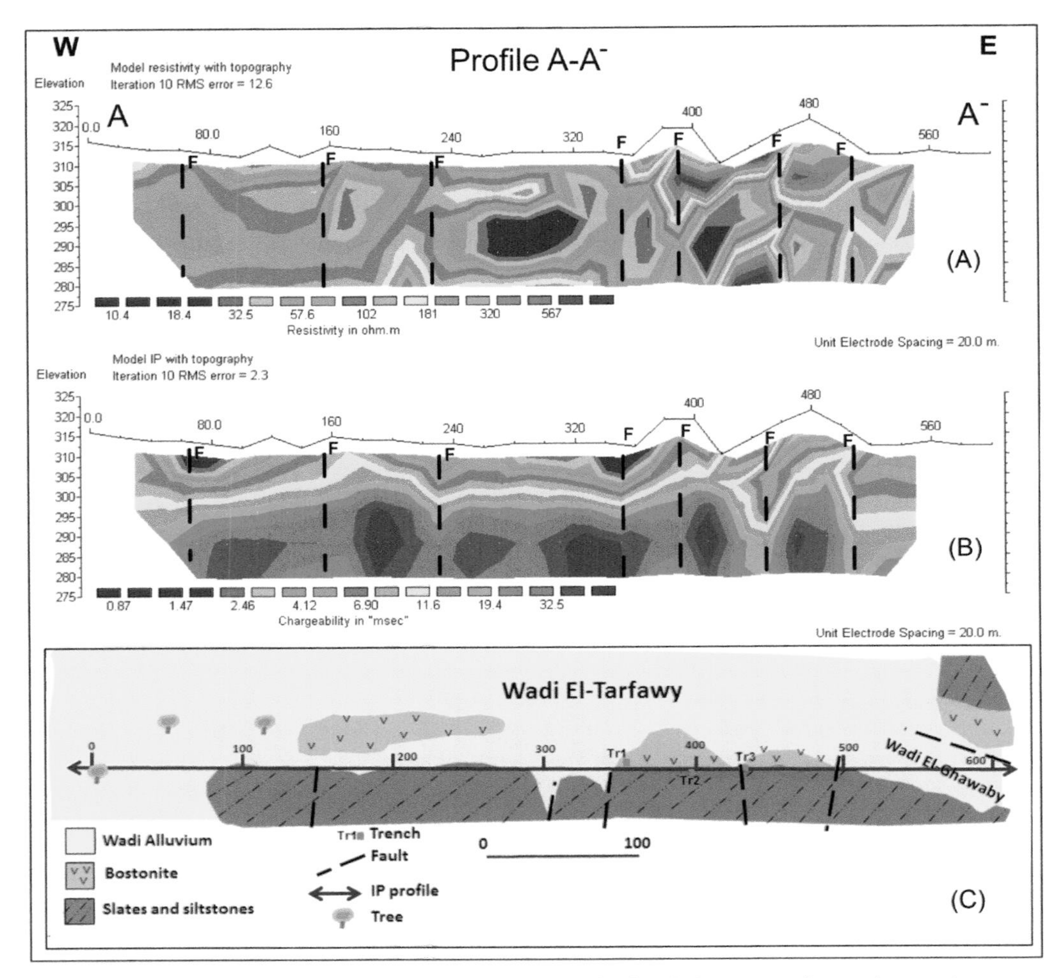

FIGURE 5.24 Resistivity model (A), chargeability model (B), and the detailed corresponding surface geology (C) along with profile A–A¯ with a dipole spacing of 20 m, Wadi El-Ghawaby area, Central Eastern Desert, Egypt (after El-Qassas, 2014).

Three trenches were excavated along this line at stations 360, 390, and 430 (Fig. 5.24C). Low chargeability values are corresponding to moderate resistivity values at the first trench, which may reflect that the surface radiometric anomalies have a limited depth and are associated with clay and altered materials (El-Qassas, 2014). Meanwhile, at the second and third trenches, relatively moderate and high resistivity values are corresponding to moderate and high chargeability values, respectively. This may attribute to the recorded radiometric anomalies having a considerable depth and are associated with disseminated metallic mineralization of relatively moderate and high grades, respectively. It is recommended to carry out a core drilling at station 440 with an angle of 45° to the west and east

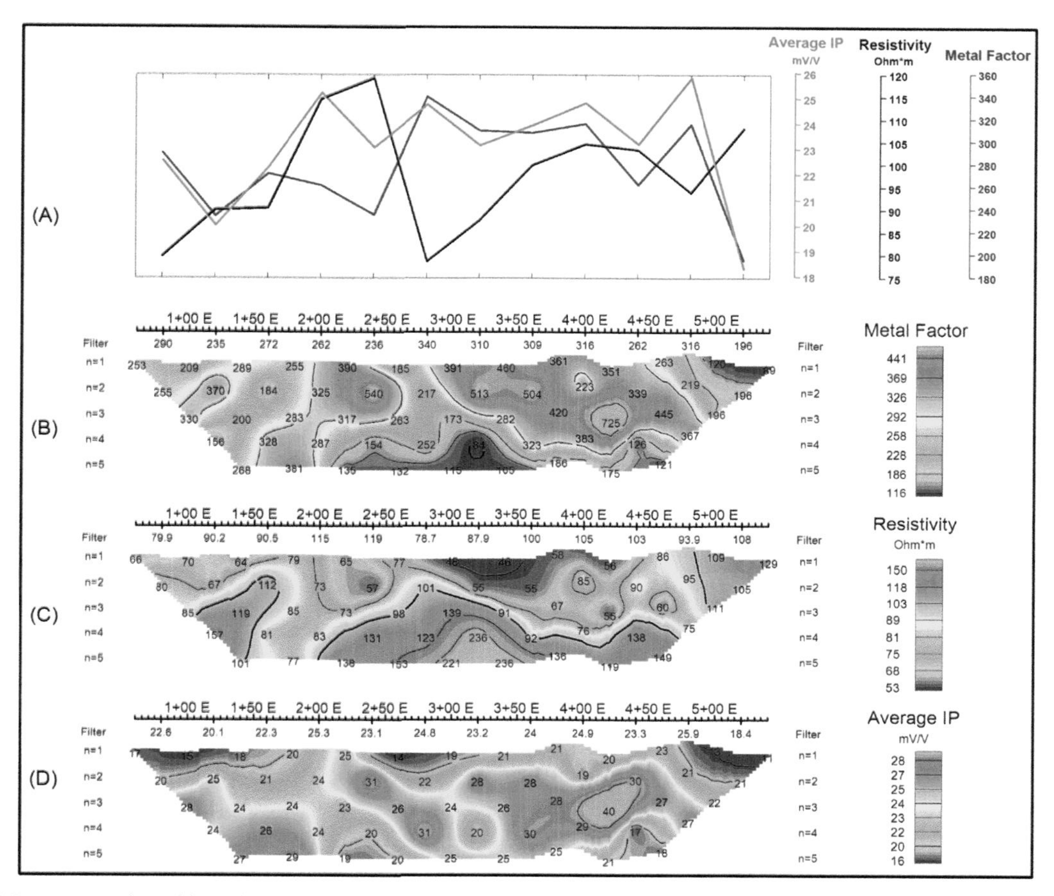

FIGURE 5.25 Filtered lines (A) and pseudosections of metal factor (B), resistivity (C), and chargeability (D) along line A–A⁻ with a dipole spacing of 40 m, Wadi El-Ghawaby area, Central Eastern Desert, Egypt.

to follow the surface radiometric anomalies at a depth of about 40 m and to determine the extensions and grades of mineralization for the fifth and sixth chargeability anomalies (El-Qassas, 2014).

Inferred faults (Marked F) can be deduced from the resistivity model, which is based on the abrupt and significant lateral changes in the resistivity values and the thickness of materials. Some of these faults are observed on the corresponding surface geology, and most of them are confirmed by the chargeability model (Fig. 5.24). These deduced inferred faults are recorded at stations 60, 160, 230, 360, 390, 460, and 510 (El-Qassas, 2014).

Fig. 5.26 shows the surface geology with the 2D resistivity and chargeability sections that provide the resistivity and chargeability distributions from the surface to a depth of about 75 m along profile A–A⁻ with a dipole length of 40 m (El-Qassas, 2014). They confirm the obtained results from resistivity and chargeability models with a dipole length of 20 m. Also,

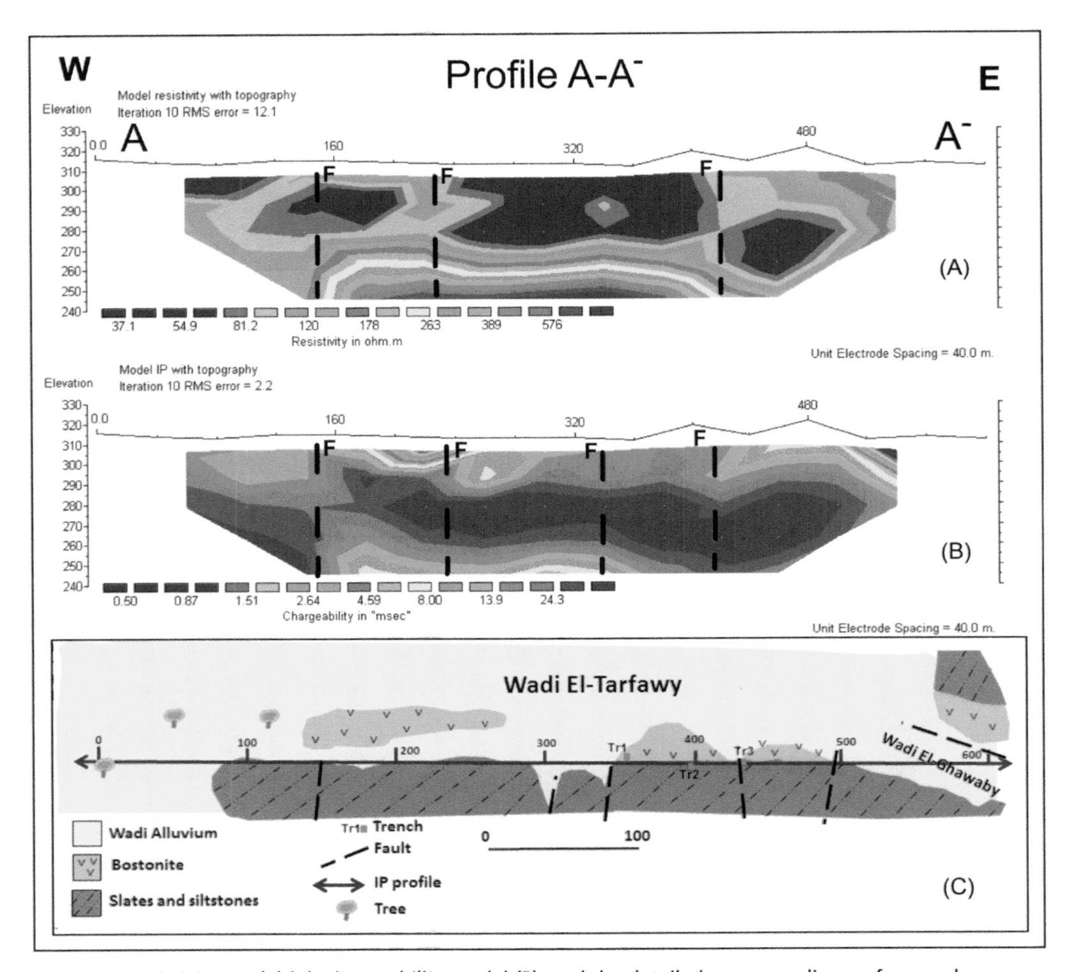

FIGURE 5.26 Resistivity model (A), chargeability model (B), and the detailed corresponding surface geology (C) along profile A–A – with a dipole spacing of 40 m, Wadi El-Ghawaby area, Central Eastern Desert, Egypt (after El-Qassas, 2014).

they determine the bottom of the mineralization zone displayed in Fig. 5.24B. The chargeability model displays a well-defined high chargeability zone (stations 140 to the end of profile), which is sandwiched between low and moderate chargeability layers. This zone is corresponding to moderate resistivity values that are also sandwiched between low- and high-resistivity values that are mainly associated with Hammamat (top) and bostonite (bottom), respectively. This reflects that the high chargeability zone is associated with the contact zone between these rocks. Meanwhile, the western part (station 0–140) is separated from the eastern part by an inferred fault, as deduced from the resistivity and chargeability data, and the direction of movement is well observed along the chargeability model. At this part, the high chargeability zone has a higher depth (~40 m), and the expected mineralization may be continued to more than 70 m as deduced from the opening of the chargeability zone downward. Also, the resistivity and chargeability models of dipole length 40 m confirm some of the deduced inferred faults from the models of dipole length 20 m (Fig. 5.24) (El-Qassas, 2014).

References

Abbas, A.M., Abdallatif, T.F., Shaaban, F.A., Salem, A., Suh, M., 2005. Archaeological investigation of the eastern extensions of the Karnak Temple using ground-penetrating radar and magnetic tools. Geoarchaeology 20 (5), 537–554. Available from: http://onlinelibrary.wiley.com/journal/10.1002/(ISSN)1520-6548; https://doi.org/10.1002/gea.20062.

Abd El Nabi, S., 2009. Airborne γ-ray spectrometric signatures – a case study, El-Bakriyah granitic plutons. Cent. East. Desert 20 (1), 19–33.

Abdallatif, T., El Emam, A.E., Suh, M., El Hemaly, I.A., Ghazala, H.H., Ibrahim, E.H., et al., 2010. Discovery of the causeway and the mortuary temple of the pyramid of Amenemhat II using near-surface magnetic investigation, Dahshour, Giza, Egypt. Geophys. Prospect. 58 (2), 307–320. Available from: https://doi.org/10.1111/j.1365-2478.2009.00814.x.

Abdel Hadi, M., 1978. Aeroradiometric data and its relation to the regional geology of Bakriya area. East. Desert (Corroborated Aeromagn. Surv.) 250.

Abdelrahman, E.S.M., Essa, K.S., 2005. Magnetic interpretation using a least-squares, depth-shape curves method. Geophysics 70 (3), L23–L30. Available from: http://library.seg.org/loi/gpysa7; https://doi.org/10.1190/1.1926575.

Aero-Service, 1984. Final operational report of airborne magnetic/ radiation survey in the Eastern Desert, Egypt. For the Egyptian General Petroleum Corporation (EGPC) and the Egyptian Geological Survey and Mining Authority (EGSMA). Aero-Service Division.

Ahmed, S.B., El Qassas, R.A.Y., El Salam, H.F.A., 2020. Mapping the possible buried archaeological targets using magnetic and ground penetrating radar data, Fayoum, Egypt. Egypt. J. Remote. Sens. Space Sci. 23 (3), 321–332. Available from: http://www.elsevier.com/wps/find/journaldescription.cws_home/723780/description#description; https://doi.org/10.1016/j.ejrs.2019.07.005.

Ammar, A., Abdelhady, H., Soliman, S., Sadek, H., 1988. Aerial radioactivity of phosphates in the Western Central Eastern Desert of Egypt. J. King Abdulaziz Univ.-Earth Sci. 1 (1), 181–203. Available from: https://doi.org/10.4197/Ear.1-1.10.

Anderson, R., Bell, B., Reynolds, R., 1990. Induced Polarization used for Highway Planning: Advances in Applications and Case Histories of Induced Polarization SEG Special Publication.

Assaf, H.S., 1966. Ground Exploration and Geological Studies on Some Radioactive Occurrences in the Area South of Kosseir.

Assran, A.S.M., Abuelnaga, H.S.O., Khalil, A.F., 2007. Application of induced polarization, self-potential and magnetic techniques for the radioactive Bostonite Dykes of Wadi Rahia Area, Central Eastern Desert III. In: Second International Conference on the Geology of the Tethys, pp. 553–560.

Assran, A.S.M., El Qassas, R.A.Y., Yousef, M.H.M., 2019. Detection of prospective areas for mineralization deposits using image analysis technique of aeromagnetic data around Marsa Alam-Idfu road, Eastern Desert, Egypt. Egypt. J. Pet. 28 (1), 61–69. Available from: https://doi.org/10.1016/j.ejpe.2018.11.002; http://www.journals.elsevier.com/egyptian-journal-of-petroleum.

Balkaya, Ç., Ekinci, Y.L., Göktürkler, G., Turan, S., 2017. 3D non-linear inversion of magnetic anomalies caused by prismatic bodies using differential evolution algorithm. J. Appl. Geophys. 136, 372–386. Available from: https://doi.org/10.1016/j.jappgeo.2016.10.040; http://www.elsevier.com/inca/publications/store/5/0/3/3/3/3/.

Baranov, V., 1957. A New method for interpretation of aeromagnetic maps: pseudo-gravimetric anomalies. Geophysics 22 (2), 359–382. Available from: https://doi.org/10.1190/1.1438369.

Barker, R.D., 1981. The offset system of electrical resistivity sounding and its use with a multicore cable* + . Geophys. Prospect. 29 (1), 128–143. Available from: https://doi.org/10.1111/j.1365-2478.1981.tb01015.x.

Barongo, J.O., 1985. Method for depth estimation on aeromagnetic vertical gradient anomalies. Geophysics 50 (6), 963–968. Available from: https://doi.org/10.1190/1.1441974; http://library.seg.org/loi/gpysa7.

Blakely, R.J., 1995. Potential Theory in Gravity and Magnetic Applications. Cambridge University Press. Available from: https://doi.org/10.1017/CBO9780511549816.

Blakely, R.J., Simpson, R.W., 1986. Approximating edges of source bodies from magnetic or gravity anomalies. Geophysics 51 (7), 1494–1498. Available from: https://doi.org/10.1190/1.1442197.

Bott, M.H.P., 1960. The use of rapid digital computing methods for direct gravity interpretation of sedimentary basins. Geophys. J. R. Astron. Soc. 3 (1), 63–67. Available from: https://doi.org/10.1111/j.1365-246X.1960.tb00065.x.

Bücker, C., Rybach, L., 1996. A simple method to determine heat production from gamma-ray logs. Mar. Pet. Geol. 13 (4), 373–375. Available from: https://doi.org/10.1016/0264-8172(95)00089-5; http://www.sciencedirect.com/science/journal/02648172.

Cagniard, L., 1953. Basic theory of the magneto-telluric method of geophysical prospecting. Geophysics 18 (3), 605–635. Available from: https://doi.org/10.1190/1.1437915.

Chakravarthi, V., Pramod Kumar, M., Ramamma, B., Rajeswara Sastry, S., 2016. Automatic gravity modeling of sedimentary basins by means of polygonal source geometry and exponential density contrast variation: two space domain based algorithms. J. Appl. Geophys. 124, 54–61. Available from: https://doi.org/10.1016/j.jappgeo.2015.11.007; http://www.elsevier.com/inca/publications/store/5/0/3/3/3/3/.

Chambers, J.E., Kuras, O., Meldrum, P.I., Ogilvy, R.D., Hollands, J., 2006. Electrical resistivity tomography applied to geologic, hydrogeologic, and engineering investigations at a former waste-disposal site. Geophysics 71 (6), B231–B239. Available from: https://doi.org/10.1190/1.2360184; http://library.seg.org/loi/gpysa7.

Clark, S.P., Peterman, Z.E., Heier, K.S., 1966. Abundances of uranium, thorium and potassium, handbook of physical constant. Geol. Soc. Am. Memoir. 47.

Conoco Coral Corporation and the Egyptian General Petroleum Corporation (EGPC), 1987a. Geological Map of Egypt 1.

Conoco Coral Corporation and the Egyptian General Petroleum Corporation (EGPC), 1987b. Geological Map of Egypt.

Constable, S.C., Parker, R.L., Constable, C.G., 1987. Occam's inversion: A practical algorithm for generating smooth models from electromagnetic sounding data. Geophysics 52, 289–300.

Cooper, G.R.J., Cowan, D.R., 2006. Enhancing potential field data using filters based on the local phase. Comput. Geosci. 32 (10), 1585–1591. Available from: https://doi.org/10.1016/j.cageo.2006.02.016.

Cooper, G.R.J., Cowan, D.R., 2008. Edge enhancement of potential-field data using normalized statistics. Geophysics 73 (3), H1–H4. Available from: https://doi.org/10.1190/1.2837309; http://library.seg.org/loi/gpysa7.

Cordell, L., 1979. Gravimetric expression of graben faulting in Santa Fe Country and the Espanola Basin, New Mexico. New Mexico Geological Society Guidebook, pp. 59–64.

Dewi, R.K., Kurniawan, A., Taqwantara, R.F., Iskandar, F.M., Naufal, T.Z., Widodo, 2017. Identification of buried victims in natural disaster with GPR method. In: AIP Conference Proceedings 1861, 15517616. American Institute of Physics Inc. Indonesia. Available from: https://doi.org/10.1063/1.4990909; http://scitation.aip.org/content/aip/proceeding/aipcp.

dGB Earth Sciences, 2015. OpendTect dGB Plugins User Documentation version 4.6. dGB Earth Sciences.

Dobrin, M.B., 1960. Introduction to Geophysical Prospecting.

Drascovits Hobot, J., Vero, L., Smith, B.D. 1990. Induced Polarization – Applications and Case Histories. SEG 379-396 Induced Polarization Surveys Applied to Evaluation of Groundwater Resources.

Eaton, D.W., Adam, E., Milkereit, B., Salisbury, M., Roberts, B., White, D., et al., 2010. Enhancing base-metal exploration with seismic imaging. Canadian Journal of Earth Sciences 47 (5), 741–760.

Ekwok, S.E., Akpan, A.E., Achadu, O.I.M., Thompson, C.E., Eldosouky, A.M., Abdelrahman, K., et al., 2022. Towards understanding the source of brine mineralization in Southeast Nigeria: evidence from high-resolution airborne magnetic and gravity data. Minerals 12 (2). Available from: https://doi.org/10.3390/min12020146; https://www.mdpi.com/2075-163X/12/2/146/pdf.

El-Amin, 1975. Radiometric and Geological Investigations of El-Bakriya Area. p. 244.

El Kassas, I.A., 1969. Comparative Geological Investigation of the Radioactive Mineralizations in the Central Eastern Desert.

El-Qassas, R.A.Y., 2014. Airborne and ground geophysical studies for the radioactive mineralized bostonite rocks in Wadi El-Tarfawy–Wadi El-Dabbah area, Central Eastern Desert.

El-Qassas, R.A.Y., 2018. Application of airborne spectral gamma-ray data for delineating surface distribution of heat production at Gebel El Bakriyah – Wadi El Batur area. Central Eastern Desert, Egypt. Egypt. Geophys. Soc. 16, 187–198.

El-Qassas, R.A.Y., Abu-Donia, A.M., 2019. Uranium mobility and its implication on favourable potentiality concentration in Gabal Um-Araka area, Northern Eastern Desert, Egypt. Arab. J. Geosci. 12 (23). Available from: https://doi.org/10.1007/s12517-019-4902-2; http://www.springer.com/geosciences/journal/12517?cm_mmc = AD-_-enews-_-PSE1892-_-0.

El-Sehamy, M., Abdel Gawad, A.M., Aggour, T.A., Orabi, O.H., Abdella, H.F., Eldosouky, A.M., 2022. Sedimentary cover determination and structural architecture from gravity data: East of Suez Area, Sinai, Egypt. Arab. J. Geosci. 15 (1). Available from: https://doi.org/10.1007/s12517-021-09348-6.

Eldosouky, A.M., Mohamed, H., 2021. Edge detection of aeromagnetic data as effective tools for structural imaging at Shilman area, South Eastern Desert, Egypt. Arab. J. Geosci. 14 (1). Available from: https://doi.org/10.1007/s12517-020-06251-4; http://www.springer.com/geosciences/journal/12517?cm_mmc = AD-_-enews-_-PSE1892-_-0.

Eldosouky, A.M., Abdelkareem, M., Elkhateeb, S.O., 2017. Integration of remote sensing and aeromagnetic data for mapping structural features and hydrothermal alteration zones in Wadi Allaqi area, South Eastern Desert of Egypt. J. Afr. Earth Sci. 130, 28–37. Available from: https://doi.org/10.1016/j.jafrearsci.2017.03.006; http://www.sciencedirect.com/science/journal/1464343X.

Eldosouky, A.M., Sehsah, H., Elkhateeb, S.O., Pour, A.B., 2020a. Integrating aeromagnetic data and Landsat-8 imagery for detection of post-accretionary shear zones controlling hydrothermal alterations: the Allaqi-Heiani Suture zone, South Eastern Desert, Egypt. Adv. Space Res. 65 (3), 1008–1024. Available from: https://doi.org/10.1016/j.asr.2019.10.030; http://www.journals.elsevier.com/advances-in-space-research/.

Eldosouky, A.M., Elkhateeb, S.O., Ali, A., Kharbish, S., 2020b. Enhancing linear features in aeromagnetic data using directional horizontal gradient at Wadi Haimur area, South Eastern Desert, Egypt. Carpathian J. Earth Environ. Sci. 15 (2), 323–326. Available from: https://doi.org/10.26471/cjees/2020/015/132; http://www.cjees.ro/actions/actionDownload.php?fileId = 1376.

Eldosouky, A.M., El-Qassas, R.A.Y., Pour, A.B., Mohamed, H., Sekandari, M., 2021a. Integration of ASTER satellite imagery and 3D inversion of aeromagnetic data for deep mineral exploration. Adv. Space Res. 68 (9), 3641–3662. Available from: https://doi.org/10.1016/j.asr.2021.07.016; http://www.journals.elsevier.com/advances-in-space-research/.

Eldosouky, A.M., Pham, L.T., Abdelrahman, K., Fnais, M.S., Gomez-Ortiz, D., 2022a. Mapping structural features of the Wadi Umm Dulfah area using aeromagnetic data. J. King Saud. Univ. - Sci. 34 (2). Available from: https://doi.org/10.1016/j.jksus.2021.101803; http://www.sciencedirect.com/science/journal/10183647.

Eldosouky, A.M., El-Qassas, R.A.Y., Pham, L.T., Abdelrahman, K., Alhumimidi, M.S., Bahrawy, A.E., et al., 2022b. Mapping main structures and related mineralization of the Arabian Shield (Saudi Arabia) using sharp edge detector of transformed gravity data. Minerals. 12 (1). Available from: https://doi.org/10.3390/min12010071; https://www.mdpi.com/2075-163X/12/1/71/pdf.

Eldosouky, A.M., Elkhateeb, S.O., Mahdy, A.M., Saad, A.A., Fnais, M.S., Abdelrahman, K., et al., 2022c. Structural analysis and basement topography of Gabal Shilman area, South Eastern Desert of Egypt, using aeromagnetic data. J. King Saud. Univ. - Sci. 34 (2). Available from: https://doi.org/10.1016/j.jksus.2021.101764; http://www.sciencedirect.com/science/journal/10183647.

Eldosouky, A.M., Pham, L.T., El-Qassas, R.A.Y., Hamimi, Z., Oksum, E., 2021b. Lithospheric structure of the Arabian–Nubian shield using satellite potential field data. Springer Sci. Bus. Media LLC 139–151. Available from: https://doi.org/10.1007/978-3-030-72995-0_6.

Elkhateeb, S.O., Eldosouky, A.M., 2016. Detection of porphyry intrusions using analytic signal (AS), Euler Deconvolution, and Center for Exploration Targeting (CET) Technique Porphyry Analysis at Wadi Allaqi Area, South Eastern Desert. Egypt. Int. J. Sci. Eng. Res. 7 (6), 471–477.

Elkhateeb, S.O., Eldosouky, A.M., Khalifa, M.O., Aboalhassan, M., 2021. Probability of mineral occurrence in the Southeast of Aswan area, Egypt, from the analysis of aeromagnetic data. Arab. J. Geosci. 14 (15). Available from: https://doi.org/10.1007/s12517-021-07997-1; http://www.springer.com/geosciences/journal/12517?cm_mmc = AD-_-enews-_-PSE1892-_-0.

El Shazly, L.M., Abd-Elkhalek, A., Hafez, H.E.-A., 1981. Geological and radiometric study of G. El-Bakriya area, Eastern Desert. J. Geol. 25, 1–20.

Evjen, H.M., 1936. The place of the vertical gradient in gravitational interpretations. Geophysics. 1 (1), 127–136. Available from: https://doi.org/10.1190/1.1437067.

Farquharson, C.G., Ash, M.R., Miller, H.G., 2008. Geologically constrained gravity inversion for the Voisey's Bay ovoid deposit. Lead. Edge (Tulsa, OK) 27 (1), 64–69. Available from: https://doi.org/10.1190/1.2831681; http://library.seg.org/tle/.

Francke, J., 2010. Applications of GPR in Mineral Resource Evaluations. In Proceedings of the 13th International Conference on Ground Penetrating Radar, GPR. Canada. Available from: https://doi.org/10.1109/ICGPR.2010.5550188.

Gharieb, A.G., 1995. Characterization of Phosphate Belts in the Central Eastern Desert of Egypt by Management of Geological and Geophysical Databases, p. 179.

Granar, L., 1982. The Impression of Computer-Made Coloured Geophysical Maps, Abstract, 8th European Geophysical Society Meeting, p. 119.

Granser, H., 1986. Convergence of iterative gravity inversion. Geophysics 51 (5), 1146–1147. Available from: https://doi.org/10.1190/1.1442169.

Granser, H., 1987. Three-dimensional interpretation of gravity data from sedimentary basins using an exponential density-depth function. Geophys. Prospect. 35 (9), 1030–1041. Available from: https://doi.org/10.1111/j.1365-2478.1987.tb00858.x.

Griffiths, D.H., Barker, R.D., 1993. Two-dimensional resistivity imaging and modelling in areas of complex geology. J. Appl. Geophys. 29 (3–4), 211–226. Available from: https://doi.org/10.1016/0926-9851(93)90005-J.

Haldar, S.K., 2018. Mineral Exploration: Principles and Applications. p. 378.

Halim, I., Asyari, Wijaksana, Alfadli, M.K. 2017. 3D Modeling from induced polarization method for identification of gold deposit exploration in North Minahasa. In: International Geophysical Conference.

Hall-Beyer, M., 2017. Practical guidelines for choosing GLCM textures to use in landscape classification tasks over a range of moderate spatial scales. Int. J. Remote. Sens. 38 (5), 1312–1338. Available from: https://doi.org/10.1080/01431161.2016.1278314; https://www.tandfonline.com/loi/tres20.

Hanafy, S.M., 2002. Contribution of airborne magnetic and spectral gamma-ray survey data to mineral exploration and geological mapping at Wadi Al Miah-Wadi Ash Shalul area, Central Eastern Desert, Egypt; aided with the application of geostatistical methods of analysis, p. 144.

Hansen, R.O., Simmonds, M., 1993. Multiple-source Werner deconvolution. Geophysics. 58 (12), 1792–1800. Available from: https://doi.org/10.1190/1.1443394.

Hinze, W., Frese, R., Saad, A., 2012. Gravity and Magnetic Exploration, Principles, Practices and Applications. Cambridge University Press.

Holden, E.J., Dentith, M., Kovesi, P., 2008. Towards the automated analysis of regional aeromagnetic data to identify regions prospective for gold deposits. Comput. Geosci. 34 (11), 1505–1513. Available from: https://doi.org/10.1016/j.cageo.2007.08.007.

Hussein, H.A., El Kassas, I.A., 1972. Occurrence of some primary uranium mineralization at el Atshan locality. Central Eastern Desert: U. A. R. J. Geol. 14, 97–110.

Ibrahim, A.B., 1968. Geology of Radioactive Occurrences in El Kereim-El Oweirsha Area, Central Eastern Desert.

International Atomic Energy Agency (IAEA), 1979. Gamma-Ray Surveys in Uranium Exploration. International Atomic Energy Agency (IAEA).

International Atomic Energy Agency (IAEA), 1991. Airborne Gamma-Ray Spectrometer Surveying, International Atomic Energy Agency (IAEA).

International Atomic Energy Agency (IAEA), 2003. Guidelines for Radioelement Mapping Using Gamma Ray Spectrometry Data. International Atomic Energy Agency (IAEA).

King, A., 2007. Review of Geophysical Technology for Ni-Cu-PGE deposits Fifth Decennial International Conference on Mineral Exploration, pp. 647–665.

Kelbert, A., Meqbel, N., Egbert, G.D., Tandon, K., 2014. ModEM: A modular system for inversion of electromagnetic geophysical data. Computers & Geosciences 66, 40–53.

Korsch, R.J., Huston, D.L., Henderson, R.A., Blewett, R.S., Withnall, I.W., Fergusson, C.L., et al., 2012. Crustal architecture and geodynamics of North Queensland, Australia: insights from deep seismic reflection profiling. Tectonophysics 572-573, 76–99. Available from: https://doi.org/10.1016/j.tecto.2012.02.022.

Lelièvre, P., Carter-McAuslan, A., Farquharson, C., Hurich, C., 2012. Unified geophysical and geological 3D Earth models. Lead. Edge 31 (3), 322–328. Available from: https://doi.org/10.1190/1.3694900.

Loke, M.H., 2004. Tutorial 2-D and 3-D Electrical Imaging Surveys.

Loke, M.H., Barker, R.D., 1996b. Practical techniques for 3D resistivity surveys and data inversion. Geophys. Prospect. 44 (3), 499–523. Available from: https://doi.org/10.1111/j.1365-2478.1996.tb00162.x.

Lyatsky, H.V., 2010. Magnetic and Gravity Methods in Mineral Exploration: The Value of Well-Rounded Geophysical Skills, Recorder (Canadian Society of Exploration Geophysics), pp. 30–35.

Maag, E., Li, Y., 2018. Discrete-valued gravity inversion using the guided fuzzy c-means clustering technique. Geophysics 83 (4), G59–G77. Available from: https://doi.org/10.1190/geo2017-0594.1; http://library.seg.org/loi/gpysa7.

Malehmir, A., Koivisto, E., Manzi, M., Cheraghi, S., Durrheim, R.J., Bellefleur, G., et al., 2014. A review of reflection seismic investigations in three major metallogenic regions: The Kevitsa Ni-Cu-PGE district (Finland), Witwatersrand goldfields (South Africa), and the Bathurst Mining Camp (Canada). Ore Geol. Rev. 56, 423–441. Available from: https://doi.org/10.1016/j.oregeorev.2013.01.003.

Melouah, O., Eldosouky, A.M., Ebong, E.D., 2021. Crustal architecture, heat transfer modes and geothermal energy potentials of the Algerian Triassic provinces. Geothermics 96, 102211. Available from: https://doi.org/10.1016/j.geothermics.2021.102211.

Mierczak, M., Karczewski, J., 2021. Location of agate geodes in Permian deposits of Simota gully using the GPR. Acta Geophys. 69 (2), 655–664. Available from: https://doi.org/10.1007/s11600-021-00537-1; http://www.springer.com/11600.

Millegan, P.S., Bird, D.E., 1998. How basement lithology changes affect magnetic interpretation. In: Geologic applications of Gravity and Magnetic: case Histories. SEG Geophysical Reference, pp. 40–48.

Miller, H.G., Singh, V., 1994. Potential field tilt-a new concept for location of potential field sources. J. Appl. Geophys. 32 (2–3), 213–217. Available from: https://doi.org/10.1016/0926-9851(94)90022-1.

Mohamed, H., Saibi, H., Bersi, M., Abdelnabi, S., Geith, B., Ismaeil, H., et al., 2018. 3-D magnetic inversion and satellite imagery for the Um Salatit gold occurrence, Central Eastern Desert, Egypt. Arab. J. Geosci. 11 (21). Available from: https://doi.org/10.1007/s12517-018-4020-6; http://www.springer.com/geosciences/journal/12517?cm_mmc = AD-_-enews-_-PSE1892-_-0.

Murty, B.V.S., Haricharan, P., 1985. Nomogram for the complete interpretation of spontaneous potential profiles over sheet-like and cylindrical two-dimensional sources. Geophysics 50 (7), 1127–1135. Available from: https://doi.org/10.1190/1.1441986.

Nabighian, M.N., 1972. The analytic signal of two-dimensional magnetic bodies with polygonal cross-section: its properties and use for automated anomaly interpretation. Geophysics 37 (3), 507–517. Available from: https://doi.org/10.1190/1.1440276; http://library.seg.org/loi/gpysa7.

Nabighian, M.N., Ander, M.E., Grauch, V.J.S., Hansen, R.O., LaFehr, T.R., Li, Y., et al., 2005. Historical development of the gravity method in exploration. Geophysics 70 (6). Available from: https://doi.org/10.1190/1.2133785; http://library.seg.org/loi/gpysa7.

Nabighian, M.N., 1974. Additional comments on the analytic signal of two-dimensional magnetic bodies with polygonal cross-section. Geophysics 39 (1), 85–92. Available from: https://doi.org/10.1190/1.1440416.

Odah, H., Ismail, A., Elhemaly, I., Anderson, N., Abbas, A.M., Shaaban, F., 2013. Archaeological exploration using magnetic and GPR methods at the first court of Hatshepsut Temple in Luxor, Egypt. Arab. J. Geosci. 6 (3), 865–871. Available from: https://doi.org/10.1007/s12517-011-0380-x.

Okiwelu, A.A., Ude, I.A., 2012. 3D Modelling and Basement Tectonics of the Niger Delta Basin from Aeromagnetic Data. InTech, 10.5772/48158.

Oksum, E., Le, D.V., Vu, M.D., Nguyen, T.H.T., Pham, L.T., 2021. A novel approach based on the fast sigmoid function for interpretation of potential field data. Bull. Geophys. Oceanogr. 62 (3), 543–556. Available from: https://doi.org/10.4430/bgta0348; http://www3.inogs.it/bgo/index.php.

Oksum, E., 2021. Grav3CH_inv: A GUI-based MATLAB code for estimating the 3-D basement depth structure of sedimentary basins with vertical and horizontal density variation. Comput. Geosci. 155, 104856. Available from: https://doi.org/10.1016/j.cageo.2021.104856.

Oldenburg, D., Pratt, W.A., 2007. Geophysical inversion for mineral exploration: a decade of progress in theory and practice. In: Proceedings of Exploration 07: Fifth Decennial International Conference on Mineral Exploration, pp. 61–95.

Oldenburg, D.W., 1974. Inversion and interpretation of gravity anomalies. Geophysics 39 (4), 526–536. Available from: https://doi.org/10.1190/1.1440444.

Oldenburg, D.W., Li, Y., 1994. Inversion of induced polarization data. Geophysics 59 (9), 1327–1341. Available from: https://doi.org/10.1190/1.1443692.

Oruç, B., 2011. Edge detection and depth estimation using a tilt angle map from gravity gradient data of the Kozaklı-central Anatolia region, Turkey. Pure Appl. Geophys. 168 (10), 1769–1780. Available from: https://doi.org/10.1007/s00024-010-0211-0; http://www.springer.com/birkhauser/geo + science/journal/24.

O'Brien, D.P., 1972. CompuDepth, A New Method for Depth-to-Basement Computation: Presented at the 42nd Annual International Meeting, SEG.

Parasnis, D.S., 1986. Principles of Applied Geophysics. Chapman and Hall.

Parker, R.L., 1973. The rapid calculation of potential anomalies. Geophys. J. R. Astron. Soc. 31 (4), 447–455. Available from: https://doi.org/10.1111/j.1365-246X.1973.tb06513.x.

Pelton, W.H., Rijo, L., Swift, C.M., 1978. Inversion of two-dimensional resistivity and induced-polarization data. Geophysics 43 (4), 788–803. Available from: https://doi.org/10.1190/1.1440854.

Pham, L.T., 2021. A high-resolution edge detector for interpreting potential field data: a case study from the Witwatersrand basin, South Africa. J. Afr. Earth Sci. 178.

Pham, L.T., Oksum, E., Do, T.D., 2018. GCH_gravinv: A MATLAB-based program for inverting gravity anomalies over sedimentary basins. Comput. Geosci. 120, 40–47. Available from: https://doi.org/10.1016/j.cageo.2018.07.009; http://www.elsevier.com/inca/publications/store/3/9/8/.

Pham, L.T., Oksum, E., Do, T.D., 2019. Edge enhancement of potential field data using the logistic function and the total horizontal gradient. Acta Geodaetica et Geophysica. 54 (1), 143–155. Available from: https://doi.org/10.1007/s40328-019-00248-6; http://rd.springer.com/journal/40328.

Pham, L.T., VanVu, T., Le Thi, S., Trinh, P.T., 2020a. Enhancement of potential field source boundaries using an improved logistic filter. Pure Appl. Geophys. 177 (11), 5237–5249. Available from: https://doi.org/10.1007/s00024-020-02542-9; http://www.springer.com/birkhauser/geo + science/journal/24.

Pham, L.T., Eldosouky, A.M., Oksum, E., Saada, S.A., 2020b. A new high resolution filter for source edge detection of potential field data. Geocarto Int. Available from: https://doi.org/10.1080/10106049.2020.1849414; http://www.tandfonline.com/toc/tgei20/current.

Pham, L.T., Oksum, E., Le, D.V., Ferreira, F.J.F., Le, S.T., 2021. Edge detection of potential field sources using the softsign function. Geocarto Int. Available from: https://doi.org/10.1080/10106049.2021.1882007; http://www.tandfonline.com/toc/tgei20/current.

Pham, L.T., Oksum, E., Eldosouky, A.M., Gomez-Ortiz, D., Abdelrahman, K., Altinoğlu, F.F., et al., 2022. Determining the Moho interface using a modified algorithm based on the combination of the spatial and frequency domain techniques: a case study from the Arabian Shield. Geocarto Int. Available from: https://doi.org/10.1080/10106049.2022.2037733; http://www.tandfonline.com/toc/tgei20/current.

Porsani, J.L., de Jesus, F.A.N., Stangari, M.C., 2019. GPR survey on an iron mining area after the collapse of the tailings Dam I at the Córrego do Feijão mine in Brumadinho-MG, Brazil. Remote. Sens. 11 (7). Available from: https://doi.org/10.3390/RS11070860; https://res.mdpi.com/remotesensing/remotesensing-11-00860/article_deploy/remotesensing-11-00860-v2.pdf?filename = &attachment = 1.

Ram Babu, H.V., Atchuta Rao, D., 1988. A rapid graphical method for the interpretation of the self-potential anomaly over a two-dimensional inclined sheet of finite depth extent. Geophysics. 53 (8), 1126–1128. Available from: https://doi.org/10.1190/1.1442551.

Reid, A.B., Allsop, J.M., Granser, H., Millett, A.J., Somerton, I.W., 1990. Magnetic interpretation in three dimensions using Euler deconvolution. Geophysics. 55 (1), 80–91. Available from: https://doi.org/10.1190/1.1442774.

Reynolds, R.L., Webring, M., Grauch, V.J.S., Tuttle, M., 1990. Magnetic forward models of Cement oil field, Oklahoma, based on rock magnetic, geochemical and petrologic constraints. Geophysics 55, 344–353.

Richardson, K.A., Killeen, P.G., 1980. Regional radiogenic heat production mapping by airborne gamma-ray spectrometry. Curr. Res., B: Geol. Surv. Can. 227–232.

Robert, F., Brommecker, R., Bourne, B.T., Dobak, P.J., McEwan, C.J., Rowe, R.R., et al., 2007. Models and exploration methods for major gold deposit types. Proceedings of Exploration 07: Fifth Decennial International Conference on Mineral Exploration, pp. 691–711.

Roest, W.R., Verhoef, J., Pilkington, M., 1992. Magnetic interpretation using the 3-D analytic signal. Geophysics. 57 (1), 116–125. Available from: https://doi.org/10.1190/1.1443174.

Saada, S.A., Mickus, K., Eldosouky, A.M., Ibrahim, A., 2021a. Insights on the tectonic styles of the Red Sea rift using gravity and magnetic data. Mar. Pet. Geol. 133. Available from: https://doi.org/10.1016/j.marpetgeo.2021.105253; http://www.sciencedirect.com/science/journal/02648172.

Saada, S.A., Eldosouky, A.M., Abdelrahman, k, Al-Otaibi, N., Ibrahim, E., Ibrahim, A., 2021b. New insights into the contribution of gravity data for mapping the lithospheric architecture. J. King Saud. Univ. - Sci. 33 (3). Available from: https://doi.org/10.1016/j.jksus.2021.101400; http://www.sciencedirect.com/science/journal/10183647.

Sadek, H.S., 1972. Regional Aeroradiometric Survey of East Luxor Area and Its Relation to Stratigraphy and Regional Geology (M.Sc. thesis Fac). p. 131.

Salem, A., Ravat, D., 2003. A combined analytic signal and Euler method (AN-EUL) for automatic interpretation of magnetic data. Geophysics 68 (6), 1952–1961. Available from: https://doi.org/10.1190/1.1635049; http://library.seg.org/loi/gpysa7.

Salem, A., ElSirafy, A., Aref, A., Ismail, A., Ehara, S., Ushijima, K., 2005. Mapping radioactive heat production from airborne spectral gamma-ray data of Gebel Duwi Area, Egypt. In: Proceedings World Geothermal Congress, pp. 24–29.

Salman, A.B., 1974. Structures and radioactivity of some phosphate deposits. Fac. Sci. Ain Shams Univ. 322.

Santos, D.F., Silva, J.B.C., Martins, C.M., dos Santos, R.D.C.S., Ramos, L.C., de Araújo, A.C.M., 2015. Efficient gravity inversion of discontinuous basement relief. Geophysics 80 (4), G95–G106. Available from: https://doi.org/10.1190/GEO2014-0513.1; http://library.seg.org/loi/gpysa7.

Sarma, D.D., Koch, G.S., 1980. A statistical analysis of exploration geochemical data for uranium. J. Int. Assoc. Math. Geol. 12 (2), 99–114. Available from: https://doi.org/10.1007/BF01035242.

Sasaki, Y., 1992. Resolution of resistivity tomography inferred from numerical simulation. Geophys. Prospecting 40 (4), 453–463. Available from: https://doi.org/10.1111/j.1365-2478.1992.tb00536.x.

Sato, M., 2015. Near range radar and its application to near surface geophysics and disaster mitigation. J. Earth Sci. 26 (6), 858–863. Available from: https://doi.org/10.1007/s12583-015-0595-y; http://www.springer.com/earth + sciences/journal/12583.

Sato, M., Mooney, H.M., 1960. The electrochemical mechanism of sulfide self-potentials. Geophysics 25, 47–58.

Saunders, D.F., Potts, M.J., 1976. Interpretation and application of high sensitivity airborne gamma ray spectrometric data. In: IAEA Symp. Exploration for Uranium Ore Deposits, pp. 107–124.

Sehsah, H., Eldosouky, A.M., 2022. Neoproterozoic hybrid forearc – MOR ophiolite belts in the northern Arabian-Nubian Shield: no evidence for back-arc tectonic setting. Int. Geol. Rev. 64 (2), 151–163. Available from: https://doi.org/10.1080/00206814.2020.1836523; http://www.tandfonline.com/loi/tigr20.

Sehsah, H., Eldosouky, A.M., El Afandy, A.H., 2019. Unpaired ophiolite belts in the Neoproterozoic Allaqi-Heiani Suture, the Arabian-Nubian Shield: Evidences from magnetic data. J. Afr. Earth Sci. 156, 26–34. Available from: https://doi.org/10.1016/j.jafrearsci.2019.05.002; http://www.sciencedirect.com/science/journal/1464343X.

Seigel, H.O., 1979. Geophysics and geochemistry in the search for metallic ores. Geol. Surv. Can. Econ. Geol. Rep. 31, 7–23.

Sharma, P.V., 1997. Environmental and Engineering Geophysics. Cambridge University Press.

Silva, J.B.C., Santos, D.F., 2017. Efficient gravity inversion of basement relief using a versatile modeling algorithm. Geophysics 82 (2), G23–G34. Available from: https://doi.org/10.1190/GEO2015-0627.1; http://library.seg.org/loi/gpysa7.

Simpson, F., Bahr, K., 2005. Practical magnetotellurics. Practical Magnetotellurics 9780521817271. Available from: https://doi.org/10.1017/CBO9780511614095.

Smith, N.C., Vozoff, K., 1984. Two-dimensional DC resistivity inversion for dipole-dipole data. IEEE Trans. Geosci. Remote. Sens. GE-22 (1), 21–28. Available from: https://doi.org/10.1109/TGRS.1984.350575.

Spector, A., Grant, F.S., 1970. Statistical models for interpreting aeromagnetic data. Geophysics 35 (2), 293–302. Available from: https://doi.org/10.1190/1.1440092; http://library.seg.org/loi/gpysa7.

Sumner, J.S., 1976. Principles of Induced Polarization for Geophysical Exploration. Elsevier.

Telford, W.M., Geldart, L.P., Sheriff, R.E., 1990. second edApplied Geophysics, 770.

Thanassoulas, C., Xanthopoulos, N., 1991. Location of possibly productive geothermal fracture zones/faults using integrated geophysicals methods over Lesvos island geothermal field, Greece. Geothermics 20 (5–6), 355–368. Available from: https://doi.org/10.1016/0375-6505(91)90026-R.

Thompson, D.T., 1982. EULDPH: a new technique for making computer-assisted depth estimates from magnetic data. Geophysics 47 (1), 31–37. Available from: https://doi.org/10.1190/1.1441278.

Tikhonov, A.N., 1950. On determining electrical characteristics of the deep layers of the Earth's crust. Doklady 73, 295–297.

Tsourlos, P.I., Szymanski, J.E., Tsokas, G.N., 1999. The effect of terrain topography on commonly used resistivity arrays. Geophysics 64 (5), 1357–1363. Available from: https://doi.org/10.1190/1.1444640.

Ulrich, C., Slater, L., 2004. Induced polarization measurements on unsaturated, unconsolidated sands. Geophysics 69 (3), 762–771. Available from: https://doi.org/10.1190/1.1759462; http://library.seg.org/loi/gpysa7.

Ustra, A.T., Elis, V.R., Mondelli, G., Zuquette, L.V., Giacheti, H.L., 2012. Case study: a 3D resistivity and induced polarization imaging from downstream a waste disposal site in Brazil. Environ. Earth Sci. 66 (3), 763–772. Available from: https://doi.org/10.1007/s12665-011-1284-5.

Vieira, L.B., Moreira, C.A., Côrtes, A.R.P., Luvizotto, G.L., 2016. Geophysical modeling of the manganese deposit for Induced Polarization method in Itapira (Brazil). Geofisica Int. 55 (2), 107–117. Available from: http://www.geofisica.unam.mx/unid_apoyo/editorial/publicaciones/investigacion/geofisica_internacional/anteriores/2016/02/2vieira.pdf.

Werner, S., 1949. Interpretation of magnetic anomalies at sheet-like bodies. Sver. Geol. Underso. Ser. C. 43.

Wijns, C., Perez, C., Kowalczyk, P., 2005. Theta map: Edge detection in magnetic data. Geophysics 70 (4), L39–L43. Available from: https://doi.org/10.1190/1.1988184; http://library.seg.org/loi/gpysa7.

6

Geological data for mineral exploration

Ahmed M. Eldosouky[1], Hatem Mohamed El-Desoky[2], Ahmed Henaish[3], Ahmed Moustafa Abdel-Rahman[2], Wael Fahmy[2], Hamada El-Awny[2], Amin Beiranvand Pour[4]

[1]GEOLOGY DEPARTMENT, FACULTY OF SCIENCE, SUEZ UNIVERSITY, SUEZ, EGYPT [2]GEOLOGY DEPARTMENT, FACULTY OF SCIENCE, AL-AZHAR UNIVERSITY, CAIRO, EGYPT [3]DEPARTMENT OF GEOLOGY, FACULTY OF SCIENCE, ZAGAZIG UNIVERSITY, ZAGAZIG, EGYPT [4]INSTITUTE OF OCEANOGRAPHY AND ENVIRONMENT (INOS), UNIVERSITY MALAYSIA TERENGGANU (UMT), KUALA NERUS, TERENGGANU, MALAYSIA

6.1 Introduction

Exploration and prospecting of mineral deposits is a branch of geology that explores the origins of mineral resources and the methods for estimating economic deposits, or those that can be recovered under current conditions. It is a part of applied geology, together with economic geology, engineering geology, and hydrogeology. The duty of exploration concludes with the gathering of data, while the development and exploitation of resources fall under the purview of mining and mining geology (Pohl, 2011). Mining has supplied resources for fire, housing, and food acquisition from prehistory to the present. From the dawn of man until the dawn of civilization, the earth's goods obtained via mining were vital for the people's needs in the creation of implements, buildings, trade, jewelry, cosmetics, and treasure (Gandhi and Sarkar, 2016). There would be no mining, processing, or industry if mineral exploration did not exist, as well as a lack of cultural amenities and creature comforts. It is unavoidable for every country's economy to have an appropriate supply of mineral resources. Mineral exploration is distinct from other resource sectors in that it has its own set of features (Gandhi and Sarkar, 2016). Today, the world's people's level of living is measured by their per capita consumption of various metals. Transportation, communication, and building have all evolved over time as a result of the materials gained from the earth's resources. Metal and mineral products will continue to play an important part in satisfying society's demands for the foreseeable future. This chapter is intended to serve as a field guide for geologists who are concerned with ore and mineral exploration.

Geospatial Analysis Applied to Mineral Exploration. DOI: https://doi.org/10.1016/B978-0-323-95608-6.00006-8

6.1.1 Definitions

It's important to establish a few keywords that will serve as the foundation for the book in question. This chapter has been written as a field guide for those geologists interested in ore and minerals exploration.

6.1.2 Mineral and rock

The International Mineralogical Association defines a **mineral** as a naturally occurring, inorganic, homogeneous element or compound with a specified chemical structure and an organized atomic arrangement that has been formed as a result of geological processes (Gandhi and Sarkar, 2016). It's also important to remember that a **rock** is a naturally occurring assemblage of several varieties of crystals or mineral particles (Fig. 6.1A).

Ore is another important word in mining and is only used to describe the material that is removed for treatment purposes. The ore is described solely as a concentration of mineralization, with no economic context; however, this is a somewhat uncommon term (Bateman, 1950). **Ore** is a naturally occurring assemblage of one or more minerals that may be mined, processed, and profitably sold (Fig. 6.1B; Guilbert and Park, 1986). Ore deposit is a mineral deposit that has been evaluated and confirmed to be of sufficient size, grade, and accessibility to be profitable to mine. Surface mapping and sampling, as well as drilling through the deposit, are popular methods of testing (Gandhi and Sarkar, 2016). In general, ore deposits of metal are rocks with considerably greater metal concentrations than the typical crust.

FIGURE 6.1 (A) Hand specimen of granite sample composed of large; (B) interlocking mineral crystals; (C) gold flakes associated with a hydrothermal vein in gray quartz, Greenland. Deep open pit at El-Sukari gold mine, Egypt; (D) underground (subsurface) gold mine; (E) reclamation stage around mining sites.

They are geochemical anomalies or natural enrichments of metal in the Earth's crust. Deposits of industrial minerals have a similar definition: They are natural concentrations of the material of interest (Ridley, 2013).

Gangue minerals mean the valueless mineral particles or crystals within an ore, while waste is while; waste is the material that must be mined to obtain the ore. G**angue minerals** must be separated through milling, followed, typically, by flotation to separate the minerals of economic interest from the **gangue** (Ridley, 2013).

Grade of ore: In terms of the ore idea, it's important to remember that the ore (or mineralization) has a **grade**, which is defined as the average concentration of a valuable substance (e.g., gold or tin) in a sample or a mineral deposit. Metal ores are often graded as a percentage or grams per ton, which is equivalent to parts per million (ppm). **Prospect** is a phrase that is commonly used in the mining industry. It may be described generically as a small patch of land with the potential to contain a mineral deposit. It is frequently given the name of a specific locality. The Geological Survey of Western Australia uses the definitions given out by Cooper et al. (1998), which define prospect as any working or exploratory operation that has discovered subeconomic mineral deposits but no documented production.

Ore body is a solid and relatively continuous body of ore containing gangue minerals that differ in form and character from host rocks. A single ore body or multiple ore bodies can make up an ore deposit. The ore bodies appear in a range of forms, including isometric, flat, veins, lenses, and pipes (Gandhi and Sarkar, 2016). **Mineral exploration**, or simply exploration, is the process of looking for ore. Exploration is a multistep process that can take years, if not decades, to identify and define an ore body that could be mined. **Syngenetic deposits** form at the same time as the rock in which they are found. They developed in the same way as the surrounding rocks and at the same period in geological history. On the other place the **epigenetic deposits** are believed to have come much later than the host rocks in which they occur. They were introduced into the preexisting country rock by the movement of metal-bearing fluids after its formation. A good example is a mineral vein.

6.1.3 Stages of mineral exploration

1. Exploration and prospecting stage

   The exploration or prospective stage, known as the search for mineral resources, is considered the first phase in the mining life and involves a diverse set of operations (Stevens, 2010). To start a mine, businesses must first locate a deposit that is economically viable (an amount of ore or mineral that makes exploitation worthwhile).
2. Discovery and exploitation

   Mining will be the next step if the economic evaluation of a mining operation shows that there is a good possibility that exploiting the mineral deposit will provide benefits. The extraction and recovery of mineral and waste rock from the Earth's crust is known as exploitation or mining. Surface (Fig. 6.1C) and subsurface (Fig. 6.1D) mining methods are classified according to the distance from the surface.

3. Development

Feasibility, geology, and engineering studies are all part of the mine development stage. If all of these outcomes are positive and all necessary approvals have been obtained, the firm will determine whether or not to proceed with the project. At this point the firm is raising funds to commence the construction and development of a mine. This is the most costly stage of the mining process.

4. Production and mineral processing

The extraction, grinding, and processing of raw materials, such as coal, metals, industrial minerals, and aggregate, are all part of the production phase. The volume and grade of mineral or metal in the deposit, as well as the profitability of the operation, determine how long a mine will be in operation. Mineral processing removes valuable minerals from waste rock or gangue during extraction, resulting in a more concentrated substance for subsequent processing.

5. Closure and reclamation

Because mining is a transitory industry with an operational life varying from a few years to many decades, mine closure is the final phase of the mining cycle. The land that was mined will be rehabilitated and parts of this objective include ensuring public safety and health, minimizing environmental impact, preserving water quality, establishing new landforms and vegetation, stabilizing land to protect against erosion, and removing waste and hazardous material (Fig. 6.1E). Fig. 6.2 shows the different stages of mineral exploration.

6.1.4 Mineral system

A mineral system, according to Wyborn et al. (1994), is "all the geological processes that regulate the creation and maintenance of mineral deposits and is similar to petroleum systems." According to them, there are major geological variables to consider, as indicated in Fig. 6.3.

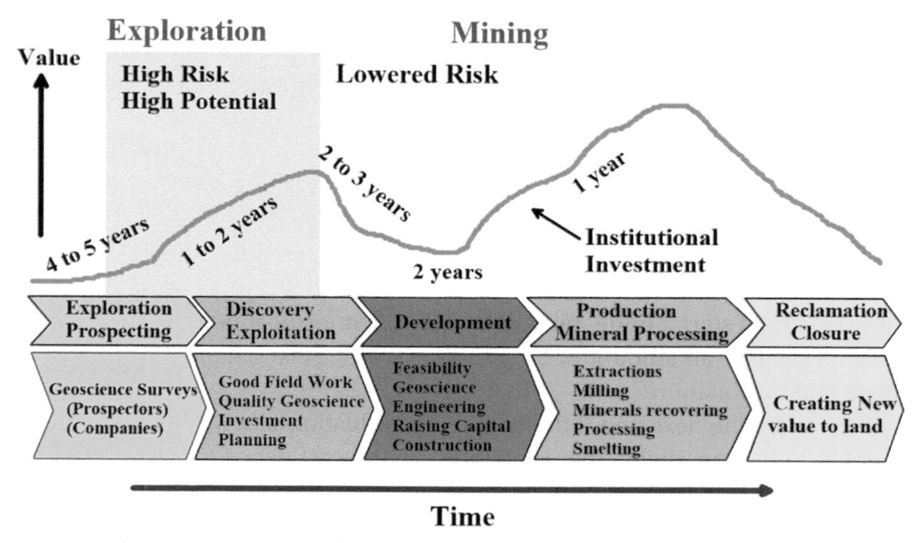

FIGURE 6.2 Stages of mineral exploration (lifecycle of mine).

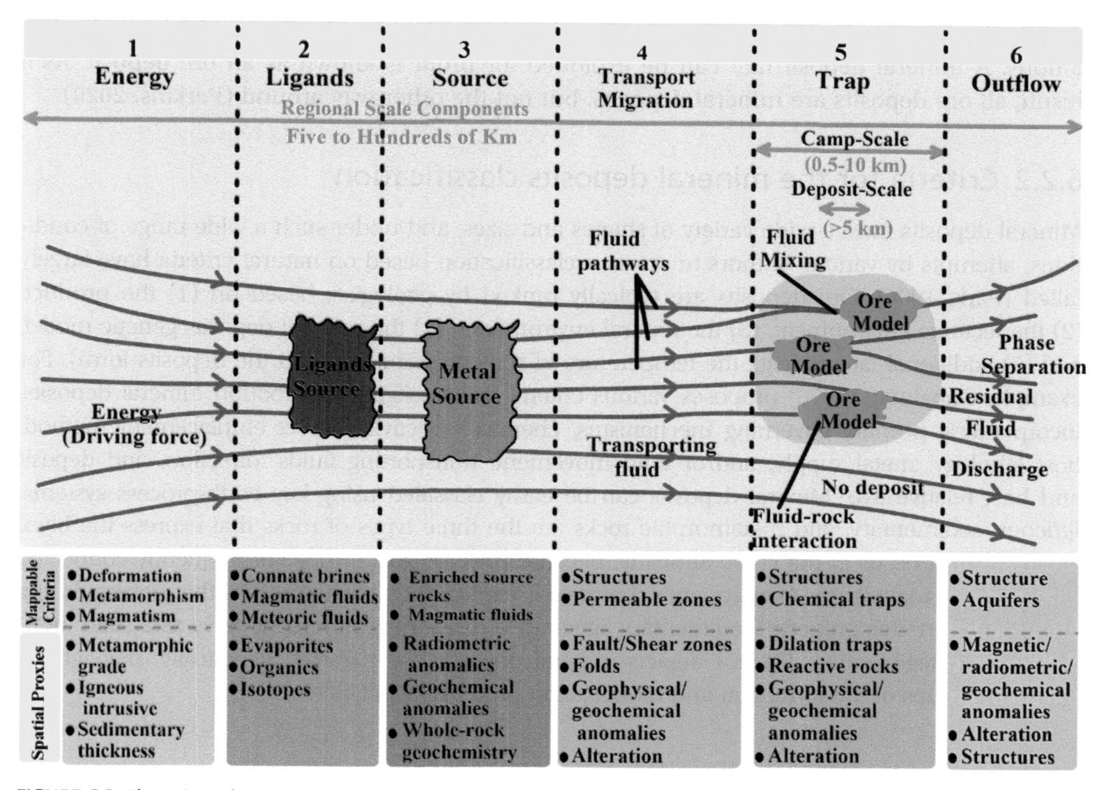

FIGURE 6.3 The mineral systems concept of ore-body formation (Knox-Robinson and Wyborn, 1997). *From Knox-Robinson, C.M., Wyborn, L.A.I., 1997. Towards a holistic exploration strategy: using geographic information systems as a tool to enhance exploration. Austr. J. Earth Sci., 44(4), 453–463. https://doi.org/10.1080/08120099708728326.*

The mineral system idea is a useful framework for mineral exploration because it provides more flexibility in interpreting their data than typical ore deposit models depending on geological information systems (GIS) (Brauhart, 2019; Duuring, 2020). The assumption of this concept is that mineral deposits will only form and remain preserved where critical earth processes (geodynamic setting, lithosphere architecture, fluid, ligand, and ore component reservoir(s), fluid flow drivers, and pathways, depositional mechanisms, and postdepositional processes) have occurred in a spatial and temporal coincidence and that the occurrence of these critical processes can be recognized from mappable geological features expected to reappear (Fig. 6.3).

6.2 Mineral deposits and occurrence

6.2.1 Definition and formation of mineral deposits

Mineral deposit: Any precious lump of ore a mineral deposit is what it's called. It is a concentration of valuable minerals or metals found in the crust of the Earth of sufficient size and grade that may be commercially exploited under current technology and economic conditions. A mineral deposit that can be exploited for profit is known as an ore deposit. As a result, all ore deposits are mineral deposits, but not the other way around (Perkins, 2020).

6.2.2 Criteria for the mineral deposits classification

Mineral deposits have a wide variety of shapes and sizes, and under such a wide range of conditions, attempts by various authors to create a classification based on natural criteria have largely failed (Park, 1906). Ore deposits are typically ranked by geologists based on (1) the product, (2) the tectonic environment, (3) the natural environment, (4) the mineral deposits genetic model, and (5) additional factors (e.g., the temperature of mineral formation and the deposits form). For example, Gabelman (1976) proposes various criteria for classifying stratabound mineral deposits, incorporating primary governing mechanisms, chemical reactivity, direct emplacement method, host lithology, metal supply, and/or fluid movement, transporting fluids' direction, and deposit and host relative age. Mineral deposits can be easily classified using key earth-process systems. Igneous, sedimentary, and metamorphic rocks are the three types of rocks that express the basic processes that occur in the crust. Similarly, because ores are rocks, they are frequently connected with rocks of many sorts. Consequently, because it depicts the genetic process that occurs during ore production, this property (sedimentary, igneous, or metamorphic) might serve as a solid foundation for categorization. Fig. 6.4 depicts mineral deposits are classified genetically, highlighting the main groups of ore formation and modification processes (McQueen, 2005).

6.2.3 Mineral deposits classification

6.2.3.1 *Industrial classification*

Mineral deposits are classified into three categories according to their industrial classification: **metallic** minerals, **nonmetallic** (industrial) minerals, and **combustible** (energy) minerals. **Metallic** minerals include gold, silver, iron, titanium, manganese, nickel, chromium, copper, tin, lead, molybdenum, zinc, while **nonmetallic** (industrial) minerals are represented by bentonites, kaolin, phosphate, clay, graphite, gravel, mica, fluorite, sulfur, asbestos, and

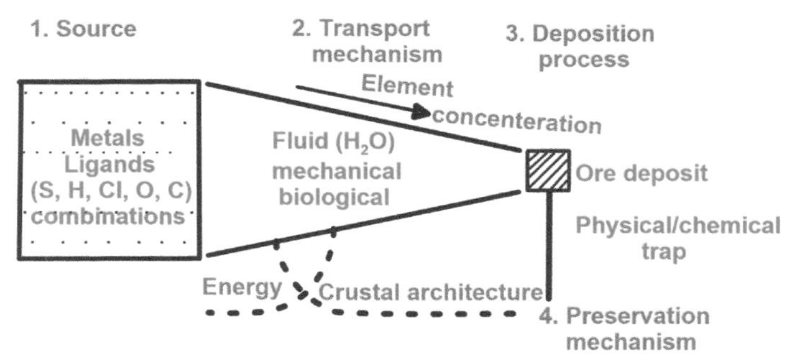

FIGURE 6.4 The four basic geological conditions for the formation of every mineral deposit (McQueen, 2009). *From McQueen, K.G., 2009. Ore Deposit Types and Their Primary Expressions, 14.*

sands. On the other hand, **combustible** (energy) minerals include oil and gas, bituminous shales, coal, tar sands, tight gas, and uranium.

6.2.3.2 Geological classification based on origin or genetic and occurrence

Fig. 6.5 illustrates mineral deposits' genetic categorization system, highlighting the primary aggregates of ore formation and modification procedures (McQueen, 2005). To summarize, classifying deposits based on their relationship to the processes of ore formation and origin is potentially the best method (Herrington, 2011). Internal, hydrothermal, metamorphic, and surficial processes can all be categorized as ore-forming processes (Evans, 1993). The first three processes are concerned with subterranean phenomena, whereas the fourth process is concerned with those occurring at the surface of the Earth. Thus the first method of ore formation can be summarized as follows: magmatic, metamorphic, sedimentary, and hydrothermal procedures.

1. Magmatic ore deposits

   Magma is a molten material (i.e., melt, including dissolved gases and any floating crystals in it). Magma may cool down and harden at depth to create a huge rock body called a pluton, erupt as lava at a volcano, or ascend as a finely fragmented ash cloud. Corresponding igneous (magmatic) rocks are known as plutonic (or intrusive) and volcanic (or extrusive) rocks. Ortho-magmatic deposits (also known as magmatic igneous-related deposits) are generated directly from the melt by magmatic operations. Certain ore deposits are so closely related to igneous rocks that a shared common ancestor may be deduced.

   There are mainly three types of magmatic deposits. The first type relates to primordial magmas that are directly formed from the basalt of the Earth's mantle. Cr, Ni, Mn, platinum (Pt), iron (Fe), Ti, and V deposits can occur in such basic and ultrabasic

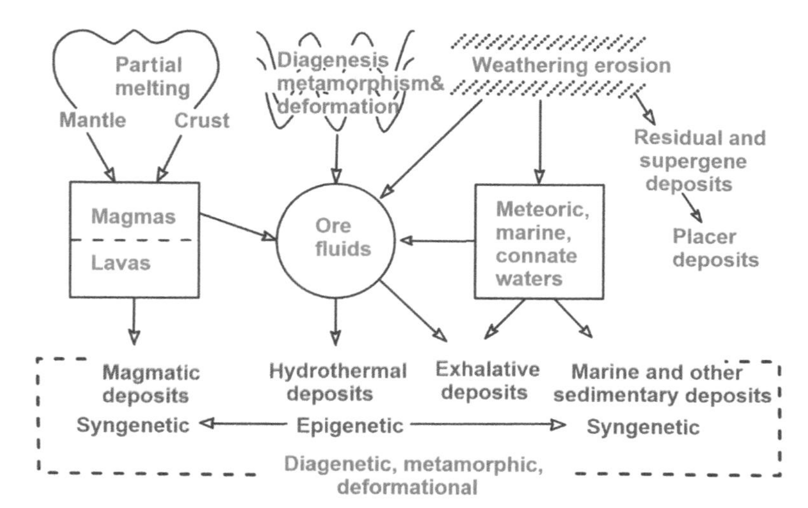

FIGURE 6.5 Ore deposit classification scheme based on genetics (McQueen, 2005). *From McQueen, K.G. (2009). Ore Deposit Types and Their Primary Expressions, 14.*

magmas. The second type is associated with granitic rocks. They could be generated by severe fractionation from a primitive melt that originated in the mantle or through melt production in the crust. The third kind of magma is alkali-rich magma (also known as alkaline magma and alkaline rocks), which is found near continental rifts and hotspots Fig. 6.6A–D.

FIGURE 6.6 Photograph showing magmatic and hydrothermal deposits: (A) a hand specimen of the chromite deposit from Barramiya area, Egypt; (B and C) massive ilmenite ore deposit from Abu Ghalga area, Egypt; (D) a hand specimen of the pyrolusite from Um Bogma area, Egypt; (E) gold deposits disseminated in quartz vein from Umm El-Russ, Egypt; (F) gold deposits disseminated in quartz vein from Gabal Elba, Egypt; (G) a hand specimen of the galena of Um Gheig area, Egypt; (H) cubic pyrite crystals disseminated in quartz vein from Eqat, Egypt; (I) cubic and triangular pyrite crystals disseminated in quartz vein Colorado, USA; (J and K) quartz vein bearing chalcopyrite from El-Fawakhir mine, Egypt.

6.2.3.3 Classification of chromite deposits

These chromites are thought to have developed during the development of the new oceanic crust by early crystallization followed by crystal settling from basic magmas at spreading centers. Chromite ore is frequently discovered in ophiolite complexes. Exploration for novel chromite deposits interstratified with the lowermost areas of layered mafic-ultramafic intrusions, along with serpentinite ophiolite belt segments, must be prioritized (Hussein and El Sharkawi, 2017). Most of the chromite globally comes from mafic and ultramafic rocks (Fig. 6.6A). Chromite ores are usually assigned to one of the following two classes based on the mode of occurrences, petrologic character, and tectonic setting of their host rocks (Thayer, 1970).

Stratiform ores are sheetlike chromite agglomerations connected with mafic rocks. Precambrian ultramafic layered intrusions in an intracranial context. The majority of the chromite attributed with layered intrusions is excavated from South Africa at the Bushveld Complex and the Great 'Dyke' in Zimbabwe, which is usually observed within the southern African Proterozoic mega-fracture system.

Podiform ores are chromite-rich rocks that are irregular but lenticular, podlike, and tabular. They are found mostly in tectonic peridotites and basal dunite cumulates of ophiolites and Alpine-type peridotites (Naldrett and Gruenewaldt, 1989; Zhou et al., 1996; Shi et al., 2012). They are mostly Phanerozoic in age, with very irregular shapes and associations with ophiolites. The combination of podiform deposits with ophiolites limits their presence to Palaeozoic or relatively young island arcs and dynamic orogenic belts (Thayer, 1970). Despite size or shape, podiform chromites are often surrounded by a dunite sheath or halo (Ahmed and Arai, 2002).

2. Hydrothermal ore deposits

   Hydrothermal ore deposits are those generated by hydrothermal action in conjunction with igneous rocks deposited at higher elevations in the crust of the Earth (Fig. 6.6E–K). A hydrothermal fluid is a hot (more than 50°C) aqueous solution that contains solutes frequently precipitate as characteristics of the solution change over time and distance. It is critical to recognize that the terms fluid and solution are interchangeable. Hot water is an excellent carrier and enrichment medium for certain elements, which are then precipitated from the water. This can happen in several sites, including a cave, small cracks above a granite pluton, a hot spring on the seabed, a fault, or rock pores.

   Saltwater, meteoric, connate, metamorphic, juvenile, and magmatic water can all be encountered within hydrothermal solutions. Connate water sometimes called formation water (is water that is caught during sediment deposition and formed during diagenetic processes). It is widely known that hydrothermal fluids may arise during burial diagenesis and achieve high salinities and temperatures. During diagenesis, temperatures can fluctuate at a few degrees below 8°C to 30°C–250°C; however, water ejection is carried out at temperature points ranging between 90°C and 1208°C throughout diagenetic operations (Hanor, 1979).

   The existence of volatile-rich fluids liberated through metamorphism is well known, and these fluids may be regarded as diluted brines containing H_2O, CH_4, and CO_2. The presence and quantity of these volatiles in magma are frequently linked to their chemical composition and origin. Generally, hydrothermal solutions are often supposed to be liquid combinations with water as the solvent. The fluids that compose ore could be colloidal or molecular in origin. The movement of hydrothermal fluids in the crust of Earth causes the genesis and maintenance of hydrothermal systems. Because they help in the separation of components from solution, boiling and solubility are significant in the deposition of ores from hydrothermal fluids. The boiling of hydrothermal solutions is important since it causes ore constituents to precipitate (e.g., Au, As, Sb, Ag).

Hydrothermal activity is defined by the existence and transportation of these fluids, whether they drain at the surface or not. A hydrothermal system is composed of two main parts: a heat source (magmatic, geothermal gradient, radiogenic decay, or metamorphism) and a fluid phase (solutions formed from magmatic/juvenile fluids, metamorphic fluids, meteoric, connate waters, or saltwater). Hydrothermal change is crucial for mineral exploration due to its extension further than the boundaries of the ore, facilitating exploration work to be concentrated in fewer locations. Porphyry deposits and volcanic-associated sulfide deposits are considered two of the most prevalent kinds of deposits produced by magmatic–hydrothermal operations (pyrite and chalcopyrite, gold) (Fig. 6.6G–K), (gold) (Fig. 6.6E and F). The best evidence of hydrothermal fluid movement is hydrothermal veins, which can also be thought of as a manifestation of the conduits or cracks through which fluids circulate.

3. Sedimentary ore deposits

Mineral deposits, developed by sedimentary practices, are known as sedimentary mineral deposits. At or near the ground surface, economic mineral deposits can be formed because of low-temperature surface processes, under certain conditions. Sedimentation can result in the production of mineral deposits by clastic accumulation (e.g., diamond or gold placer deposits) and biochemical precipitation and/or chemistry of economically significant elements in coastal settings, lakes, or shallow to deep seas, including evaporation.

By forming metal–organic complexes, organic compounds generated as a result of bacterial breakdown of more complex organic materials may facilitate the metal transfer. Similarly, biogenic H2S may form persistent aquatic complexes with metal ions, enabling the transportation of specific metals, such as Ag at low temperatures (Kyle and Saunders 1996). Weathering can also reveal the reformation of a residual concentration of the parent rock's weathering-resistant minerals or relatively insoluble components being reconstructed into stable minerals (Misra, 2000). Weathering is an essential ore-forming process that involves migrating solutions causing chemical changes and redeployment of components in surface rocks. Chemical differences between minerals at the surface–crustal boundary and at the Earth's surface can result in chemical dissolution, residue upgrading, and reprecipitation operations that concentrate the desired mineral/metal.

The circulation of mostly meteorically produced water near the Earth's surface drives ore production in these settings; however, identical equivalent processes can occur on the seafloor. At favorable mineral locations or surface contacts, these subterranean waters can dissolve and reprecipitate components (Herrington, 2011). Numerous raw minerals, such as iron, manganese, and aluminum ores, are often produced by supergene processes. The process known as supergene enrichment comprises leaching ore-forming components (e.g., copper) off the surface of a low-grade sulfide deposit, and reprecipitation under the water table is a kind of weathering.

The ore metals are released from weak sulfide minerals into descending meteoric water, whereas, in the subsurface ecosystem, more stable secondary oxide and sulfide mineral assemblages are precipitated. These deposits are commonly referred to as (gossan) (Fig. 6.7A). Gossans were significant guides for prospectors looking for underground mineral resources in the 19th and 20th centuries.

6.2.3.4 Placer deposits

They comprise deposits connected to supergene modification, water infiltration, and diagenetic processes, as well as placer deposits formed by erosion, transportation, and sedimentation processes. A placer ore body is a sedimentary deposit of soil, gravel, or sand that

FIGURE 6.7 Photograph showing: (A) goethite (iron-bearing mineral) of Abu Ghalaga, Egypt; (B and C) gold production from placer deposits by vibrating from Eqat, Egypt; (D–F) beach and black sand deposits at Rashid and Damietta, Egypt; (G) banded iron ore from Abu Ghalaga, Egypt; and (H and I) greisen from Abu Dabbab, Egypt.

includes eroded valuable mineral particles. Minerals have the ability to withstand erosion and become concentrated in the surface environment due to their physical and chemical characteristics. Platinum metals, gold (Fig. 6.7B and C) in its metallic or native form, ilmenite, rutile, monazite, and zircon, and gemstones, such as diamond, ruby, garnet, are all common minerals found in placer deposits (Revuelta, 2017). Furthermore, the precious minerals contained in this sort of deposit are far denser than those carried at the Earth's surface.

Minerals may then be recovered and concentrated in ore bodies from the detrital elements or rock fragments that compose the majority of the sediment load. Numerous categorization schemes have been suggested by Wells (1969), Smirnov (1976), and Macdonald (1983), which are founded principally on geological environment. The most important classifications for placer deposits of economic relevance are residual, fluviatile, colluvial, eluvial, and coastal; coastal placer deposits are often referred to as marine and beach placers.

Other processes that lead to successful fractionation include weathering, transport, and sedimentation. When weather-resistant minerals with higher densities are enriched, placer deposits occur. Palaeo-placers are the lithified counterparts to placer deposits. Nevertheless, high mineral concentrations retained in sedimentary layers are often of scholarly relevance exclusively (Zimmerle, 1973). Because placer deposits are created by natural surface processes, black sands, they can be found all over the world. Most of the World's placer deposits are of Tertiary and Quaternary age. In Egypt, black sand can be found along the Mediterranean Sea shoreline from Rasheed to Rafah, 400 km. In those places, it is disseminated by sea currents and waves, and it is found in sand dunes. Ilmenite, magnetite, zircon, rutile, garnet, and monazite, which contain radioactive minerals, are among the key minerals found in black sand (Fig. 6.7D–F).

4. Metamorphic and metamorphosed mineral deposits

Ore deposits can occur before, during, or after metamorphic processes in metamorphosed rocks. The first is the class of metamorphosed ore deposits, which are of premetamorphic origin and are unaffected by later metamorphic overprinting, graphite veins, orogenic gold, and many significant talc deposits are among the ore deposits formed by regional metamorphism, according to Pohl (2011). Contact or regional metamorphism causes metamorphic deposits, which entail recrystallization and the mobilization of dispersed ore components by metamorphic fluids (Misra, 2000). Numerous metal deposits are found in metamorphic rocks, and it is hypothesized that metamorphic fluids are the source of a variety of mineral deposits. For example, gold ore is a kind of mineralization that is frequently associated with metamorphic fluids. As a result, according to chemistry, it is conceivable to assert that metamorphic fluids can contain high concentrations of metals and hence constitute prospective ore fluids under specific scenarios (Banks et al., 1994). Ore deposits are generated in three distinct ways as a result of metamorphic fluid processes, according to Yardley and Cleverley (2015): (1) where metal-rich metamorphic fluids act as a segregation medium, (2) decarbonation processes cause concentrated fluid flow and the creation of skarns, and (3) the pace of fluid generation is not restricted by heat flow because rapid uplift drives dehydration processes despite dropping temperatures. Given the prevalence of magmatic activity in certain metamorphic environments, it is reasonable to assume that some mineral deposits in metamorphic rocks are magmatic (Fig. 6.7G), which were created by a combination of magmatic and metamorphic processes.

Skarn deposits are without a doubt the most important ore deposit type. Thermal aureoles of magmatic intrusions are intimately linked to skarn and contact metamorphism ore deposits. They can be thought of as contact metamorphism products, but the interaction with magmatic fluids is the causal agent, not just heating (Pohl, 2011). Skarn deposits, on the other hand, have enough distinguishing properties to be considered a separate class (Misra, 2000). Skarn is an ancient Swedish mining word that refers to a broad spectrum of mostly coarse-grained calc–silicate rocks that are rich in iron, aluminum, calcium, manganese, and magnesium, regardless of their association with potentially important minerals. They were created by metasomatic processes replacing formerly carbonate-rich rocks (Einaudi et al., 1981). Mineral assemblages are a diagnostic feature of typical skarns; the main assembly varies in structure according to the constitution of the skarn-forming fluids and invaded rocks but is dominated by anhydrous Ca–Fe–Mg silicates and pyroxenes (including pyroxenoids), with garnets playing a significant role. They are the world's leading source of tungsten as well as key sources of copper, molybdenum, iron, and zinc and range in age from Precambrian to late Cenozoic. Between porphyry and skarn ore deposits, there is a continuum, and at least several skarn deposits appear to be mined within porphyry systems' carbonate wall rocks. The composition, metallogenic affinities, and oxidation states of the igneous intrusion all contribute to the variety of metals found in skarn deposits. Au and Fe skarn deposits, for example, are frequently linked with mafic to intermediate-composition intrusions. The bulk of large and commercially viable skarn deposits are associated with calcic exoskarns, which are composed of limestone (calcic) host rock and a metasomatic assemblage located outside the intruding pluton (exo—prefix).

Greisen is a type of endoskarn created by self-produced modification of granite. Greisen is a granitic rock that has been hydrothermally metamorphosed. It's largely made up of quartz and light-colored mica (muscovite, zinnwaldite, and lepidolite) (Fig. 6.7H and I). Greisen develops near the ceiling of several granite plutons, sometimes capped by pegmatite, and frequently hosts workable—though not necessarily economically viable—W–Sn–Mo ore deposits.

6.3 Geological mapping

6.3.1 Lithological mapping

Lithological mapping is considered one of the vital steps in mineral exploration which deals with determining the spatial distribution of various rock types, including igneous, metamorphic, and sedimentary rocks. It represents a challenging step as it requires integrative field, laboratory, and remote sensing techniques. Moreover, lithological mapping techniques of sedimentary rocks differ from those used in igneous as well as metamorphic rocks. Consequently, the next paragraphs deal with clarifying the different procedures used for mapping each rock type.

6.3.1.1 Petrography

The geological environment is divided into two parts (Hawkes, 1957; Hawkes and Webb, 1963): (1) the primary environment, including the igneous differentiation and metamorphism; and (2) the secondary environment comprises weathering and sedimentation processes at the Earth's surface.

6.3.1.1.1 Igneous Rocks

Magma (magma = molten materials) is the more general term that embraces mixtures of melt and any crystals that may be suspended in it inside the surface of the Earth. Magmatic rocks are classified into four main groups based on their chemistry: **ultramafic/ultrabasic, mafic/basic, intermediate, and felsic/acid** categories.

Ultrabasic igneous rocks Although ultramafic and ultrabasic rocks are not widespread on the surface of the Earth, they give important information on basalt magma generation and mantle source zones (Gill, 2011). Igneous rocks classified as ultrabasic or ultramafic have SiO_2 (45%), MgO (> 18%), high FeO, and low potassium. In ultrabasic rocks, ferromagnesian minerals, including olivine, clinopyroxenes, orthopyroxenes, hornblende, and opaque minerals, comprise more than 90% of the mineral composition. These rocks are containing iron ores, Cu–Ni–Co sulfides, chromite–ilmenite, talc, magnesite deposits, and gold mineralization, associated with other mineral deposits. Petrographical characterization of distinct kinds in ultrabasic rocks is shown in Fig. 6.8. There are well-known examples of rocks in this group: peridotite, lherzolite, harzburgite, hornblendite, websterite, dunite, and pyroxenite, kimberlite, and komatiites.

FIGURE 6.8 Petrographic characteristics of ultrabasic rocks (A–D) and basic rocks (E–H): coarse grains of dominant Ol crystals in dunite, Crossed Polarized Light (XPL) image, 2× (7 mm). Well-developed Hbl crystals in a hornblendite rock. Plain Polarized Light (PPL) image, 2× (7 mm). Hypidiomorphic texture in pyroxenite rocks in Antarctica. XPL image. 2× (7 mm). Spinifex texture in Komatiite rock showing skeletal pyroxene (Cpx) at Alexo, Canada. PPL image, 10× (2 mm). Coarse grains of Pl and Ol altered to Idd in gabbroic rocks at Wadi Dungash and Wadi Shait. CN, 25× Ol-pyroxene ferrogabbros rocks showing brown Hbl contain opaques as inclusions at Wadi Khashir and Ras Al-Khair. CN, 40×. Diabasic and subophitic textures between Pl and Aug crystals. XPL image, 2× (7 mm). Pl and Ol crystals in Ol basalt rocks. XPL image, 10× (2 mm). *Aug*, augite; calcite; *Cpx*, clinopyroxene; *Hbl*, hornblende; *Idd*, iddingsite; iron oxides; *Ol*, olivine; *Op*, orthopyroxene; *Pl*, plagioclase; spinel.

Basic rocks igneous rocks Basic rocks have a low-content silica concentration, ranging from 45% to 52% SiO_2. It is mostly made up of calcic plagioclase (labradorite–bytownite) with high ferromagnesian minerals such as orthopyroxene, clinopyroxene, olivine, and opaque minerals. Pyroxene (diopside, augite, and hypersthene), hornblende, and olivine are ferromagnesian minerals that can be found together or separately. These families include the gabbro, norite, gabbronorite, troctolite, anorthosite, dolerite, and basalt (Fig. 6.8). Gabbros are common examples of coarse-grained igneous rocks that are similar in composition to basalts and reflect basalt magma that has slowly crystallized at deep. The widespread textures in these rocks are allotriomorphic, ophitic–subophitic, and diabasic texture (Fig. 6.8).
Intermediate igneous rocks Intermediate rocks have an SiO_2 content of 52%–63% and are divided into two types based on chemical calcic-intermediate. The sodic-intermediate includes the syenite, monzonite, trachyte, while the calcic-intermediate comprises andesite and diorite analysis. The sodic-intermediate rocks have a similar overall composition to granite, but quartz is either missing or present in minor levels (less than 5%). It is essentially composed of alkali feldspar (60%–80%, orthoclase–sanidine) and a ferromagnesian mineral (20%–40% hornblende, biotite, and pyroxene). The alkaline feldspar component (typically orthoclase or sanidine) is the most common. Plagioclase feldspars (oligoclase) can be found in small amounts, less than 10%. Feldspathoids (nepheline), titanite, apatite, zircon, magnetite, and pyrites are among the accessory minerals.

Syenite is a plutonic intermediate rock with a coarse-grained texture that is panidiomorphic and hypidiomorphic (Fig. 6.9A). On the other place the trachytic rocks are the volcanic equivalent of the plutonic (intrusive) rock syenite, which is distinguished by directive texture (Fig. 6.9B).

Regarding, calcic-intermediate rocks are mostly composed of 75% of plagioclase feldspar and 25% of mafic minerals (amphiboles, biotite, augite, and orthopyroxene). They contain a low of quartz, zircon, apatite, sphene, magnetite, ilmenite, and sulfides that occur as accessory minerals. The prominent rocks of this category are diorite and andesite, which have hypidiomorphic (granular) and porphyritic textures, respectively (Fig. 6.9C and D).

The intermediate rocks are containing the most abundant wall rock alterations such as carbonatization, kaolinitization, sericitization, chloritization, pyritization, and propylitization. The mineralization is represented by the occurrence of banded iron formation (BIF), uranium deposits, Zn–Pb–Cu sulfide minerals associated with syenite and andesite rocks.
Acidic igneous rocks **Felsic/acidic** igneous rocks define as light-color, low specific gravity, and high-silicate minerals ($>65\%$ SiO_2). The most prominent felsic minerals are quartz, orthoclase and sodic plagioclase, and mica (biotite and muscovite). Granitic rocks (alkali feldspar granites, granodiorite, tonalite, adamellite) are representing plutonic felsic rocks, whereas rhyolite and dacite indicate volcanic felsic rocks (Fig. 6.9). The primary mineral composition of granitic rocks is K-feldspar (microcline and/or orthoclase: 50%–80%), quartz (20%–40%), Na-plagioclase, and micas (biotite and rare muscovite). It is characterized by granitic texture such as hypidiomorphic, perthitic, myrmekitic, rapakivi, graphic, poikilitic, and porphyritic texture.

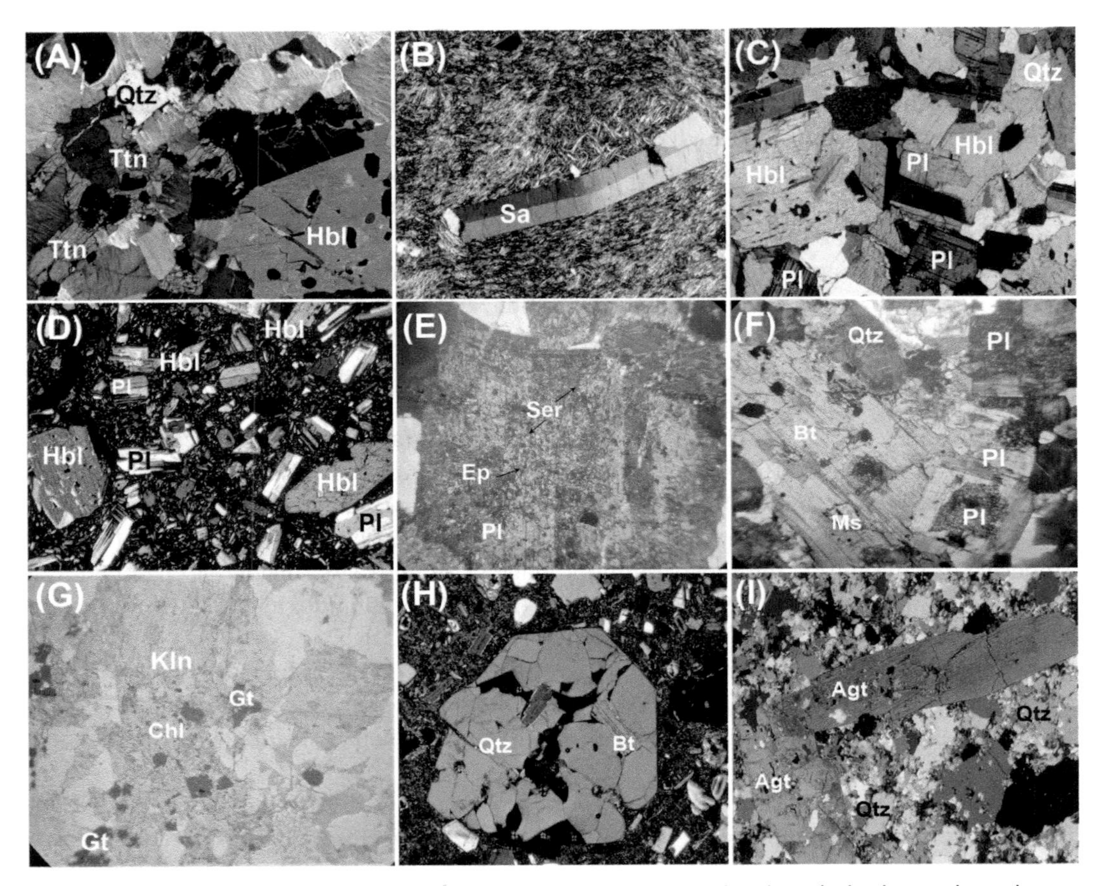

FIGURE 6.9 Petrographical characterization of intermediate rocks (A–D) and acidic rocks (E–I); granular and perthite texture in alkali feldspar syenite as well as hornblende and titanite. XPL image, 2 × (7 mm). Phenocryst of sanidine crystal within fine groundmass in trachyte rocks. XPL image, 2 × (7 mm). Hypidiomorphic texture in diorite showing plagioclase and brown hornblende. XPL image, 2 × (7 mm). Large phenocryst of hornblende and plagioclase crystals in andesite. XPL image, 2 × (7 mm). Plagioclase altered to saussuritization (epidote and sericite) in granodiorite, Hamash mine. XPL image, 4 × (7 mm). Large muscovite–biotite crystal and saussurite in core of plagioclase in biotite granites, Samadai-Um-Tunduba area. XPL image, 4 × (7 mm). Kaolinite and chlorite alteration associated with goethite after pyrite and granophyric texture in alkali feldspar granite, Samadai-Um-Tunduba area. PPL image, 4 × (7 mm). Phenocryst of quartz crystal in rhyolite with felsic groundmass at Vincenzo (Italy). XPL image, 2 × (field of view = 7 mm). Aegirine crystals in fine-grained quartz and feldspar at Brandberg Complex (Namibia). XPL image, 2 × (7 mm). *Agt*, Augite; *Bt*, biotite; *Chl*, chlorite; *Ep*, epidote; *Gt*, goethite; *Hbl*, hornblende; *Kln*, kaolinite; *Ms*, muscovite; *Pl*, plagioclase; *PPL*, plain polarized light; *Qtz*, quartz; *Sa*, sanadine; *Ser*, sericite; *Ttn*, titanite; *XPL*, crossed polarized light.

On the other side, rhyolitic rocks are similar to granite in mineral composition, but it has a porphyritic texture. For acidic rocks, the most prevalent types of alteration are sericitization, epidotization, argillic, and silicification. The common mineralizations recorded in acidic rocks are Ta–Nb–Sn–W, uranium, and gold.

6.3.1.1.2 Sedimentary rocks

Sediments and sedimentary rocks represent more than 85% of Earth's crust. Detrital sediments account for more than 75% of all sediments on Earth's surface (Hefferan and Brien, 2010). Sedimentary rocks are considered a host of a variety of metallic and nonmetallic minerals, including iron ore and a small amount of copper, manganese, uranium, and magnesium.

Classification of sedimentary rocks Sedimentary rocks are divided into two groups: (1) exogenous (clastic) and (2) chemical and biological (endogenous) sediments. This classification is based on the sorts of physical, chemical, biological, and geological processes they undergo. Between these two separate and different groups, there are mixed sediments and sedimentary rocks.

1. Exogenous (clastic = terrigenous)sediments

   The weathering of rocks produces clastic (terrigenous) materials. These materials are also known as siliciclastic because they are essentially siliceous (quartz-rich) clastic sediments transported by detrital mechanisms. Except for those erupted from volcanoes, clastic deposits, and sedimentary rocks are classified according to the size of the clasts, independent of their origin: rudaceous, arenaceous, and argillaceous.

   A sedimentary rock composed mainly of rounded grains and boulder-sized fragments welded together in a matrix is known as a conglomerate (Fig. 6.10A). Grain rounding occurs when grains are moved a long distance from their original source (e.g., by a river or glacier). By contrast, breccias are angular aggregates consisting of angular to subangular, randomly oriented clasts from different sedimentary rocks (Flint et al., 1960). It is classified into cataclastic breccia, talus breccia, and volcanic breccia (Fig. 6.10B).

   Sandstones are a common and significant sedimentary rock type. Sandstones are composed essentially of quartz, feldspar (which can be partly or completely converted to clay minerals), micas, and clay minerals (Fig. 6.10C). Two main groups of sandstone are recorded: pure sandstone (arenites) and impure sandstones (graywacke). Graywacke is a variety of sandstone and composed of quartz, fine-grained clay, while many of the cloudy grains are feldspar (Fig. 6.10D).
2. Chemical and biochemical sedimentary rocks

   This type is classified into two essential types: allochemical carbonates and orthochemical rocks. **Allochemical** carbonates are divided into three groups: the calcite, dolomite, and aragonite group, that is, rocks composed constituted mostly more than 50% of $CaCO_3$ composition such as calcite, Mg–calcite, and aragonite, or dolomite. The grains that make up the structure of limestones and dolomitic limestones are known as allochems (Haldar, 2020). They appear in a range of forms, including fossils, skeletal, ooids, peloids, and intraclasts (Fig. 6.10E). According to Haldar (2020), dolomite can originate in two ways: a primary origin and secondary replacement of calcite–aragonite (Fig. 6.10F).

   Orthochemical rocks comprise evaporite deposits of halite, gypsum, anhydrite, and BIFs (Fig. 6.10G–I). Minerals that precipitate straight from salty fluids because of evaporation make up evaporites. Sulfates, halides, and carbonates are the three most

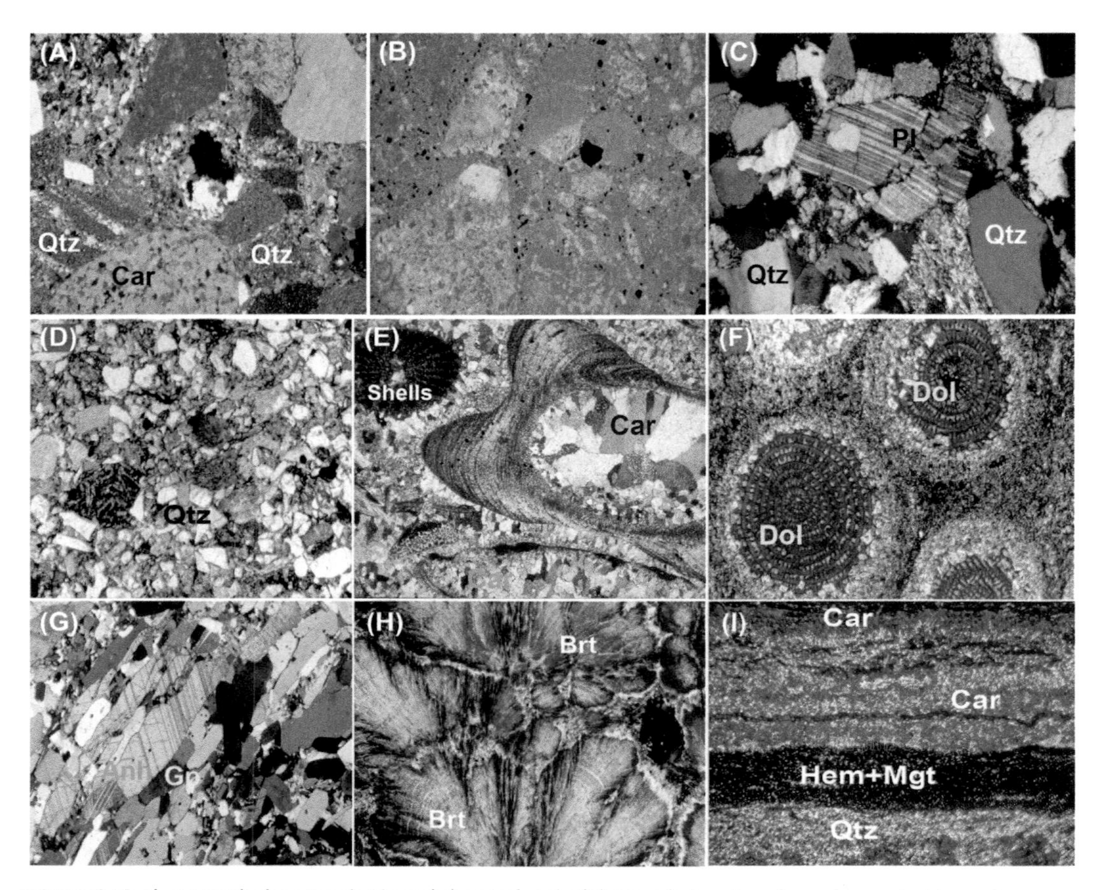

FIGURE 6.10 Photograph showing clastic and chemical rocks (A) Rounded grains of conglomerate. Crossed polarized light (XPL), 2 × (7 mm); (B) angular habit of breeccia grains. XPL, 2 × (7 mm); (C) quartz and feldspar fragments in sandstone, XPL, 10 × (2 mm); (D) medium grains of quartz and iron oxides in greywacke, plain polarized light (PPL); (E) small particles of carbonate (Car) with tiny fragments of shells fossil in Limestone. XPL 3.5 mm; (F) medium rhombohedral crystals of dolomitized (Dol) fossils in fossiliferous limestone, Tunisia, PPL, 1 × (9 mm); (G) high interference colors of anhydrite (Anh) and gray color of gypsum (Gp) crystals, Tuscany, Italy. XPL, 10 × (2 mm); (H) plumose texture and well-developed crystals of barite crystals. XPL, 2 × (7 mm); (I) alternative mode between Fe oxide-rich bands (Hematite [Hem] + Magnetite [Mgt]) and silica (Qtz) + carbonate (Car) band thickly banded in BIF, Precambrian age from Australia.

frequent types of evaporite minerals. **Gypsum** ($CaSO_4 \cdot 2H_2O$), **anhydrite** ($CaSO_4$), **sylvite** (KCl), **barite** ($BaSO_4$), and **halite** (rock salt) are the most common evaporite minerals among the sulfates and halide minerals (NaCl). Gypsum is the most common sulfate mineral and may be found in limestones, shales, and evaporite deposits (Fig. 6.10G and E).

Iron is present in most sedimentary rocks, but those that contain more than 15% iron are known as iron-rich sedimentary rocks. Iron stones and BIFs (Fig. 6.10I) are the two forms of deposits that can be found (BIFs).

6.3.1.1.3 Metamorphic rocks

The metamorphic rocks were classified into foliated rocks and nonfoliated (granular) rocks. The foliated rocks include slate, phyllite, schist, gneiss, and migmatite. Meanwhile, the non-foliated rocks comprise amphibolite, marble, quartzite, eclogite, serpentinite, soapstone, skarn, greisen, jasperoid, fenite, rodingite, spilite, mylonite, phyllonite, buchite, argillite, hornfels, granulite, granofels, metabasite, charnockites, and cataclasite.

6.4 The foliated rocks

The foliated rocks were originated from regional metamorphic terranes. They readily break along a foliation that is usually characterized by lepidoblastic phyllosilicate minerals; however, flattened felsic and carbonate grains are less prevalent (Best, 2003).

6.4.1 Schists

6.4.1.1 Tourmaline-bearing schists

Tourmaline-bearing schists are forming a large exposure of low-to-moderate relief extending between the muscovite granites and mélange rocks at Wadi Sikait, Um El-Debbaa, Wadi Abu Rusheid, and Wadi El-Gemal, South Eastern Desert, Egypt (Fig. 6.11A and B). Beryl and

FIGURE 6.11 Photograph showing (A) muscovite granites intruded within tourmaline-bearing mélange schists at Wadi Um El-Debbaa, Egypt (Looking NW); (B) tourmaline-bearing mélange schists at Wadi Um El-Debbaa, Egypt (Looking NW); (C and D) hand specimen of tourmaline minerals-bearing mélange schists at Sikiat area, Egypt.

tourmaline occur as elongated crystals disseminated in the tremolite–graphite schists and tremolite–actinolite schists (Saleh, 1997), many of ancient beryl mines related to that unit are located near the contact with peraluminous granitic rocks.

The tourmaline-bearing schists and tourmalinites occur as layers within tremolite–graphite schists and were interpreted as being deposited under pneumatolytic metasomatic conditions (El-Rahmany et al., 2015; El-Awny, 2015). The elongate tourmaline crystals of foliated tourmaline-bearing schists are commonly parallel to bedding and minor crenulations (Fig. 6.11C and D). Tourmaline (60.47%), tremolite (21.62%), and graphite (16.18%) were the most abundant minerals in these rocks discovered in Wadi Um El-Debbaa. Opaque mineral (1.15%) is a widely distributed accessory constituent, whereas rutile (0.58%) is a minor accessory. They have a schistose structure and a brownish color and are medium to coarse-grained.

Petrographically, tourmaline is found as idiomorphic prismatic crystals with two sets of cleavage, hexagonal crystals (Fig. 6.12A). They have a brownish color, complicated twinning, and irregular zoning, and range in length from 2 to 13 mm and width from 2 to 3 mm. It comes in a variety of pale brown colors that is pleochroic. Most tourmaline grains are optically unzoned, although some tourmalines have dark brown cores and lighter rims in narrow parts (Fig. 6.12B). The peripheral parts of some tourmaline porphyroblasts contain subidiomorphic crystals of tremolite, graphite, rutile, and opaque minerals conforming to the schistosity (Fig. 6.12C). This feature appears to be due to later growth or recrystallization and this indication on poikiloblastic texture. In equidimensional tourmaline crystals and porphyroblastic texture (Fig. 6.12D).

6.4.1.2 Tourmalinites

Dismembered fragments of tourmalinites are known in the mélange schists around granitic rocks encountered at Wadi Sikait, Um El-Debbaa, and Wadi Abu Rusheid (El-Desoky, 2010). The essential components of tourmalinites are tourmaline (65.23%) and quartz (26.98%). Rutile (1.72%), graphite (3.54%), actinolite (1.03%), and opaque minerals (1.5%) comprise the accessory minerals. It is surrounded by numerous slender radiating aggregates of tourmaline (Fig. 6.12E; El-Rahmany et al., 2015; El-Awny, 2015).

There are two generations of tourmaline encountered in the tourmalinites. The first generation of tourmaline is usually massive prismatic and much embayed by corrosion (Fig. 6.12F). The distribution of the color zones is very irregular (Fig. 6.12G). The core is generally nearly colorless to greenish-brown or yellowish-brown (Fig. 6.12H). The second generation of tourmaline is idiomorphic prismatic crystals, which shows two sets of cleavage, pleochroic from pale yellow to yellowish-brown, coarse- to medium-grained from 1 to 13 mm in length and from 1 to 2 mm in width (Fig. 6.12F; El-Awny, 2015).

6.4.1.3 Sillimanite-bearing schists

Sillimanite-bearing schists are strongly foliated, fine- to medium-grained, and characterized by their grayish-white color and resistance to weathering. These schists are encountered in Gabal Hafafit, and Wadi Allaqi, South Eastern Desert, Egypt. The major constituents are sillimanite (more than 25%), quartz, and plagioclase (El-Desoky, 2010; Abdel-Rahman, 2020). Sillimanite differs in color from white, yellowish-brown, grayish-green to dark gray (Fig. 6.13A). These

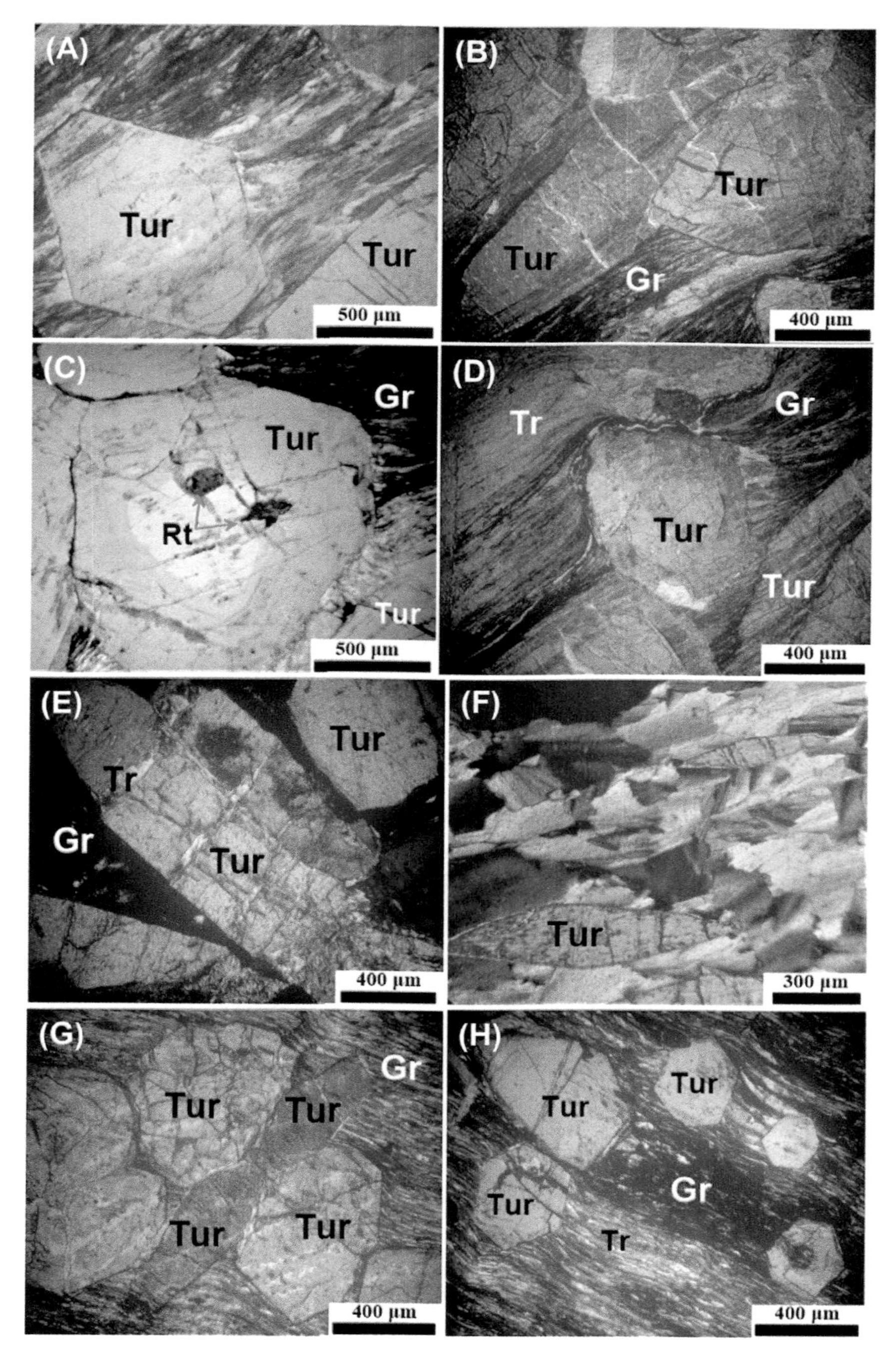

FIGURE 6.12 Photomicrograph showing (A) hexagonal tourmaline (Tur) porphyroblast and well-developed schistosity at wadi Nugrus area, Egypt (CN & TD); (B) well-developed dentations of tourmaline (Tur) crystals (CN & TD); (C) zonation in tourmaline (Tur) crystals associated graphite (Gr) and rutile (Rt) at wadi Nugrus, Egypt (PPL & TD); (D) wrapping on tourmaline (Tur) crystals (CN & TD); (E) tourmaline (Qtz) filling the tourmaline crystal (Tur; CN & TD); (F) toothed shape of tourmaline (Tur) crystal parallel to the schistosity (Tur; CN & TK); (G) pressure has done the teeth of well-developed tourmaline (Tur) crystals with graphite (Gr) (CN & TS); (H) Hexagonal tourmaline (Tur) crystals and kinking shape of tremolite (Tr), graphite (Gr), and opaque minerals (Op. Mi; CN & TS).

FIGURE 6.13 Photomicrograph showing (A) hand specimen of fibrous sillimanite aggregate from Jeseniky Czech. Photo: Zbyněk Buřival. https://mineralexpert.org/article/sillimanite-aluminosilicate-mineral-overview; (B) the structural contact between the foliated metagabbro and sillimanite-bearing schists at Hafafit area, Egypt; (C) sillimanite (Sil) crystals altered to muscovite (Ms) at wadi Allaqi, Egypt, C.N; (D) sillimanite aggregates of hair-like shape and fibrous bundles, parallel to the foliation in the sillimanite-bearing schists. C.N; (E) very coarse almandine crystals in garnet-bearing schists, Passo Rombo, Alto Adige, Italy; (F) well-developed garnet crystals (Grt) in the same direction of the foliation within garnet-bearing schists; (G) large porphyroblast of garnet (Grt) perforated by quartz (Qtz) and replaced along its cracks by iron oxides. Plain polarized light (PPL), × 2.5/0.08; (H) Garnet (Grt) porphyroblast and muscovite (Ms) infiltered by the later medium quartz crystals (Qtz) in micaschist. Val Lanterna area (Valmalenco), Italy. Crossed polarized light (XPL), 2 × (7 mm).

rocks are extending between the foliated metagabbros (Fig. 6.13B) and/or in between the foliated metagabbros and Hafafit psammitic gneisses (El-Desoky, 2010).

Sillimanite occurs often in sheaves of fine fibers with cross-fractures (Fig. 6.13C and D) or as aggregates of hairlike needles and fibrous bundles, such as Quartzforms' granoblastic aggregates as well as individual crystals. Plagioclase occurs as hypidiomorphic crystals, 0.3 mm long, oftentimes untwined, and highly altered and saussurite. Muscovite forms irregular streaks as well as small flakes, up to 0.06 mm long. Iron oxide forms allotriomorphic crystals, up to 0.2 mm long, frequently coating biotite and rarely titanite (El-Desoky, 2010).

6.4.1.4 Garnetiferous schists

Garnetiferous schists are predominant and related to amphibolite facies of medium-grade metamorphism and are to be considered of sedimentary origin. They show graded bedding, to be foliated, fine- to medium-grained, moderately hard to compact, change in color (from yellow, brownish-yellow to green), and containing variable amounts of garnet, mica, potash feldspar, and plagioclase (El-Desoky, 2010). Garnetiferous schists are crop out mainly in the eastern part of the Gabal Hafafit, South Eastern Desert, Egypt, as a low hilly to the moderately elevated country with gentle slopes.

The outer inclusions show that the garnet has well-developed crystals between quartz grains (Fig. 6.13E). Although the garnet is poikiloblastic, its grain boundaries show a very strong tendency to run parallel to the edges of a hexagon with the quartz and feldspar crystals, which represents a cross section through the rhombic dodecahedron that is a common crystal habit of members of the garnet group of minerals (Fig. 6.13F).

Petrographically, garnet-bearing schists are fine- to medium-grained, foliated, and characterized by porphyroblast texture (El-Desoky, 2010). Garnet occurs either as allotriomorphic porphyroblasts, or as fragmented crystals, up to 2.0 cm across almandine and spessartine composition. It is highly corroded and sieved by quartz blebs and replaced along its cracks by biotite and slightly altered to chlorite (Fig. 6.13G). Garnet occurs as pretectonic crystals that form shattered (Fig. 6.13H).

6.5 The nonfoliated rocks

6.5.1 Greisens

Trace and rare-earth-element-bearing greisens of the Eastern Desert Province of Egypt can be recorded in the following regions: Abu Dabbab, Igla, Nuweibi, Mueilha, Homrit Waggat, Um Naggat, Abu Rusheid, Wadi Zarieb, Ras Barud, and Wadi El-Miyah (El-Desoky, 2020). The **Abu Dabbab** greisen consists of a series of sheeted quartz and quartz greisen veins and zones-bearing cassiterite-rare-metal mineralization. Abu Dabbab greisenization displays an intensive greisenization, producing pocket-like bodies of lithionite granites, where fine disseminations of cassiterite and tantalite–columbite are recorded (Figs. 6.14 and 6.15). Three types of ore veins at Abu Dabbab are composed mainly either of quartz, or of quartz greisen with, rare-metal mineralization, topaz, and fluorite or of quartz–microcline–amazonite

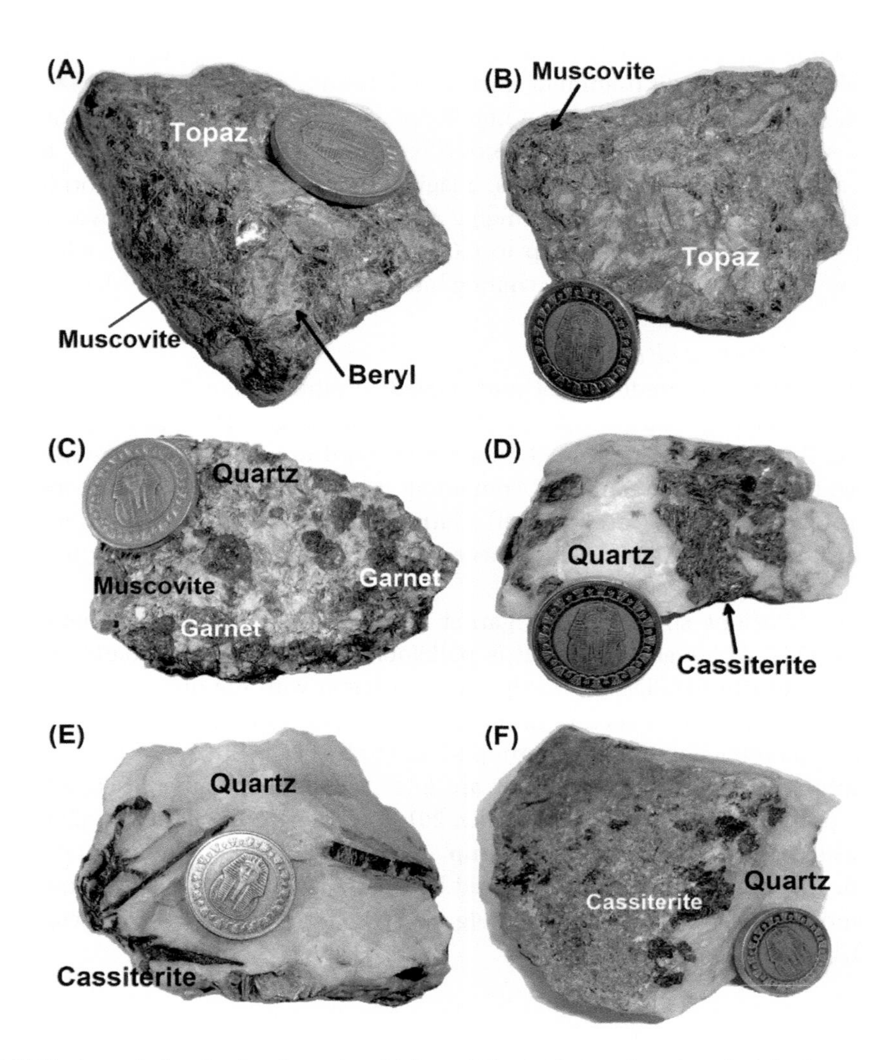

FIGURE 6.14 Photograph showing hand samples of Igla and Abu Dabbab greisen deposits (A) Igla muscovite-topaz-beryl greisen veins; (B) Abu Dabbab muscovite-topaz greisen veins; (C) Nuweibi muscovite-garnet-quartz greisen veins; (D–F) Abu Dabbab cassiterite-quartz greisen veins.

(Fig. 6.14). According to El-Desoky (2020), the quartz greisen veins are composed occasionally of muscovite, phlogopite, and cassiterite (Fig. 6.15).

The **Igla** greisen bodies are represented by several greisens-bearing tin–beryllium mineralizations (Figs. 6.14 and 6.15). The mineralized fluids are found as quartz stockworks and veins and also as greisen zones-bearing Sn, W, Nb–Ta, and Be elements. These deposits seem to be genetically related to the albitized granites (Mohamed and Bishara, 1998). The greisen samples bearing cassiterite, quartz, and phlogopite minerals come from the edges of fissures in the altered granites (Fig. 6.15). **Nuweibi** greisen is represented by albite granite

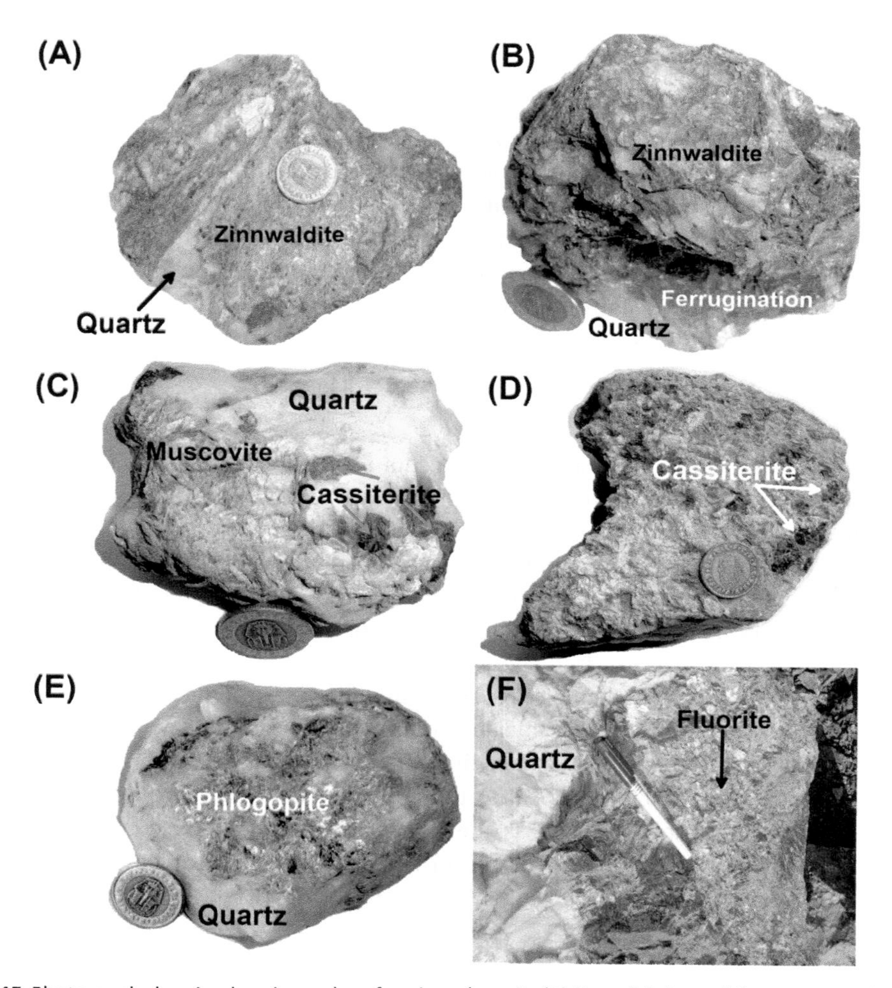

FIGURE 6.15 Photograph showing hand samples of greisen deposits (A) Nuweibi zinnwaldite-quartz greisen veins; (B) Nuweibi zinnwaldite-ferrugination-quartz greisen veins; (C & D) Abu Dabbab greisen-hosted Sn deposits; (E) Igla phlogopite-quartz greisen veins; (F) Abu Dabbab quartz-fluorite greisen veins.

masses and quartz–cassiterite veins. These veins display extremely erratic dissemination of cassiterite. Almandine and spessartine occur as red to brownish-red cubic euhedral crystals distributed in Nuweibi greisens. Remarkably, and unlike the Abu Dabbab granite, the Nuweibi intrusion does not contain Sn mineralization.

6.5.2 Chromite in serpentinites

The serpentinite forms N–S ridge covering about 40 km^2 encountered at El-Rubshi, Central Eastern Desert, Egypt. El-Rubshi serpentinite–talc carbonate range forms a mountainous ridge with sharply irregular peaks with dark black to pale violet colors (El-Desoky et al.,

2015). Petrographically, the serpentinites are mainly composed of serpentine minerals (antigorite, chrysotile as well as lizardite) associated with variable amounts of talc, carbonates, magnetite, chromite, and tremolite with relics of the parent olivine and pyroxene. Serpentine crystals replacing the parent olivine, particularly at the grain boundaries, form a mesh texture (Fig. 6.16A). Magnetite, of primary origin, is fine-grained with euhedral to subhedral form and is generally dispersed throughout the rocks. Talc forms minute colorless aggregates replacing antigorite and may occur as microveinlets (Fig. 6.16B). Huge veins of carbonate minerals (magnesite and calcite) disseminated the serpentinite rocks (Fig. 6.16C and D). Chromite occurs interstitially as rounded and fine-sized grains with a deep red to reddish-brown color (Fig. 6.16E; Abdel-Rahman, 2020).

6.5.3 Talc

Talc is a common metamorphic mineral that occurs in metamorphic belts, which are derived from ultramafic rocks, such as soapstone (a high-talc rock) and within metavolcanics and blueschist metamorphic terranes (Deer et al., 1992). The majority of talc occurrences in Egypt are derived from or hosted within the ultramafic rocks, mainly by serpentinite and mafic rocks as highly sheared metabasalts that are encountered in Umm Rilan, South Eastern Desert. Talc occurs as a secondary mineral formed by the alteration of magnesium carbonates (magnesite and dolomite) and magnesium and/or iron silicates such as serpentine (El-Desoky and Khalil, 2011). Umm Rilan area is of low-to-moderate relief, made up of two suites, metamorphic talc carbonate and talc serpentinite rocks of ophiolitic mélange (Fig. 6.16F and G), metavolcanics (metabasalts), and granitic rocks. Talc carbonate rocks contain euhedral cubes of goethite pseudomorphs most probably after pyrite or hematite (Fig. 6.16H); these goethite cubes are well developed and show colloform texture. Petrographically, talc is the most abundant mineral constituent that forms more than 90% of the rock volume. It occurs in coarse to fine platy or fibrous aggregates that often have a more or less parallel arrangement (Fig. 6.16I).

6.5.3.1 *Stratigraphy*

Stratigraphic studies, including sequence stratigraphy, are considered the most important steps when dealing with sedimentary ore deposits. This can help in sedimentary basin analysis and classify the facies changes relative to different depositional environments. Hence, it is possible to predict and understand mineralization associated with different sedimentary facies (e.g., red-beds, evaporites, organic-rich shales, and carbonates). Several factors can control the ending locality of sedimentary ore deposit, including the type of depositional environment, regional structural setting of the basin, and the geometry of sedimentary sequence. Tracing and mapping various stratigraphic units and sequences depend mainly on surface investigation and rock sampling to trace the boundaries of different sequences and judge the depositional environment. Also, this is verified by laboratory investigation of the collected rock samples. Additionally, the geometry and structure of a basin can be determined using the seismic stratigraphy method integrated with good data. Particularly, many examples of sedimentary-ore deposits can be provided worldwide. For, instance, in Egypt,

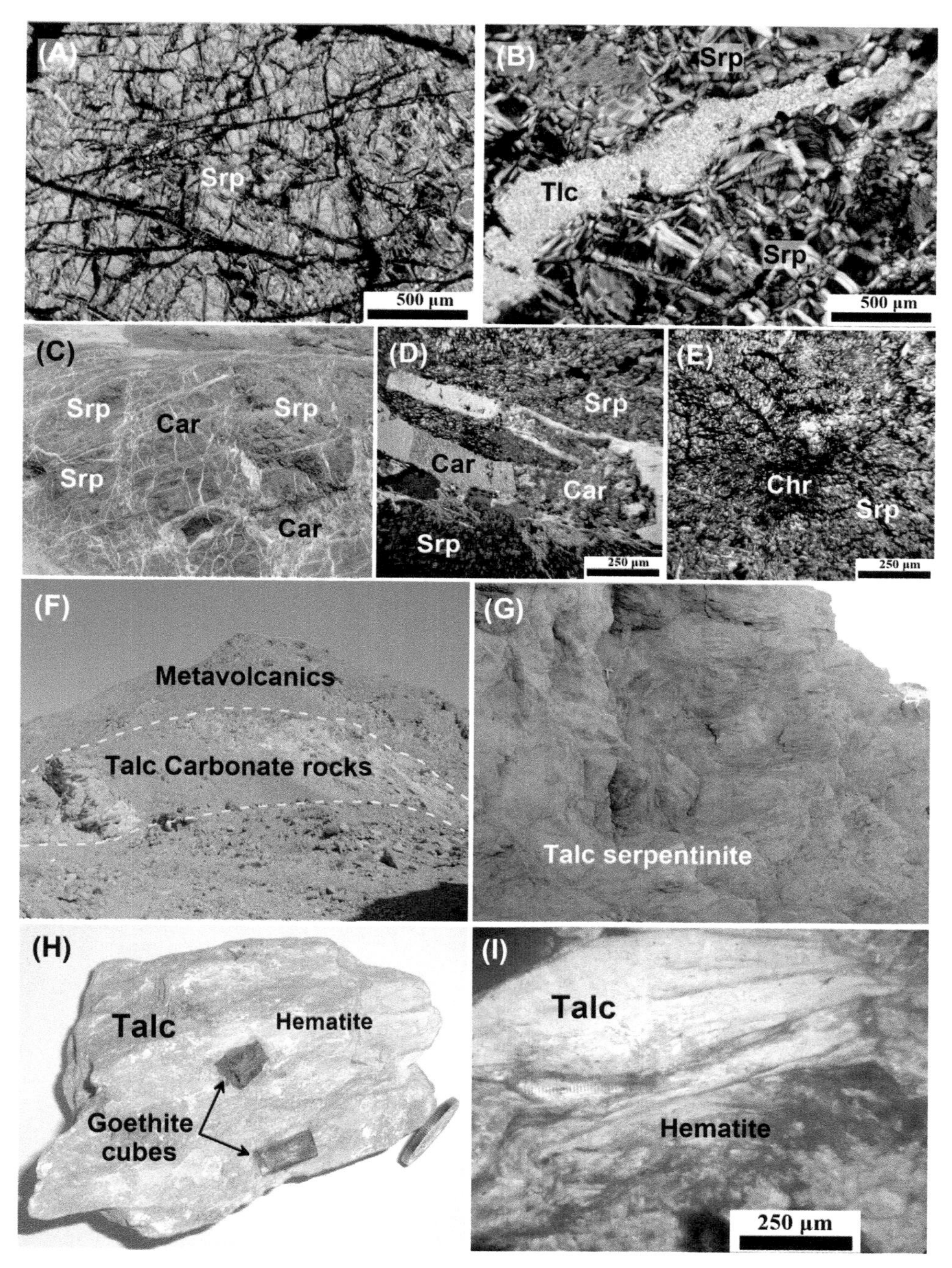

FIGURE 6.16 Photomicrograph showing (A) Serpentine minerals exhibits mesh texture and olivine grains is surrounded by fibrous serpentine and black magnetite. Crossed Polarized Light (XPL), 2 × (7 mm); (B) Talc microveinlets replacing antigorite and hourglass texture in the serpentinite matrix. XPL, Germany. 10 × (2 mm); (C) Carbonate veins (white color) in partially serpentinized peridotites, Ophiolite, Oman; (D) Two veins of carbonate within serpentinite, XPL, Wadi Allaqi, Egypt; (E) Chromite associated with secondary veins of carbonates in serpentinite, XPL, Wadi Allaqi, Egypt; (F) Talc-carbonate lens within the metavolcanics, Umm Rilan area, Egypt; (G) Talc serpentinite rocks of ophiolitic mélange, Umm Rilan area, Egypt; (H) Goethite cubes after pyrite within the talc-carbonate rocks; (I) Talc fibrous associated to hematite, C.N.

the Duwi Formation (Red Sea Coast) represents important phosphate-bearing strata; also, the Um Bogma Formation (Sinai) is an example of manganese-bearing strata. Many studies dealt with the origin and depositional environments as well as the digenetic factors controlling the hosting of the manganese in Sinai (e.g., Abou El-Anwar et al., 2017) and the phosphate in the Eastern Desert (e.g., Kora et al., 1994).

6.5.4 Structural mapping

6.5.4.1 Relation between geological structures and ore deposition

Worldwide, mineral deposits are controlled by different tectonic settings, including subduction-related settings, continental margins, strike–slip settings, collision-related settings, spreading centers, and extensional settings (Fig. 6.17). Additionally, the spatial distribution of the major mineral deposits all over the world is highly correlated to the major structural elements and plate boundaries (Fig. 6.17). For instance, copper can be formed accompanied by compressional

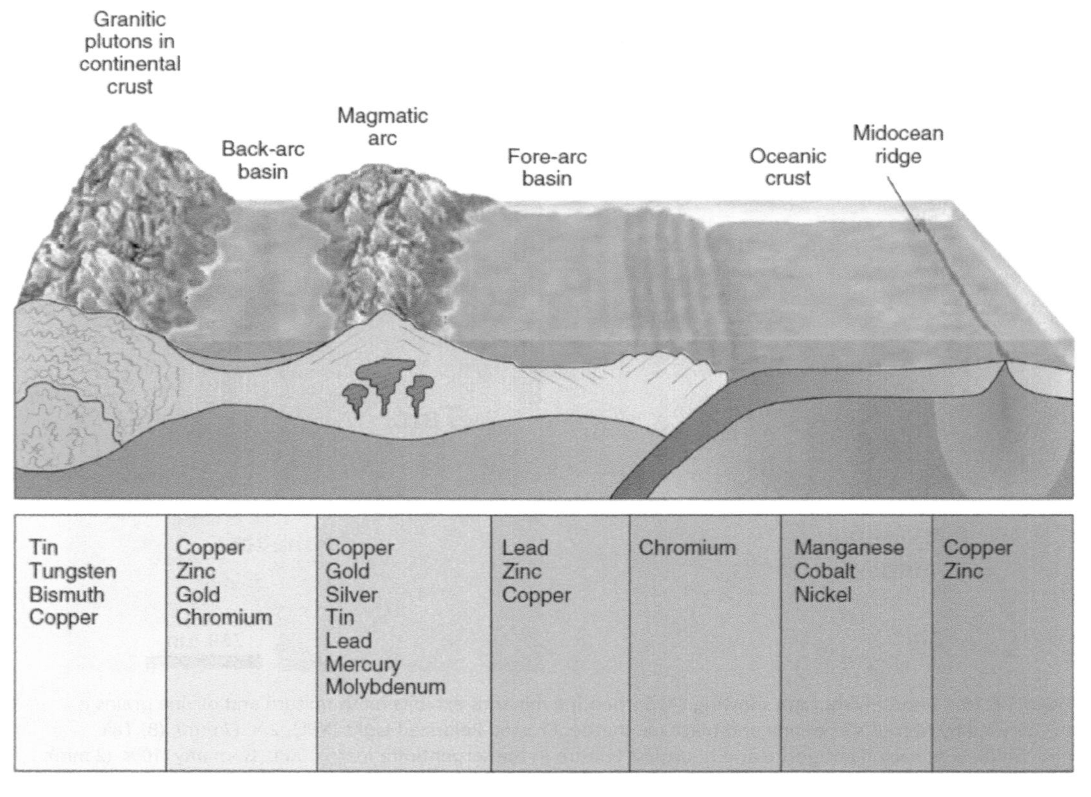

FIGURE 6.17 The role of plate tectonics controlling the distribution of ore deposits (after, Murck and Skinner, 2012). *From Murck, B., Skinner, B., 2012. Visualizing Geology. Wiley Visualizing.*

tectonics at magmatic arcs, back-arc, and fore-arc basins (e.g., Sun et al., 2018) as well as extensional tectonics through mid-ocean ridges (e.g., Wang et al., 2006). Also, gold is favorable at magmatic arcs and back-arc basins (e.g., Groves et al., 2018).

Particularly, for several decades, geologists registered many attempts to relate different types of mineralization and ore deposits to geological structures affecting the Earth's crust. In fact, geological structure plays a chief role in controlling ore deposits in different tectonic regimes (e.g., Blenkinsop and Doyle, 2014; Funedda et al., 2018; Qin et al., 2022). Almost, all hydrothermal deposits display a few degrees of structural control on mineralization. Such control is reflected through various types of geological structures, including fractures, folds and faults, and/or shear zones.

Faults and shear zones constitute types of the most famous structures controlling mineralization. They are closely related structures where both are strain localization structures. Also, both occupy displacement parallel to the walls, and both are likely to develop in width and length throughout displacement accumulation. Hence, the term shear zone can be simply defined as a tabular zone where strain is remarkably greater than in the surrounding rock. Commonly, a single shear zone is bordered by two boundaries (i.e., shear zone walls) that divide the main shear zone from its wall rock (Fig. 6.18A). Generally, shear zones can be categorized according to ductility and plasticity (Fig. 6.18B). Usually, field measurements deal with defining the thickness, extension, and displacement of the shear zone. Where markers are found around both walls of the shear zone, the displacement can be honestly measured. Also, the width and extension of a shear zone can be defined from field investigation or with the aid of remote sensing data regarding the scale of the shear zone. Shear zones enclose internal complex patterns of deformation second-order structures, including veins, dykes, and fracture systems. Such structures influence localizing ore deposits as it provides pathways for hydrothermal solutions. Also, fault zones and their associated fractures and fault breccias due to the grinding action of the rocks nearby to the fault plane increase the structural permeability of hydrothermal solutions (e.g., Grare et al., 2018).

The structure controlling ore bodies is commonly complex to observe and its relations to the regional structural framework are often absent. Folds and domal structures represent one of the challenge structures where mineralization can be with (e.g., Cugerone et al., 2018; Jacques et al., 2018). Cleavage and veins are planar structures that are usually abundant in folded metamorphic rocks. Also, they play a major role as indicators for mineralization zones and deformation processes (Fig. 6.19). Veins represent planar structures that can be described as extensional structures filled with minerals. Geometrically, a cleavage splits the fold relatively along the axial surface, near the hinge zone. It is necessary to measure and record geometric parameters of cleavages and veins, also, structural analysis using stereographic projection techniques is crucial. As a result, structural studies focusing on the interaction between mineralization and geometry of veins are vital. Also, determining locations where structures such as cleavage are secant, as well as interpreting textural associations between metamorphic minerals and mineralization, can guide to a well again considerate of the ore-body genesis.

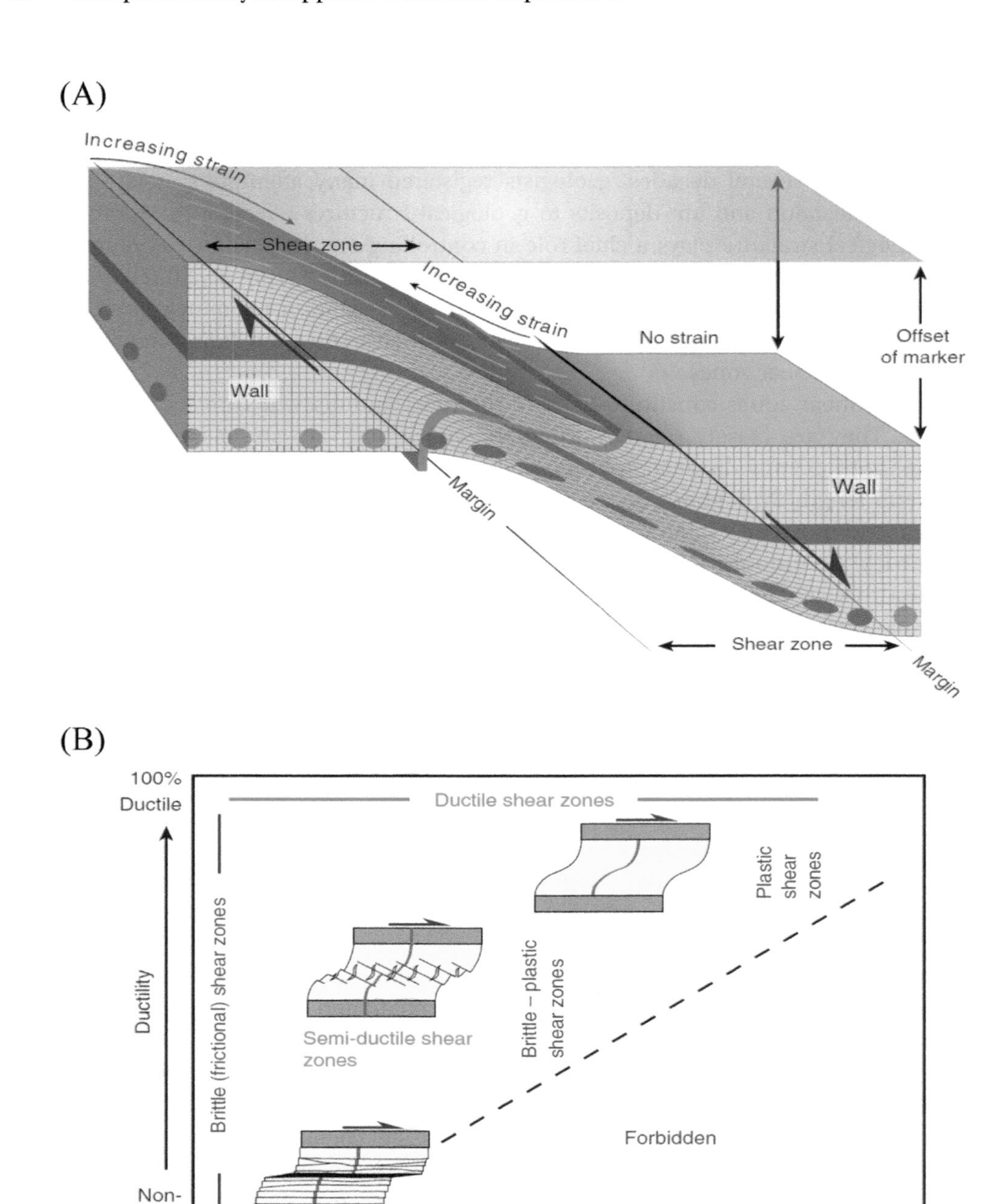

FIGURE 6.18 (A) Ideal shear zone deforming a grid with two planar markers and circular strain markers and (B) simple classification of shear zones based on the deformation mechanism and mesoscopic ductility (after, Fossen, 2010). *From Fossen, H., 2010. Structural Geology. Cambridge University Press.*

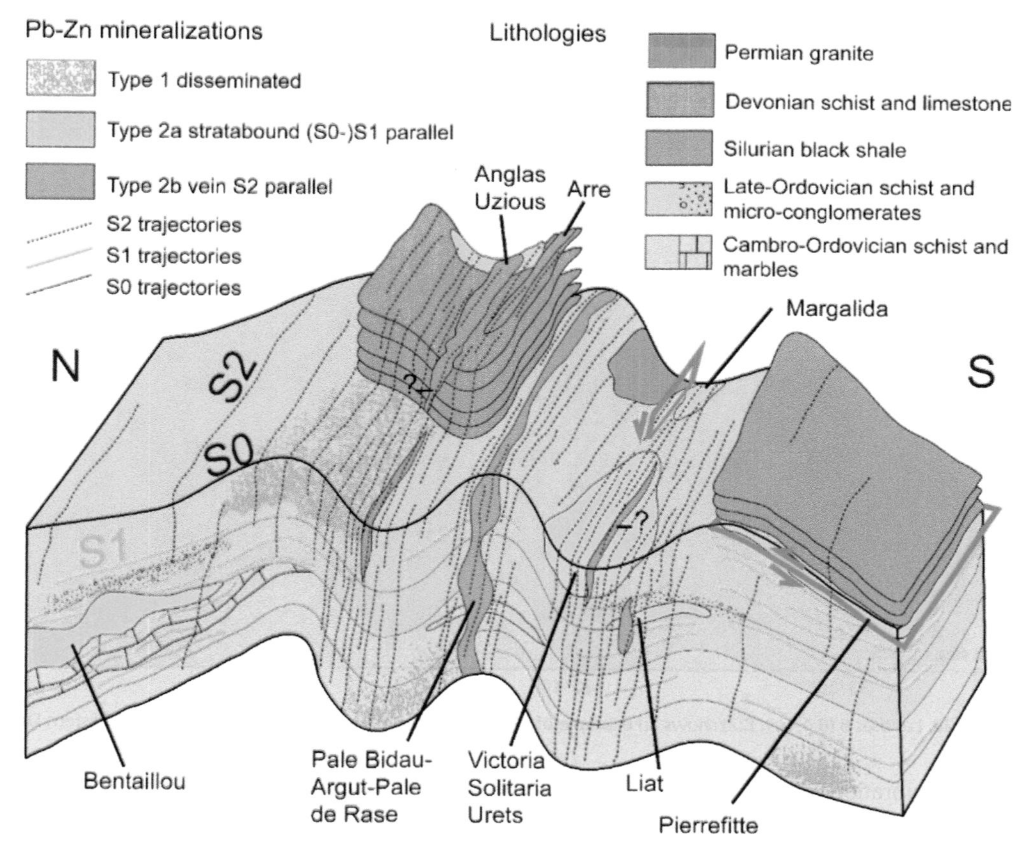

FIGURE 6.19 Schematic 3D sketch displaying the three main mineralization types of veins associated with folded rocks (after, Cugerone et al., 2018). *From Cugerone, A., Oliot, E., Chauvet, A., Gavaldà Bordes, J., Laurent, A., Le Goff, E., Cenki-Tok, B., 2018. Structural control on the formation of Pb–Zn deposits: an example from the Pyrenean axial zone. Minerals, 8(11), 489. https://doi.org/10.3390/min8110489.*

6.5.4.2 Mapping techniques of structural controlled mineralization

Exploring structural-controlled ore deposits is directly linked to delineating and mapping various geological structures via remote sensing data, field geology investigation, and geophysical data. Hence, defining the type, geometry, and timing of crustal deformation is a critical step for mineral exploration. Field data acquisition of structural geometric parameters is usually achieved through recording strike and dip of planar structures, including fault planes, fold limbs, and fractures as well as trend and plunge of linear structures. Also, the width and extension of fault zones are an important demand in structural mapping. The importance of geological structures comes from the deformation characteristics of each structural type and its controls on the nature of forming ore deposits. The nature and characteristics of the most common types of geological structures controlling ore deposits are explained in the next paragraphs. Remote sensing techniques have been widely used as an initial step for geological mapping. Particularity, structural delineation, and

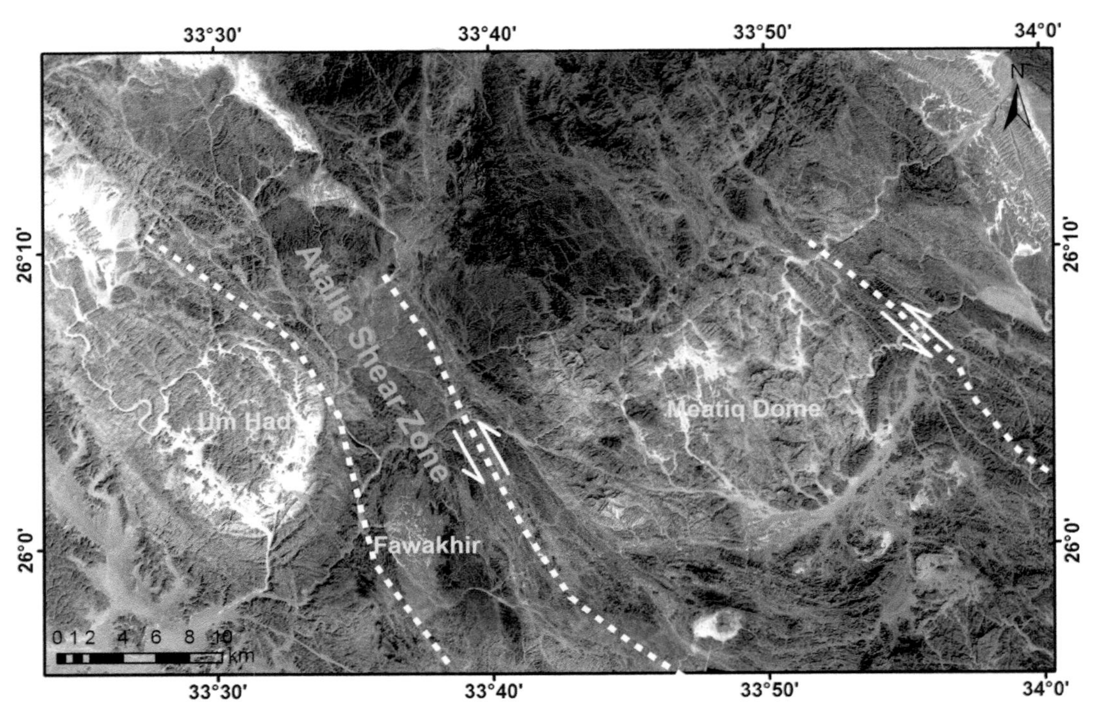

FIGURE 6.20 Landsat 8 (R:4, G:1, B:2) shows an example of a mineral-bearing shear zone from the Egyptian Eastern Desert.

mineral exploration were among the most prominent applications (e.g., Eldosouky et al., 2017). Multispectral and hyperspectral remote sensing sensors were used for geological applications, ranging from a few spectral bands to more than 100 contiguous bands, covering the visible to the shortwave infrared regions of the electromagnetic spectrum (Pour and Hashim, 2014). Passive and active remote sensing systems are widely used for mineral exploration for two applications: (1) extraction of faults and shear zones that localize ore deposits; and (2) distinguishing hydrothermally altered rocks by their spectral signatures (Fig. 6.20).

Generally, geophysical methods are usually used in interpreting geological structures because of their more spatial resolution. Particularly, geophysical potential data can provide an image of the major structural setting where deep faults and shear zones can be delineated (Faruwa et al., 2021; Zhang et al., 2022). Also, it gives an image of faults intersection, which is an important locality for mineral exploration (Fig. 6.21A). Nowadays, it represents a crucial step for defining mineralization zones according to the interpreted structural pattern. Different advanced processing methods provide tools for understanding both the regional distribution and the trends in geophysical properties to be related to near-surface and subsurface geology. The selection between different enhancement and transformation techniques depends on the geological question of interest. Processing procedures, including enhancement techniques, can provide a precise structural image of the subsurface (Eldosouky et al., 2022). The integration of remote sensing techniques, geophysical methods,

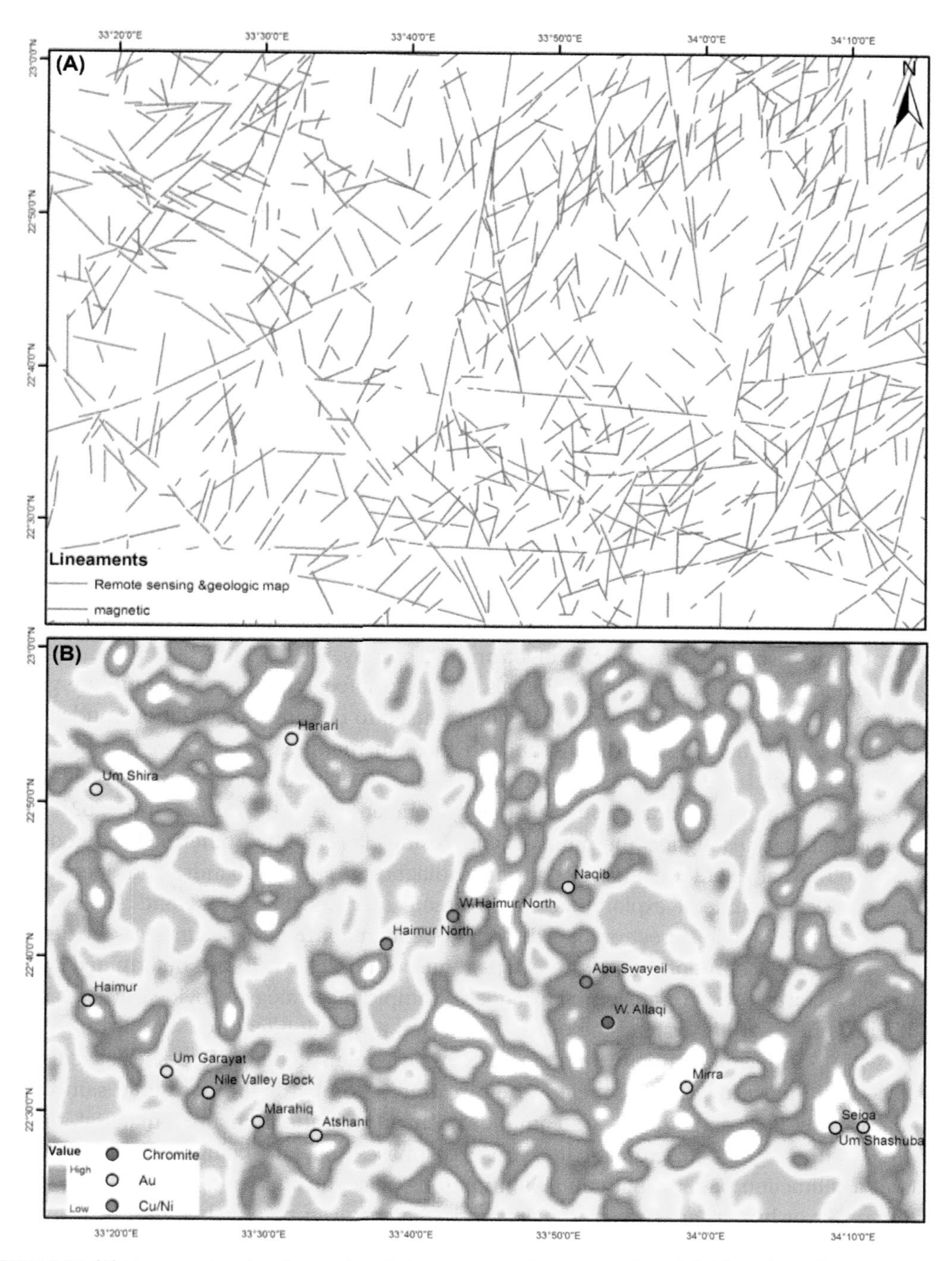

FIGURE 6.21 (A) Lineaments extraction and analysis remote sensing, magnetic and field geologic data at Wadi Allaqi area, Eastern Desert, Egypt. (B) Density map of lineaments of Wadi Allaqi overlain by known 277 sites of mineralization (after, Eldosouky et al., 2017).

and field-measured data is a very important step for determining alteration and/or mineralization zones (Fig. 6.21B).

Currently, development in mapping software and enhancement algorithms (e.g., Arc GIS) provides a platform for correlating and integrating multiple datasets. This is considered a final stage for delineating the optimum zones of mineralization occurrences. Also, machine-learning and statistical analysis techniques are widely used in the last decade, which, in turn, provide more knowledge and reliability about the constructed models. Hence, the uncertainties about mineral exploration can be reduced, which makes decision-makers more confident.

6.6 Laboratory analysis

6.6.1 X-ray diffraction analysis

X-ray diffraction (XRD) is a conventional technique to pinpoint minerals in a laboratory environment. It is an adaptable and nondestructive analytical technique to obtain detailed structural and phase information of materials, quickly. XRD is a significant device in mineralogy for identifying, quantifying, and characterizing minerals in complex mineral assemblages (Stanjek and Häusler, 2004; Ali et al., 2022). It is typically useful for detecting fine-grained minerals and mixtures or intergrowths of minerals. In many geologic investigations, XRD complements other mineralogical methods, including optical light microscopy, electron microprobe microscopy, and scanning electron microscopy.

Mineralogical identification of ore minerals and associated minerals, including fine-grained hydrothermal alteration minerals, provides evidence used to deduce the conditions under which ore deposits formed and the conditions under which, in many cases, they were subsequently altered (Mathieu, 2018). Mineralogical analysis by XRD is utilized along with remotely sensed data in several mineral exploration investigations (Pour et al., 2021; Yousefi et al., 2021, 2022). XRD is applicable to distinguish the minerals composing clay-rich, hydrothermally altered rocks. Many hydrothermal alteration minerals produce diagnostic spectral bands, and spectral data provide a basis for mineral exploration using remote sensing data (Hunt and Ashley, 1979). Analysis of remote imaging spectrometer data can openly map mineral occurrences by detecting diagnostic spectral absorption bands, the shape, and position of which are determined by individual mineral structures. A detailed knowledge of sample mineralogy, provided at least in part by XRD, is required to understand the observed spectral absorption features.

For X-ray powder diffraction, a micronized sample is constrained into a specialized holder and placed into the X-ray diffractometer. The instrument bombards the powdered sample with X-rays at varying angles. As the X-rays come into contact with the particulate material, they are diffracted by the crystal structure of the phase or phases that are being analyzed. The scan results in a diffraction pattern that contains numerous peaks or humps (Gräfe et al., 2014). The resulting peaks or humps are similar to a fingerprint, as they can be matched to specific mineral or crystalline phases. The interpretation of the diffraction pattern is a comparative method where the data are matched to an exhaustive list of reference patterns. By using this interpretation technique, it is possible to identify well over 10 different minerals or crystalline phases within an

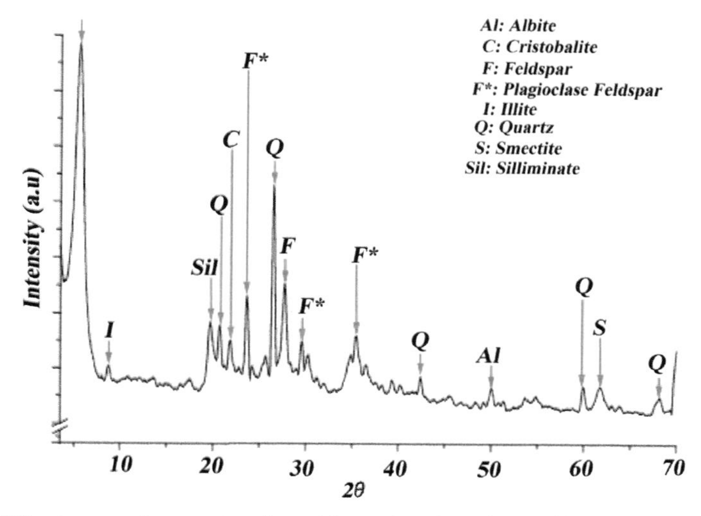

FIGURE 6.22 X-ray diffractogram of a sample collected from clay mineral assemblages.

unknown sample (Artioli, 2013; Ameh, 2019). Fig. 6.22 shows the XRD spectrum of a clay sample. A variety of minerals present in the sample were identified. For example, five Q (quartz) peaks are observed with varying peak positions and intensities, indicating a variety of atomic positions, the size, and the shape of the quartz unit cell.

6.6.2 Analytical spectral devices

Field measurements of surface reflectance are widely utilized for the identification of altered minerals and assemblages and verification of satellite remotely sensed results. Spectral data are characteristically acquired for small areas at ground level or from laboratory conditions in a variety of configurations using Analytical Spectral Devices (ASDs) field portable spectroradiometers (Deering, 1989; Milton and Goetz, 1997). The ASDs improve remote sensing field capabilities and accurately measure reflectance, transmittance, radiance, or irradiance in the full spectrum range of 350–2500 nm (UV/Vis/NIR/SWIR) with a spectral resolution of 10 nm. They deliver comprehensive spectral information of target minerals in the field or laboratory conditions.

Typically, the amount of energy reflected from a ground area or object of interest over different wavelengths is measured by the ASDs (Milton et al., 1995). Consequently, the measurements are transformed to spectral radiance values using equipment calibration factors (Peddle et al., 2001). Key spectral signatures derived from ASD data countenance the detection of iron minerals in the VNIR and clays, carbonates, micas, sulfates, and other minerals in the SWIR wavelengths (Kruse, 2012; Calvin and Pace, 2016). The USGS spectral library version 7 (https://speclab.cr.usgs.gov/spectral-lib.html), the ASTER spectral library version 1.2 (http://speclib.jpl.nasa.gov), and ECOSTRESS spectral library version 1.0 (http://speclib.jpl.nasa.gov) contain a comprehensive collection of spectra for a wide variety of materials especially minerals, rocks, lunar and terrestrial soils acquired by the ASDs (Baldridge et al., 2009; Kokaly et al., 2017; Meerdink et al., 2019).

6.7 Case studies

Title Case Study 1: Gold-mineralization of Wadi Hodein Belt, Egypt.

Audience:

Researchers and investors interested in prospecting for gold ore benefit from this study; several authors participated in this study as Mohamed Abd El-Wahed, Basem Zoheir, Amin Beiranvand Pour, and Samir Kamh.

Rational:

The main reasons for these activations:

1. Knowing the structural factors controlling the formation of gold mineralization in the study area
2. The link between the results of remote sensing analysis, field studies, and the structure data in the possibility of the presence of gold ore

Expected results and deliverables:

1. The Wadi Khashab shear zone hosts several occurrences of gold-bearing quartz veins that are commonly associated with sericitized and silicified rocks.
2. The most important conclusion of this study is that gold mineralization in the Wadi Hodein—Wadi Beitan shear belt was mostly confined to steeply dipping strike—slip shear zones in the marginal parts of the shear belt (Fig. 6.23).
3. The gold mineralization event was evidently epigenetic in the metamorphic rocks and was likely attributed to rejuvenated tectonism and the circulation of hot fluids during transpressional deformation.
4. The gold mineralization event was evidently epigenetic in the metamorphic rocks was clearly epigenetic, and it was most likely caused by revitalized tectonism and hot fluids circulation during transpressional deformation.
5. The results of image processing complying with field observations and structural analysis suggest the coincidence of shear zones, hydrothermal alteration, and crosscutting dikes in the study area.
6. In the South Eastern Desert the Wadi Hodein shear belt is a high-angle NW-oriented transcurrent shear zone. Imbricate thrust-bounded sheets and slices of serpentinized ultramafic rocks, amphibolite, and metabasalt are embedded in a strongly tectonized matrix of talc carbonate, schists, and metasiltstones in the Wadi Rahaba—Gabal Abu Dahr and Gabal Arais ophiolites (Fig. 6.24).
7. The concurrence of shear zones, hydrothermal alteration, and crosscutting dikes in the research area might be used as a model criterion in the search for new gold targets, according to the results of image processing in accordance with field observations and structural analysis.
8. Neoproterozoic rocks in Hodein shear belt include variably deformed ophiolitic mélange rocks, and island-arc metavolcanic and metavolcaniclastic successions.
9. Gold-bearing quartz veins intersect sheared metavolcanics, volcaniclastic metasediments, and gabbro—diorite complexes in the studied area.

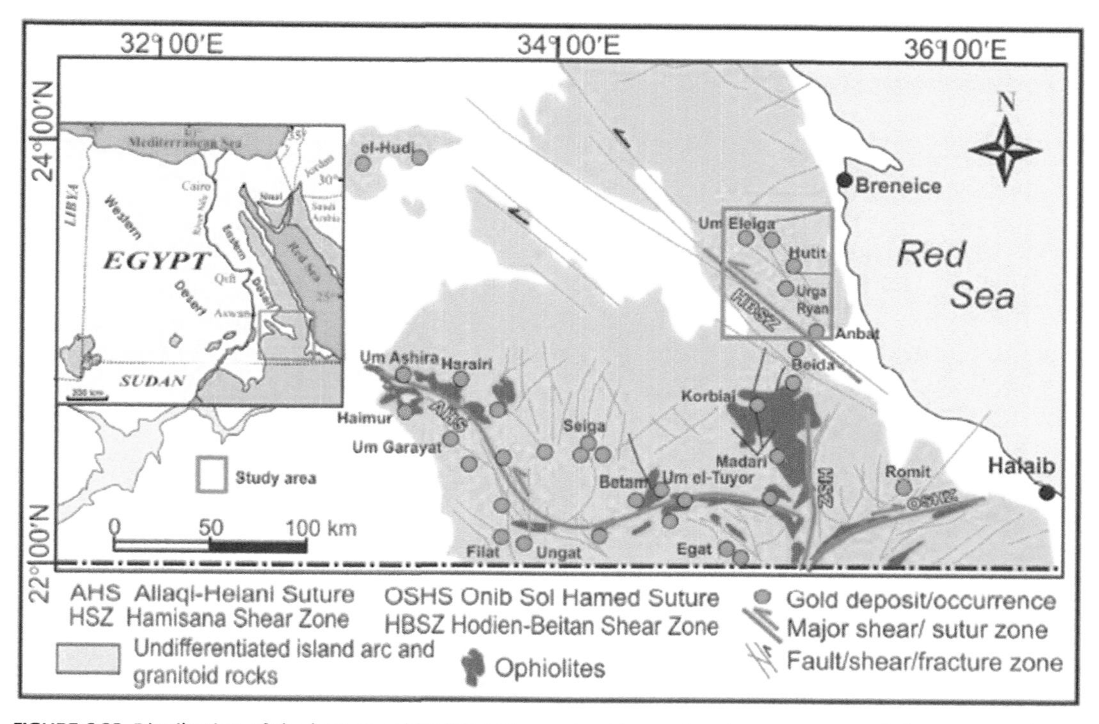

FIGURE 6.23 Distribution of the known gold mining sites and gold–quartz vein occurrences relative to the major fault/shear structures and ophiolitic masses in the South Eastern Desert terrane (**Zoheir et al., 2019**). *Modified after Zoheir, B., Emam, A., Abd El-Wahed, M., Soliman, N., 2019. Gold endowment in the evolution of the Allaqi-Heiani suture, Egypt: a synthesis of geological, structural, and space-borne imagery data. Ore Geol. Rev. 110, 102938. https://doi.org/10.1016/j.oregeorev.2019.102938. Inset shows a location map of the study.*

10. Mineralized quartz veins are constantly regulated by NW- and NNW-trending shear zones, and they are linked with widespread hydrothermal alteration.

El-Beida, Khashab, El-Anbat, Um Teneidab, Urga Ryan, Hutit, and Um Eleiga are all gold occurrences in the research region.

Gold-mineralized shear zones penetrate across deformed metavolcaniclastic and ophiolites rocks in and around granodioritic intrusions, attenuating the signal.

Albitization of the gabbroic host rocks is widespread in the Um Eleiga gold mine area.

The high strain zones are mostly connected with a variety of gold occurrences in the Wadi Hodein shear band.

The Urga Ryan gold deposit is located 3 km south of the Wadi Urga Ryan/Wadi Hutib intersection. Gneisses, volcaniclastic metasediments, and mafic to intermediate metavolcanic rocks underpin the area around Wadi Urga El-Ryan.

The serpentinite, mafic metavolcanics, volcaniclastic metasediments, ophiolitic metagabbros, syn-orogenic metagabbros, and postorogenic granites underpin the Hutit mine area.

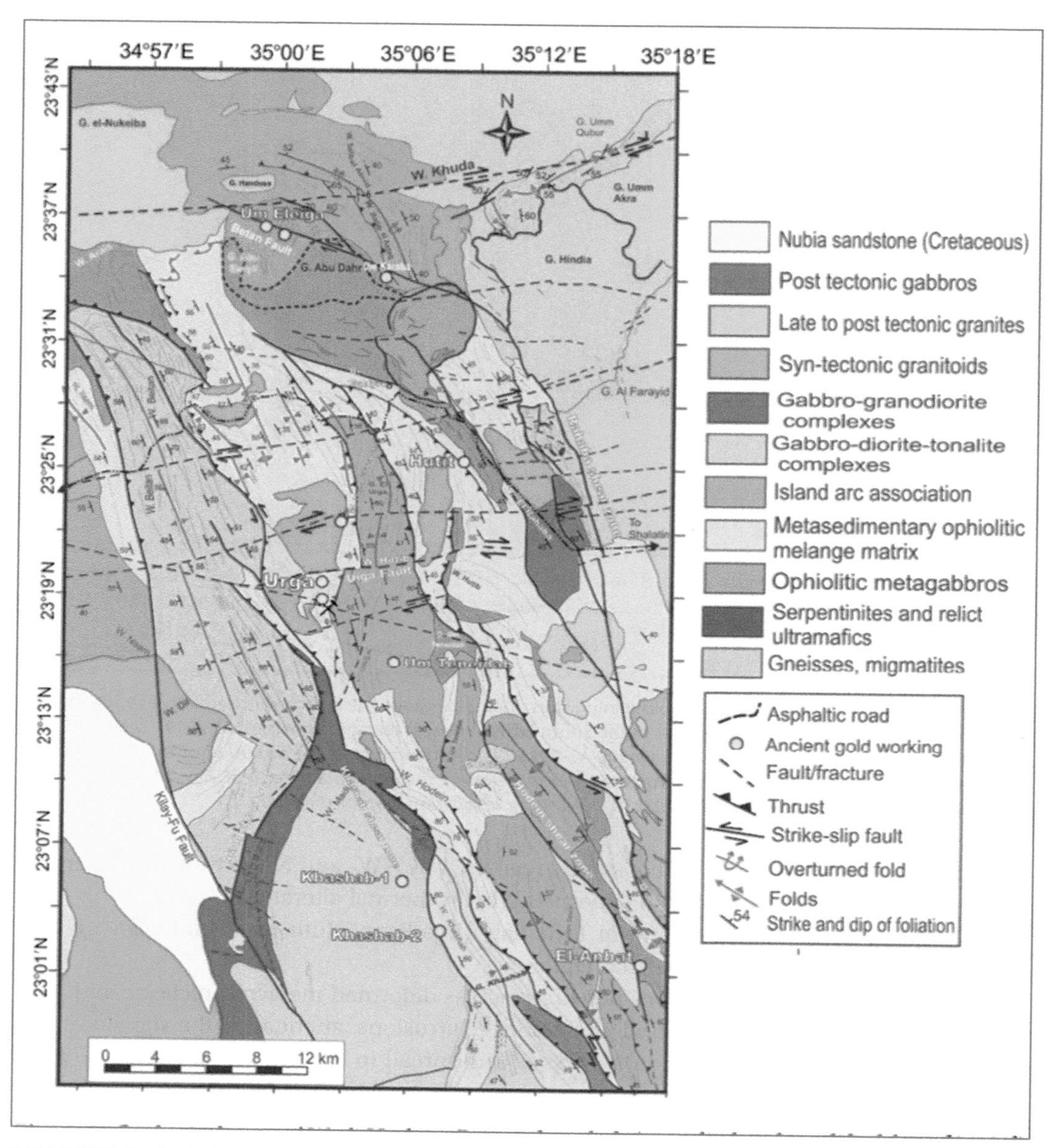

FIGURE 6.24 Geological map of Wadi Hodein–Beitan shear belt (**El Amawy et al., 2000**; **Zoheir, 2012**). *Complied and modified from El-Amawy, M.A., Wetait, M.A., Alfy, Shweel, A.S., 2000. Geology, geochemistry and structural evolution of Wadi Beida area, South Eastern Desert, Egypt. Egypt J. Geol. 44, 65–84; Zoheir, B.A., 2012. Lode-gold mineralization in convergent wrench structures: examples from South Eastern Desert, Egypt. J. Geochem. Explor. 114, 82–97. https://doi.org/10.1016/j.gexplo.2011.12.005.*

Results:

1. The Wadi Hodein–Wadi Beitan shear belt is thought to have a preferential distribution of gold mineralization.
2. The study's most noteworthy finding is that gold mineralization in the Wadi Hodein–Wadi Beitan shear band is primarily restricted to steeply dipping strike–slip shear zones at the shear belt's fringes.
3. So far, no gold-bearing quartz veins have been discovered in the center shear zone.
4. The coincidence of shear zones, alteration, and acid dikes in the research area is translated into highly prospective zones for new and perhaps important gold occurrences in a conceptual model for evaluating new and unexplored targets of gold–quartz veins in the region.

Title Case Study 2: Lithium-bearing pegmatite deposits in Dahongliutan area, China.

Audience:

Several researchers contributed to this work, including Yongbao Gao, Leon Bagas, Kan, Moushun Jin, Yuegao Liu, and Jiaxin Teng, who are all interested in exploring lithium-bearing pegmatite deposits.

Rational:

The following are the key reasons for these activations:

1. This work used a mix of geochemical methodologies, geological mapping, and high-resolution remotely sensed multispectral images to determine prospective areas of pegmatite-hosted Li deposits.
2. Our findings reveal that the Li mineralization in the Dahongliutan area is similar in age and genesis to the pegmatite-hosted deposits in the Jiajika area in western Sichuan province, implying that the Dahongliutan area is rich in pegmatite-hosted mineral deposits (Fig. 6.25).

Expected results and deliverables:

1. The value of rare metals used in rechargeable batteries, such as lithium and cobalt, has climbed as demand for them has grown.
2. Lithium compounds are widely employed in industries such as aerospace, chemical, pharmaceutical, and new energy.
3. Lithium has become an important commodity as a result of continuing discoveries in the research and development of rechargeable batteries.
4. Although many pegmatite veins and dykes in the vicinity of the deposit are anomalous in Li–Be–Ta–Nb, only 10 have been proven to be economically viable, with lengths ranging from 180 to 300 m and widths up to 40 m.
5. The process of using satellite photos for mineral exploration starts with data collection, which is often done in a geographic information system (GIS) context. This comprises exploration data, georeferenced geological maps, and preliminary interpretations of satellite pictures and orthorectified aerial photographs, among other things.

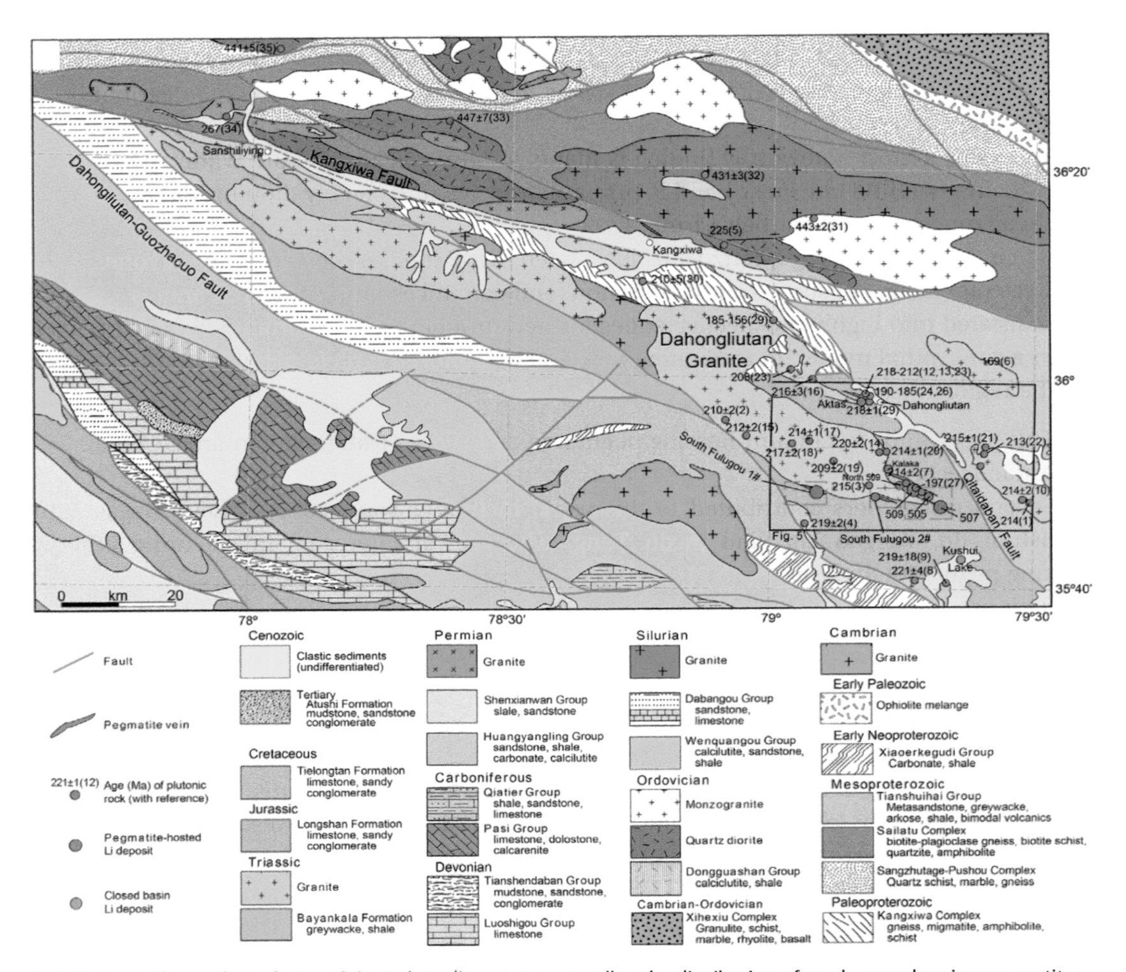

FIGURE 6.25 The geological map of the Dahongliutan area, as well as the distribution of spodumene-bearing pegmatites.

6. Lithium-bearing minerals can be found in closed-basin brines, granitic pegmatites, and associated granitic rocks that include spodumene ($LiAl(SiO_3)_2$) and other economically valuable minerals.
7. A pegmatite dyke swarm in NW China hosts the recently discovered Dahongliutan Li mineral deposits, which are also potential for Be, Rb, Nb, and Ta mineralization.
8. Landsat 5, Landsat 7, and ASTER are examples of low-resolution remote sensing satellites with high spectral resolution.

Results:

1. The main topic of this publication is remote sensing in conjunction with fieldwork and geochemistry. This resulted in the discovery of mineralized pegmatites at a high altitude in the Dahongliutan region.

2. Joints and shears within the Bayankala Formation and Dahongliutan Granite structurally limit the distribution of pegmatites in the area. Between tourmaline-bearing pegmatites near granites and feldspar-rich pegmatites and quartz veining outside the contact hornfels zone, spodumene-bearing pegmatite dykes and veins can be found. Spodumene-bearing pegmatite veins and dykes are associated with higher Be, Rb, Nb, and Ta assays.
3. Recognizing Li-bearing pegmatites in sedimentary layers is part of the exploratory method devised for discovering spodumene-bearing pegmatites in the difficult Bayankala Fold Belt. Following that, pegmatite veins were identified on high-resolution hyperspectral remote sensing pictures, as well as anomalies connected to pegmatite-type Li deposits using multispectral remote sensing.
4. The approach developed for the Dahongliutan area was successful in locating many important Li and Be deposits in the area.

References

Abdel-Rahman, A.M., 2020. Petrology and Geochemistry of The Basement Rocks and Their Suitability for Some Industrial Application Around Wadi Umm Ashira and Wadi Tilal Al Qulieb North Wadi Allaqi.

Abou El-Anwar, E.A., Mekky, H.S., Abd El Rahim, S.H., Aita, S.K., 2017. Mineralogical, geochemical characteristics and origin of Late Cretaceous phosphorite in Duwi Formation (Geble Duwi Mine), Red Sea region, Egypt. Egypt. J. Pet. 26 (1), 157–169. Available from: http://www.journals.elsevier.com/egyptian-journal-of-petroleum.10.1016/j.ejpe.2016.01.004.

Ahmed, A.H., Arai, S., 2002. Eastern Desert, Egypt. 9th Inter. Platinum Symposium 1–4 Platinum-Group Element Geochemistry in Podiform Chromitites and Associated Peridotites of the Precambrian Ophiolite.

Ali, A., Chiang, Y.W., Santos, R.M., 2022. X-ray diffraction techniques for mineral characterization: a review for engineers of the fundamentals, applications, and research directions. Minerals 12 (2). Available from: https://www.mdpi.com/2075-163X/12/2/205/pdf.10.3390/min12020205.

Ameh, E.S., 2019. A review of basic crystallography and x-ray diffraction applications. Int. J. Adv. Manuf. Technol. 105 (7–8), 3289–3302. Available from: https://doi.org/10.1007/s00170-019-04508-1.

Artioli, G., 2013. Science for the cultural heritage: the contribution of X-ray diffraction. Rendiconti Lincei 24 (S1), 55–62. Available from: https://doi.org/10.1007/s12210-012-0207-z.

Baldridge, A.M., Hook, S.J., Grove, C.I., Rivera, G., 2009. The ASTER spectral library version 2.0. Remote. Sens. Environ. 113 (4), 711–715. Available from: https://doi.org/10.1016/j.rse.2008.11.007.

Banks, D.A., Yardley, B.W.D., Campbell, A.R., Jarvis, K.E., 1994. REE composition of an aqueous magmatic fluid: a fluid inclusion study from the Capitan Pluton, New Mexico, U.S.A. Chem. Geol. 113 (3–4), 259–272. Available from: https://doi.org/10.1016/0009-2541(94)90070-1.

Bateman, A.M., 1950. Economic Mineral Deposits., 918.

Best, M.G., 2003. Igneous and Metamorphic Petrology. pp. 576–582.

Blenkinsop, T.G., Doyle, M.G., 2014. Structural controls on gold mineralization on the margin of the Yilgarn Craton, Albany-Fraser Orogen: the Tropicana deposit, Western Australia. J. Struct. Geol. 67, 189–204. Available from: http://www.sciencedirect.com/science/journal/01918141.10.1016/j.jsg.2014.01.013.

Brauhart, C.W., 2019. The role of geochemistry in understanding mineral systems. ASEG Ext. Abstr. 2019 (1), 1–5. Available from: https://doi.org/10.1080/22020586.2019.12072914.

Calvin, W.M., Pace, E.L., 2016. Mapping alteration in geothermal drill core using a field portable spectroradiometer. Geothermics 61, 12–23. Available from: http://www.elsevier.com/inca/publications/store/3/8/9/.10.1016/j.geothermics.2016.01.005.

Cooper, R.W., Langford, R.L., Pirajno, F., 1998. Mineral occurrences and exploration potential of the Bangemall Basin. Geol. Surv. West. Aust.

Cugerone, A., Oliot, E., Chauvet, A., Gavaldà Bordes, J., Laurent, A., Le Goff, E., et al., 2018. Structural control on the formation of Pb–Zn deposits: an example from the Pyrenean Axial Zone. Minerals 8 (11), 489. Available from: https://doi.org/10.3390/min8110489.

Deer, W.A., Howie, R.A., Zussman, J., 1992. An introduction to the rock-forming minerals, An Introduction to the Rock-Forming Minerals, second ed.

Deering, D.W., 1989. Theory and Applications of Optical Remote Sensing. Wiley Field Measurements of Bi-Directional Reflectance. pp. 14–65.

Duuring, P., 2020. Rare-Element Pegmatites: A Mineral Systems Analysis. Geological Survey of Western Australia.

Einaudi, M., Meinert, L.D., Newberry, R.J., 1981. Skarn deposits. Economic geology, 75th anniversary volume. Fac. Sci. Alex. Univ. 9, 135–146.

El-Amawy, M.A., Wetait, M.A., Alfy, Shweel, A.S., 2000. Geology, geochemistry and structural evolution of Wadi Beida area, South Eastern Desert, Egypt. Egypt J. Geol. 44, 65–84.

El-Awny, H.M., 2015. Petrology, geochemistry and petrogenesis of Wadi El-Gemal tourmaline deposits. South. East. Desert.

El-Desoky, H.M., 2010. Geochemical, Mineralogical and Beneficiation Studies on Some Schists Bearing Sillimanite Refractory Material.

El-Desoky, H.M., 2020. Geology, geochemistry and genetic implications of the greisen deposits. Central Eastern Desert, Egypt. Ann. Geol. Surv. Egypt. V. 37, 128–164.

El-Desoky, H.M., Khalil, A.E., 2011. Evolution of the talc-carbonate rocks in Umm Rilan Ophiolite, South Eastern Desert, Egypt: implication from mineralogy, petrography, geochemistry and p-t conditions. Al-Azhar Bull. Sci. 22 (Issue 2-D), 1–32. Available from: https://doi.org/10.21608/absb.2011.7908.

El-Desoky, H.M., Khalil, A.E., Salem, A.K.A., 2015. Ultramafic rocks in Gabal El-Rubshi, Central Eastern Desert, Egypt: petrography, mineral chemistry, and geochemistry constraints. Arab. J. Geosci. 8 (5), 2607–2631. Available from: http://www.springer.com/geosciences/journal/12517?cm_mmc = AD-_-enews-_-PSE1892-_-0.10.1007/s12517-014-1407-x.

El-Rahmany, M.M., Saleh, G.M., El-Desoky, H.M., El-Awny, H.M., 2015. Wadi El-Gemal tourmaline-bearing deposits, Southern Eastern Desert, Egypt: constraints on petrology and geochemistry. Int. J. Sci. Eng. Appl. Sci. (IJSEAS) 2395–3470.

Eldosouky, A.M., Abdelkareem, M., Elkhateeb, S.O., 2017. Integration of remote sensing and aeromagnetic data for mapping structural features and hydrothermal alteration zones in Wadi Allaqi area, South Eastern Desert of Egypt. J. Afr. Earth Sci. 130, 28–37. Available from: http://www.sciencedirect.com/science/journal/1464343X.10.1016/j.jafrearsci.2017.03.006.

Eldosouky, A.M., Pham, L.T., Henaish, A., 2022. High precision structural mapping using edge filters of potential field and remote sensing data: a case study from Wadi Umm Ghalqa area, South Eastern Desert, Egypt. Egypt. J. Remote. Sens. Space Sci. 25 (2), 501–513. Available from: https://doi.org/10.1016/j.ejrs.2022.03.001.

Evans, A.M., 1993. Ore Geology and Industrial Minerals: An Introduction. Blackwell Science.

Faruwa, A.R., Qian, W., Akinsunmade, A., Akingboye, A.S., Dusabemariya, C., 2021. Aeromagnetic and remote sensing characterization of structural elements influencing iron ore deposits and other mineralization in Kabba, southwestern Nigeria. Adv. Space Res. 68 (8), 3302–3313. Available from: https://doi.org/10.1016/j.asr.2021.06.024.

Flint, R.F., Sanders, J.E., Rodgers, J., 1960. Diamictite, a substitute term for symmictite. Geol. Soc. Am. Bull. 71 (12), 1809. Available from: https://doi.org/10.1130/0016-7606(1960)71[1809:DASTFS]2.0.CO;2.

Fossen, H., 2010. Structural Geology. Cambridge University Press.

Funedda, A., Naitza, S., Buttau, C., Cocco, F., Dini, A., 2018. Structural controls of ore mineralization in a polydeformed basement: field examples from the Variscan Baccu Locci Shear Zone (SE Sardinia, Italy). Minerals 8 (10), 456. Available from: https://doi.org/10.3390/min8100456.

Gabelman, J.W., 1976. Classifications of Strata-Bound Ore Deposits. Elsevier BV, pp. 79–110. Available from: https://doi.org/10.1016/b978-0-444-41401-4.50007-1.

Gandhi, S.M., Sarkar, B.C., 2016. Essentials of Mineral Exploration and Evaluation. Elsevier Inc, pp. 1–406. Available from: http://www.sciencedirect.com/science/book/9780128053294.

Gill, R., 2011. Igneous Rocks and Processes: A Practical Guide. John Wiley & Sons.

Grare, A., Lacombe, O., Mercadier, J., Benedicto, A., Guilcher, M., Trave, A., et al., 2018. Fault zone evolution and development of a structural and hydrological barrier: the Quartz Breccia in the Kiggavik Area (Nunavut, Canada) and its control on Uranium Mineralization. Minerals 8 (8), 319. Available from: https://doi.org/10.3390/min8080319.

Groves, D.I., Santosh, M., Goldfarb, R.J., Zhang, L., 2018. Structural geometry of orogenic gold deposits: implications for exploration of world-class and giant deposits. Geosci. Front. 9 (4), 1163–1177. Available from: https://www.sciencedirect.com/journal/geoscience-frontiers.10.1016/j.gsf.2018.01.006.

Gräfe, M., Klauber, C., Gan, B., Tappero, R.V., 2014. Synchrotron X-ray microdiffraction (μXRD) in minerals and environmental research. Powder Diffr. 128 (11). Available from: http://journals.cambridge.org/action/displayBackIssues?jid = PDJ.10.1017/S0885715614001031.

Guilbert, J.M., Park, C.F., 1986. The Geology of Ore Deposits.

Haldar, S.K., 2020. Introduction to Mineralogy and Petrology. Elsevier.

Hanor, J.S., 1979. Geochemistry of Hydrothermal Ore Deposits. The Sedimentary Genesis of Hydrothermal Fluids. John Wiley & Sons. pp. 137–168.

Hawkes, H.E., Webb, J.S., 1963. Geochemistry in mineral exploration. Soil. Sci. 95 (4), 283. Available from: https://doi.org/10.1097/00010694-196304000-00016.

Hawkes, L., 1957. Some aspects of the progress in geology in the last fifty years. I. Quarterly. J. Geol. Soc. 113 (1–4), 309–321. Available from: https://doi.org/10.1144/GSL.JGS.1957.113.01-04.13.

Hefferan, K., Brien, J., 2010. Earth Materials. John Wiley & Sons.

Herrington, R., 2011. Geological features and genetic models of mineral deposits. In: Darling, P. (Ed.), SME Mining Engineering Handbook. Society for Mining, Metallurgy, and Exploration, pp. 83–104.

Hunt, G.R., Ashley, R.P., 1979. Spectra of altered rocks in the visible and near infrared. Econ. Geol. 74 (7), 1613–1629. Available from: https://doi.org/10.2113/gsecongeo.74.7.1613.

Hussein, A.A.A., El Sharkawi, M.A., 2017. Mineral deposits. The Geology of Egypt. CRC Press, Egypt, pp. 511–566. Available from: http://www.tandfebooks.com/doi/book/10.1201/9780203736678.10.1201/9780203736678.

Jacques, D., Vieira, R., Muchez, P., Sintubin, M., 2018. Transpressional folding and associated cross-fold jointing controlling the geometry of post-orogenic vein-type W-Sn mineralization: examples from Minas da Panasqueira, Portugal. Mineralium Deposita 53 (2), 171–194. Available from: http://link.springer.de/link/service/journals/00126/index.htm.10.1007/s00126-017-0728-6.

Knox-Robinson, C.M., Wyborn, L.A.I., 1997. Towards a holistic exploration strategy: using geographic information systems as a tool to enhance exploration. Aust. J. Earth Sci. 44 (4), 453–463. Available from: https://doi.org/10.1080/08120099708728326.

Kokaly, R.F., Clark, R.N., Swayze, G.A., Livo, K.E., Hoefen, T.M., Pearson, N.C., et al., 2017. Geological Survey Data Series 1035 (61), 7. Available from: https://doi.org/10.3133/ds1035.

Kora, M., El Shahat, A., Abu Shabana, M., 1994. Lithostratigraphy of the manganese-bearing Um Bogma formation, west-central Sinai, Egypt. J. Afr. Earth Sci. 18 (2), 151–162. Available from: https://doi.org/10.1016/0899-5362(94)90027-2.

Kruse, F.A., 2012. Mapping surface mineralogy using imaging spectrometry. Geomorphology 137 (1), 41–56. Available from: https://doi.org/10.1016/j.geomorph.2010.09.032.

Macdonald, E.H., 1983. Alluvial Mining: The Geology, Technology and Economics of Placers. p. 508.

Mathieu, L., 2018. Quantifying hydrothermal alteration: a review of methods. Geosciences 8 (7), 245. Available from: https://doi.org/10.3390/geosciences8070245.

McQueen, K.G., 2005. Ore deposit types and their primary expressions. Regolith Expression of Australia Ore System: A Compilation of Exploration Case Histories with Conceptual Dispersion, Process and Exploration Models. CRC-LEME.

McQueen, K.G., 2009. Ore Deposit Types and their Primary Expressions. 14.

Meerdink, S.K., Hook, S.J., Roberts, D.A., Abbott, E.A., 2019. The ECOSTRESS Spectral Library Version 1.0. Remote Sensing of Environment. 230. Available from: http://www.elsevier.com/inca/publications/store/5/0/5/7/3/3.10.1016/j.rse.2019.05.015.

Milton, E.J., Goetz, A.F.H., 1997. Seventh International Symposium on Physical Measurements and Signatures in Remote Sensing. Atmospheric Influences on Field Spectrometry: Observed Relationships between Spectral Irradiance and the Variance in Spectral Reflectance 1, pp. 109–114.

Milton, E.J., Rollin, E.M., Emery, D.R., 1995. Advances in Environmental Remote Sensing. Advances in Field Spectroscopy. Wiley, pp. 9–32.

Misra, K.C., 2000. Understanding Mineral Deposits. p. 845.

Mohamed, M.A., Bishara, W.W., 1998. Fluid inclusions study of Sn-W mineralization at Igla area, central Eastern Desert, Egypt. Egypt. J. Geol. 207–220.

Murck, B., Skinner, B., 2012. Visualizing Geology. Wiley Visualizing.

Naldrett, A.J., Gruenewaldt, G., 1989. The association of the PGE with chromitite in layered intrusions and ophiolite complexes. Econ. Geol. 84, 180–218.

Park, J., 1906. A Text-Book of Mining Geology for the Use of Mining Students and Miners. FB &c Ltd, p. 219.

Peddle, D.R., White, H.P., Soffer, R.J., Miller, J.R., LeDrew, E.F., 2001. Reflectance processing of remote sensing spectroradiometer data. Comput. Geosci. 27 (2), 203–213. Available from: https://doi.org/10.1016/s0098-3004(00)00096-0.

Perkins, D, 2020. Mineralogy textbook. University of North Dakota, USA. Available from: https://opengeology.org/Mineralogy/.

Pohl, W.L., 2011. Economic geology: principles and practice, Metals, Minerals, Coal and Hydrocarbons – Introduction to Formation and Sustainable Exploitation of Mineral Deposits, 663.

Pour, A.B., Hashim, M., 2014. ASTER, ALI and hyperion sensors data for lithological mapping and ore minerals exploration. SpringerPlus 3 (1), 1–19. Available from: http://www.springerplus.com/archive.10.1186/2193-1801-3-130.

Pour, A.B., Sekandari, M., Rahmani, O., Crispini, L., Läufer, A., Park, Y., et al., 2021. Identification of phyllosilicates in the Antarctic environment using aster satellite data: case study from the mesa range, Campbell and Priestley glaciers, northern Victoria land. Remote. Sens. 13 (1), 1–37. Available from: https://www.mdpi.com/2072-4292/13/1/38/pdf.10.3390/rs13010038.

Qin, Q., Zhong, L., Zhong, K., He, Z., Yan, Z., Dewaele, S., et al., 2022. Structural setting of the Narusongduo Pb-Zn ore deposit in the Gangdese belt, central Tibet. Ore Geol. Rev. 143. Available from: http://www.sciencedirect.com/science/journal/01691368.10.1016/j.oregeorev.2022.104748.

Revuelta, M.B., 2017. Mineral Resources From Exploration to Sustainability Assessment. Springer Textbooks in Earth Sciences. doi: 10.1007/978-3-319-58760-8.

Ridley, J., 2013. Ore Deposit Geology. Cambridge University Press.

Saleh, G.M., 1997. The Potentiality of Uranium Occurrences in Wadi Nugrus Area, South Eastern Desert, Egypt, 15.

Shi, R., Griffin, W.L., O'Reilly, S.Y., Huang, Q., Zhang, X., Liu, D., et al., 2012. Melt/mantle mixing produces podiform chromite deposits in ophiolites: implications of Re-Os systematics in the Dongqiao Neo-tethyan ophiolite, northern Tibet. Gondwana Res. 21 (1), 194–206. Available from: https://doi.org/10.1016/j.gr.2011.05.011.

Smirnov, V.I., 1976. Geology of Mineral Deposits., 520.

Stanjek, H., Häusler, W., 2004. Basics of X-Ray Diffraction. Hyperfine Interact 154 (1–4), 107–119. Available from: https://doi.org/10.1023/B:HYPE.0000032028.60546.38.

Stevens, R., 2010. Mineral Exploration and Mining Essentials., 322.

Sun, T., Xu, Y., Yu, X., Liu, W., Li, R., Hu, Z., et al., 2018. Structural controls on copper mineralization in the Tongling Ore District, Eastern China: evidence from spatial analysis. Minerals 8 (6), 254. Available from: https://doi.org/10.3390/min8060254.

Thayer, T.P., 1970. Chromite segregations as petrogenetic indicators. Geol. Soc. S. Afr. Spec. Publ. 1, 380–390.

Wang, Q., Xu, J.F., Jian, P., Bao, Z.W., Zhao, Z.H., Li, C.F., et al., 2006. Petrogenesis of adakitic porphyries in an extensional tectonic setting, Dexing, South China: Implications for the genesis of porphyry copper mineralization. J. Petrol. 47 (1), 119–144. Available from: https://doi.org/10.1093/petrology/egi070.

Wells, J.H., 1969. Placer Examination, Principles and Practice. Dept. Int. Bureau of Land Management.

Wyborn, L.A.I., Heinrich, C., Jaques, A.L., 1994. Proceedings of the Australasian Institute of Mining and Metallurgy Annual Conference. Australian Proterozoic Mineral Systems: Essential Ingredients and Mappable Criteria. pp. 109–115.

Yardley, B.W.D., Cleverley, J.S., 2015. The role of metamorphic fluids in the formation of ore deposits. Geol. Soc. Lond. Spec. Publ. 393 (1), 117–134. Available from: https://doi.org/10.1144/sp393.5.

Yousefi, M., Tabatabaei, S.H., Rikhtehgaran, R., Pour, A.B., Pradhan, B., 2021. Application of Dirichlet process and support vector machine techniques for mapping alteration zones associated with porphyry copper deposit using aster remote sensing imagery. Minerals 11 (11). Available from: https://www.mdpi.com/2075-163X/11/11/1235/pdf.10.3390/min11111235.

Yousefi, M., Tabatabaei, S.H., Rikhtehgaran, R., Pour, A.B., Pradhan, B., 2022. Detection of alteration zones using the Dirichlet process Stick-Breaking model-based clustering algorithm to hyperion data: the case study of Kuh-Panj porphyry copper deposits, Southern Iran. Geocarto Int. Available from: http://www.tandfonline.com/toc/tgei20/current.10.1080/10106049.2022.2025917.

Zhang, P., Yu, C., Zeng, X., Tan, S., Lu, C., 2022. Ore-controlling structures of sandstone-hosted uranium deposit in the southwestern Ordos Basin: revealed from seismic and gravity data. Ore Geol. Rev. 140, 104590. Available from: https://doi.org/10.1016/j.oregeorev.2021.104590.

Zhou, M.F., Robinson, P.T., Malpas, J., Li, Z., 1996. Podiform chromitites in the Luobusa ophiolite (Southern Tibet): implications for melt-rock interaction and chromite segregation in the upper mantle. J. Petrol. 37 (1), 3–21. Available from: http://petrology.oxfordjournals.org/.10.1093/petrology/37.1.3.

Zimmerle, W., 1973. Fossil heavy mineral concentrations. Geol. Rundsch. 62 (2), 536–548. Available from: https://doi.org/10.1007/BF01840114.

Zoheir, B.A., 2012. Lode-gold mineralization in convergent wrench structures: examples from South Eastern Desert, Egypt. J. Geochem. Explor. 114, 82–97. Available from: https://doi.org/10.1016/j.gexplo.2011.12.005.

Zoheir, B., Emam, A., Abd El-Wahed, M., Soliman, N., 2019. Gold endowment in the evolution of the Allaqi-Heiani suture, Egypt: a synthesis of geological, structural, and space-borne imagery data. Ore Geol. Rev. 110, 102938. Available from: https://doi.org/10.1016/j.oregeorev.2019.102938.

7

Machine learning for analysis of geo-exploration data

Amin Beiranvand Pour[1], Jeff Harris[2], Renguang Zuo[3]

[1]*INSTITUTE OF OCEANOGRAPHY AND ENVIRONMENT (INOS), UNIVERSITY MALAYSIA TERENGGANU (UMT), KUALA NERUS, TERENGGANU, MALAYSIA* [2]*MINERAL EXPLORATION RESEARCH CENTRE, HARQUAIL SCHOOL OF EARTH SCIENCES, LAURENTIAN UNIVERSITY, SUDBURY, ON, CANADA* [3]*STATE KEY LABORATORY OF GEOLOGICAL PROCESSES AND MINERAL RESOURCES, CHINA UNIVERSITY OF GEOSCIENCES, WUHAN, P.R. CHINA*

7.1 Introduction

Modern-day exploration programs are faced with multiple challenges. First, there is an information overload given the availability of multiple datasets, requiring sophisticated tools for handling big datasets. False positives (FP) also pose certain challenges to exploration programs; anomalous contents of zinc in surficial geochemical data, for example, are not necessarily ore-related anomalies (cf. Levinson, 1974), nor are high magnetic anomalies necessarily associated with ore-bearing igneous complexes. Added to these are weak ore-related geochemical and geophysical anomalies of covered and blind mineral deposits (e.g., Zuo, 2014). In addition, to fully exploit the potential of legacy geoscientific data, one should translate the datasets into modern, digital formats. Most legacy bedrock geological maps available in hardcopy formats, for example, cannot be directly used in mineral prospectivity mapping (MPM), and manual transformation of these maps into digital formats is a time-consuming and labor-intensive process. All the previous challenges have inclined many researchers to exploit machine-learning algorithms (MLAs) for manipulating, processing, as well as translating legacy data into digital formats (Zuo, 2020; Lawley et al., 2022).

Generally, MLAs can be categorized into supervised and unsupervised algorithms. What constitutes the difference between these two categories is the use of labeled samples—samples tagged with some known features that are employed for calibrating and validating supervised algorithms. Supervised algorithms can be applied to address classification and regression problems. Some algorithms are unique to classification; there are, however, many algorithms that could be applied to both classification and regression. Predictive modeling of mineral prospectivity (also known as MPM) is a prime example of regression problems addressed by supervised algorithms. Unsupervised algorithms are commonly employed for clustering and dimensionality reduction of geoscientific data. Common applications of

Geospatial Analysis Applied to Mineral Exploration. DOI: https://doi.org/10.1016/B978-0-323-95608-6.00007-X

clustering have been demonstrated for geochemical data (Templ et al., 2008). Dimensionality reduction intends to reduce the complexity of datasets with numerous variables and has been applied to multispectral and hyperspectral remotely sensed datasets (Pour and Hashim, 2012) as well as geochemical data (Reimann et al., 2002). This chapter reviews the applications of MLAs in the geospatial analysis of geoscientific data for mineral exploration, followed by discussions on the limitations of these techniques.

7.2 Supervised algorithms

As described in Section 7.1, two common applications of supervised techniques, classification and regression, have been applied to the geospatial analysis of mineral exploration datasets. Notwithstanding the differences between these problems, applying supervised algorithms generally follows a framework summarized in Fig. 7.1. As per this figure, this procedure starts by developing two matrices—the matrix of labeled and unlabeled samples.

For predictive remote bedrock mapping (e.g., Harris et al., 2011), as an example of a classification procedure, one may pick some samples with predetermined, known bedrocks to develop a matrix of labeled samples. In a matrix of labeled samples, samples are usually represented by rows, while variables are represented by columns. The matrix constitutes the values of predictor variables (i.e., independent variables) and a target (i.e., dependent) variable, i.e., predetermined known bedrocks for some samples, in sample locations.

For MPM, as an example of a regression procedure, developing the matrix of labeled samples is more complicated. First, two types of locations should be marked for developing this matrix: positive and negative sites. Mineralized zones are usually deemed positive sites in MPM; in regional-scale studies, these are the location of mineral deposits/occurrences. Some

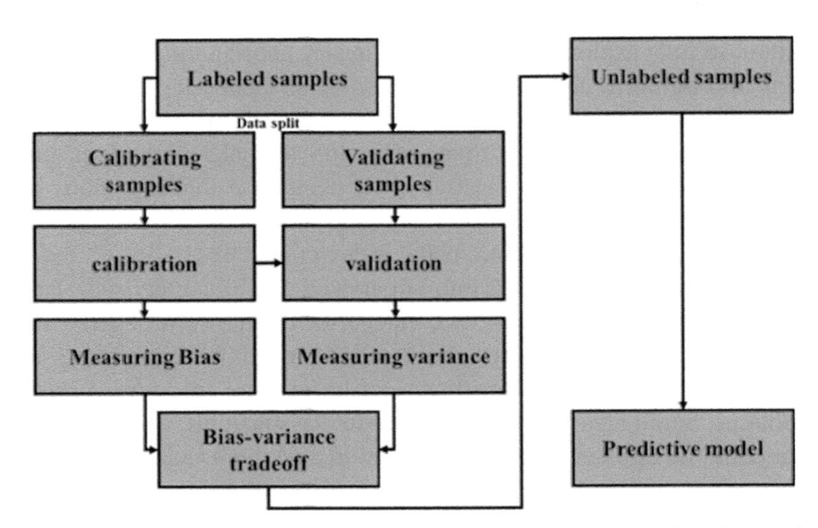

FIGURE 7.1 The general procedure employed for using supervised machine-learning algorithms, including MPM. *MPM*, Mineral prospectivity mapping.

criteria have been suggested for selecting negative sites in MPM (Nykänen et al., 2015). Foremost among these criteria is to ensure that the negative sites follow a random spatial pattern. This is because mineral deposits usually demonstrate a clustered spatial distribution (Carranza, 2009), and the spatial distribution of negative sites should be different from that of positive sites. In addition, to help maintain a data balance in the matrix of labeled samples, it has been recommended to use an equal number of positive and negative sites (Porwal et al., 2003). Furthermore, some researchers suggest selecting negative sites far from deposit locations (Carranza, 2009). This is to ensure that negative sites do not bear similarities in terms of geochemical and geophysical anomalies with positive sites. This is legit as some hydrothermal deposits, such as porphyry deposits (Sinclair, 2007), usually represent large halos of alteration. The matrix of labeled samples is then generated by assigning the values of predictor variables to positive and negative sites. Each predictor variable is considered an independent variable in the matrix, whereas a dependent variable (i.e., the target variable) is further defined by assigning the values of 1 and 0 to positive and negative sites, respectively.

Labeled samples are divided into calibrating and validating sets, which are used for the training and verification of supervised algorithms, respectively. Random splitting and k-fold cross-validation are among the methods used for dividing the labeled samples into two categories (Refaeilzadeh et al., 2009). In random splitting, usually a higher percentage of randomly selected labeled samples (usually 70%–80%) are assigned to the calibration category, leaving the remnant samples for validation (Refaeilzadeh et al., 2009). The k-fold cross-validation procedure is an iterative procedure, starting with randomly splitting the labeled samples into k groups. Each group is then considered once as the validation set, while the rest of the groups are fed into the algorithm for training. This is followed by measuring the performance of the algorithm using this training and validation set. The process is iterated k times until each group is separately considered the validation set. The average performance of the different iterations is then considered the algorithm's general performance (Refaeilzadeh et al., 2009). Selecting the number of folds depends on the number of labeled samples available, whereas fivefold and tenfold cross-validations have been commonly used for MPM (e.g., Parsa et al., 2022a,b). For a smaller number of labeled samples, k may be set to the total number of samples. This procedure is then referred to as the leave-one-out approach, which is also a common procedure in MPM given the scarcity of mineralized locations (e.g., Parsa et al., 2018a).

Bias and variance refer to the ability of a model to match the calibrating and validating samples, respectively. Receiver operating characteristic (ROC: Swets, 1988) curves, plotting the values of false-positive rate (FPR), also known as 1 − specificity, in the horizontal axis against the values of true-positive rate (TPR), also known as sensitivity, in the vertical axis, help measure the performance of a model. A prerequisite to measuring FPR and TPR is calculating the values of FP, true positives (TP), true negatives (TN), and false negatives (FN) for different threshold values. In the context of MPM, these quantities are defined by a contingency matrix as demonstrated in Fig. 7.2 (Parsa et al., 2018b). FPR and TPR are then estimated by FP/(FP + TN) and TP/(TP + FN), respectively. Once these values are quantified, the performance of a model is estimated by measuring the area under the curve (AUC). An

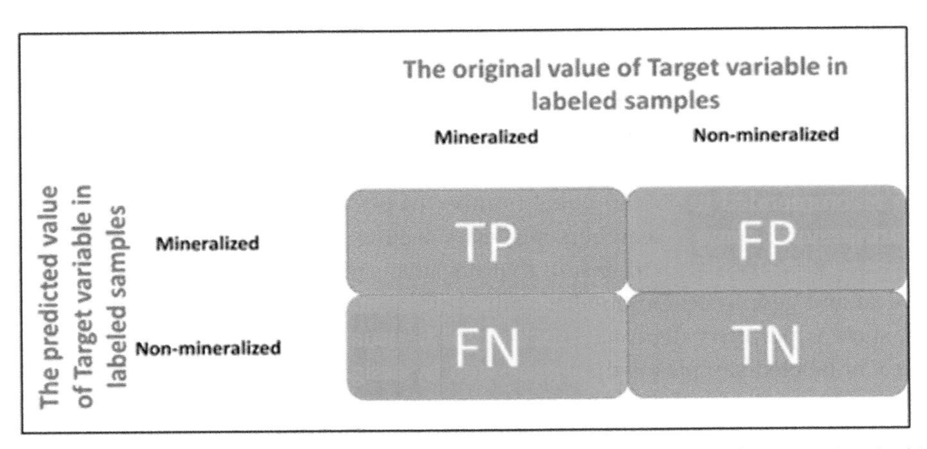

FIGURE 7.2 The contingency table is used for defining the values of TP, FP, FN, and TN for measuring the bias and variance of predictive models. *FN*, False negatives; *FP*, false positives; *TN*, true negatives; *TP*, true positives.

AUC value of 1 determines an ideal performance, whereas a random variable usually demonstrates an AUC value of ~0.5. In other words, the higher the AUC value, the better the performance of a model (Swets, 1988).

AUC values measured for calibrating and validating sets determine the values of bias and variance for a model. Higher values of variance versus low values of bias are indicators of overfitting in the model, which is further discussed in the upcoming paragraphs. If a model satisfies the bias–variance tradeoff test, it can be applied to the matrix of unlabeled samples for predictive modeling. For generating the matrix of unlabeled samples a grid of equal-sized cells is defined throughout the area being investigated. The extent of the area being investigated, the data used, and user preferences exert control over the size of cells (Carranza, 2009). In the case of using geochemical data, sample density helps determine a cell size (Zuo, 2012), as determined by Hengl (2006).

Independent variables are among the commonalities between labeled and unlabeled matrices, yet a dependent target variable is only defined for labeled samples. Perhaps the most important difference between these matrices is determined by their unbalanced size; that is, unlabeled samples largely outnumber labeled samples. The paucity of mineralized zones eventually leads to the small size of labeled samples, posing problems to the generalization of predictive models (Li et al., 2021). In other words, a majority of MLAs cannot be adequately trained with a small-sized matrix of labeled samples. Models developed with a small set of labeled data usually fit the calibration category, yet they fail to adequately correlate with the validation set. This is in fact an alert, showing that the predictive models derived from these data are probably unreliable.

Research into this problem suggests generating additional labeled samples for calibrating and validating supervised algorithms (Li et al., 2021). However, given the fact that mineralization is a rare geological phenomenon, it is intrinsically hard to define additional positive sites. One may define additional mineralized zones using the results of drilling surveys, which is especially useful while analyzing a small area. In regional-scale studies, buffering of mineral deposit location (Prado et al., 2020) and geospatial data augmentation methodologies (Li et al., 2021) have been used. These solutions assume that neighboring cells share similar geophysical and geochemical signatures to mineral deposits. This assumption, however, only makes sense when the mineral deposits of the type being investigated are marked with large halos of alteration and mineralization. For example, a window-based data augmentation procedure has been applied to generate additional positive sites in a terrain hosting supergiant porphyry copper deposits (Parsa et al., 2022a). Using augmented samples helps modulate the effects of data imbalance, thereby controlling the effects of overfitting in machine-learning-based MPM (Li et al., 2021; Parsa et al., 2022a). Geologically constrained data augmentation tools have been successfully coupled to many supervised algorithms, including convolutional neural networks (Li et al., 2021), deep group method of data handling neural networks (Parsa et al., 2022a), and extreme gradient boosting (Parsa, 2021), for MPM. Nevertheless, in most cases, the thickness of the sedimentary cover, the morphology of mineral deposits, and other contributing factors prevent the use of data augmentation tools. In these scenarios, some algorithms, such as regularized regression techniques (e.g., Parsa et al., 2022a,b), can be applied to reduce the effect of overfitting in MPM. The case studies presented in this chapter further explain the use of such algorithms. It is also imperative to note that some researchers have used positive and unlabeled algorithms to account for the data imbalance problem in MPM (Xiong and Zuo, 2021). These algorithms generally do not require negative sites, making them ideal for MPM.

An additional challenge faced by machine-learning-based MPM is the fact that given the criteria outlined for selecting negative samples, multiple sets of random negative sites can be selected for modeling mineral prospectivity. However, each set results in a different prospectivity model, questioning the reliability of some models. Research into this problem suggests selecting different sets of negative samples and developing a model with individual sets, followed by model averaging for generating the final prospectivity model (Zuo and Wang, 2020). Other researchers suggest using bootstrapping and taking subsamples from the original labeled samples instead of developing different sets of random negative samples (Parsa and Carranza, 2021). Although both approaches basically follow the same rationale, the latter appears to be less time-consuming. The latter also helps quantify a type of stochastic uncertainty linked to the discrepancies between mineral deposits used as positive sites.

A wide variety of supervised algorithms have been applied to mineral exploration, especially to MPM, making it next to impossible for the authors to delve into the details of these algorithms. Considering MPM as the focal point of this section, three categories of algorithms exceptionally work well for developing predictive models despite the intrinsic challenges mentioned earlier. These are bagging (e.g., Harris et al., 2015), boosting (e.g., Parsa, 2021), and regularized regression (e.g., Parsa et al., 2022a,b). Bagging and boosting are ensemble learning algorithms and are based on the idea of using multiple simple classifiers or regressors and combining their results (Dong et al., 2020). The difference between these two categories is that the latter employs a progressive learning algorithm, meaning that the classifier/regressor adjusts itself based on the errors of the previous classifier/regressor in the

Table 7.1 A brief literature review of studies employing supervised algorithms in mineral prospectivity mapping.

Categories	Algorithms	Examples
Bagging	Random forest	Harris et al. (2015); Carranza and Laborte (2015); Harris et al. (2022)
Boosting	AdaBoost	Brandmeier et al. (2020)
	XGBoost	Parsa (2021)
Regularized regression	LASSO, ridge regression, and Elastic Net	Parsa et al. (2022a,b)
Other common classifiers	Naïve Bayes	Parsa and Carranza (2021)
	Support vector machine	Zuo and Carranza (2011)
Deep learning and ANNs	Convolutional neural networks	Li et al. (2021)
	GMDH-type neural networks	Parsa et al. (2022a)

ANN, Artificial neural networks; *GMDH*, group method of data handling neural networks.

ensemble. The former, however, integrates the results of individual classifiers/regressors in the ensemble. Both bagging and boosting help reduce the effect of overfitting in predictive modeling (Dong et al., 2020). Many studies demonstrate that bagging and boosting usually outperform other algorithms in MPM (e.g., Harris et al., 2015). Likewise, regularized regression techniques also resist the effects of overfitting in MPM (e.g., Parsa et al., 2022a,b). The idea behind these algorithms is to use a penalty to prevent the regression from depending too much on the calibrating set of labeled samples. Although the limited number of labeled samples poses a serious challenge to the application of supervised algorithms, bagging and regularized regression have proven effective while dealing with a small number of labeled samples. In addition to these categories, supervised deep learning algorithms, such as convolutional neural networks (e.g., Zhang et al., 2021), work effectively for MPM. However, these algorithms usually need to be supplemented with additional labeled samples either with data augmentation or other procedures and, thus, have limitations for being applied to many styles of mineral deposits. Table 7.1 briefly reviews some supervised algorithms that have been applied to MPM.

From a user perspective, these algorithms either require too many or only a few parameters. Parameter tuning is an extremely important step in using MLAs, and optimization of hyperparameters usually requires extra effort. Although a majority of studies have used a trial-and-error procedure for hyperparameter optimization of these algorithms, some studies (e.g., Parsa et al., 2022a) have employed algorithm optimization procedures to optimize their parameter.

7.3 Examples of using supervised algorithms in mineral exploration

This section presents a case study centered on the use of supervised algorithms for MPM. This case study was selected to be presented herein as they highlight prime examples of challenges faced by machine-learning-based MPM, namely, unbalanced data and uncertainty.

7.3.1 Predictive modeling of volcanic-hosted massive sulfide deposits using supervised algorithms

In the first case study a geospatial database comprising geochemical, geophysical, and geological data is exploited to model the prospectivity of volcanic-hosted massive sulfide (VHMS) deposits in the Bathurst Mining Camp of New Brunswick, Canada. This area is extensively covered by glacial sediments (Parkhill and Doiron, 2003), posing challenges to the effectiveness of most geochemical and geological surveys conducted in the area. In addition, the morphology of these deposits does not warrant the use of geologically constrained data augmentation (cf. Lentz, 1999). These are all tabular ore bodies mostly overlain by an iron formation. These were historically significant producers of Pb, Zn, and Cu as prime commodities as well as Ag as a byproduct (McCutcheon et al., 2003). As presented in the bedrock map of this area (Fig. 7.3), VHMS deposits of the camp strictly occur at the uppermost contacts of the Nepisiguit Falls Formation underlying the Flat Landing Brook Formation (Lentz, 1999), both belonging to the Ordovician Tetagouche group.

The legacy geochemical, geological, and geochemical datasets were used to define a suite of predictor (independent) variables. Using these datasets, 11 independent variables describing the commonalities between VHMS mineral deposits of the study area were defined (Table 7.2). Despite the rationality employed for defining these variables, most of them are not strongly associated with the VHMS deposits, as presented by their low AUC values (see Parsa et al., 2022a,b). This is rooted in the glacial cover of the area, diluting and dislocating the concentration of elements in glacial sediment media.

Given the AUC values of these variables, the MPM of this area is faced with poor predictor variables. This problem becomes even more complicated considering the data imbalance problem, which is intrinsic to MPM. Also, despite the commonalities between the VHMS deposits employed as positive labeled sites in this study, there are discrepancies in their geochemical and geophysical signatures. Therefore a decent algorithm was required to account for all the previous challenges. An ensemble regularization technique was employed for addressing these challenges. This technique uses a bootstrapping technique for taking 10 subsamples from the original set of labeled samples. Each subsample is fed into the ridge regression model (Ahrens et al., 2020)—a regularization model that helps prevent the effects of unbalanced data in MPM. For each subset a random splitting was applied to assign 70% of the samples for calibration leaving the rest of the samples for validation. Care was taken to account for data imbalance in calibration and validation sets, meaning that an equal number of positive and negative sites were considered for each set. This procedure was somehow similar to tenfold cross-validation; however, each time a random population was selected for feeding the algorithms. For hyperparameter optimization a trial-and-error procedure was used.

Model averaging is used for generating the final predictive model, and the stochastic uncertainty is measured by calculating the standard deviation of predictive values derived per iteration. A detailed account of this hybrid framework can be found in Parsa et al. (2022a,b). The results of using this methodology are summarized in Fig. 7.4. This predictive model is marked by low bias (0.033) and low variance (0.076) and is visually correlated with

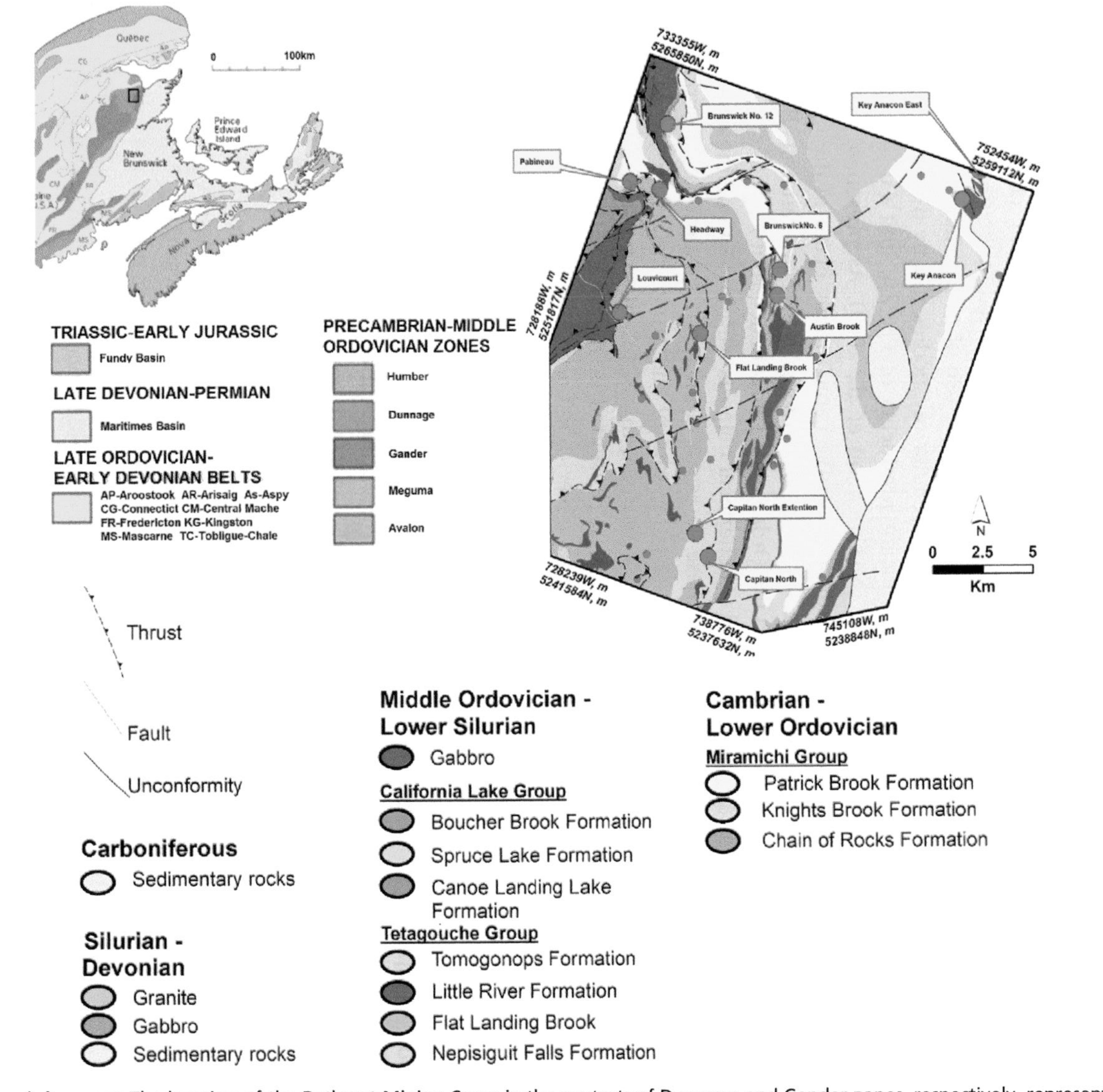

FIGURE 7.3 Upper left corner: The location of the Bathurst Mining Camp in the contacts of Dunnage and Gander zones, respectively, representing the sediments of the Miramichi group and the volcanic suites belonging to arcs, back arcs, and the Iapetus oceanic crust (Williams, 1995). Right: The bedrock map of the study area (van Staal et al., 2003). Large circles are the VHMS deposits with estimated tonnage and grade, whereas small circles are VHMS deposits that are devoid of measured tonnage and grade. *VHMS*, Volcanic-hosted massive sulfide.

Table 7.2 The list of independent, predictor variables used for mineral prospectivity mapping in the Bathurst Mining Camp.

Id	Predictor variable	Rationale behind defining the predictor variable	AUC value
V1	Measured Euclidean distance from the contacts of volcanic rocks of Nepisiguit Falls Formation and the sedimentary rocks of Miramichi group	The metal content of VHMS deposits probably originates from the underlying volcanic and sedimentary rocks. The circulating hydrothermal and meteoric waters become acidic in nature, helping absorb cations from the surrounding rocks. This is supported by the fact that the sediments of Miramichi group and the rocks of Nepisiguit Falls Formation are depleted from Ag, Pb, Zn, and Cu.	0.88
V2	Measured Euclidean distance from the contacts of Nepisiguit Falls Formation and the Flat Landing Brook Formation	This contact represents the mineralized horizon. Also, the rhyolitic rocks of the Flat Landing Brook Formation have preserved the mineralized horizon from interacting with oxidizing sea water (Goodfellow, 2007).	0.87
V3	Map of equilibrium Th	Mineralized horizons have been affected by a phase of potassium depletion (Lentz, 1999). Radiometric data can thus help narrow down these horizons (Shives et al., 2003).	0.62
V4	Map of total K		0.69
V5	Map of equilibrium U		0.72
V6	Reduced to pole total magnetic intensity	Most VHMS deposits of the area are capped by an iron formation (Peter, 2003), which is traceable by magnetic anomalies	0.9
V7	First vertical derivative map of magnetic data		0.75
V8	Pb anomalies in till samples	Pb, Zn, and Cu are the prime commodities of VHMS deposits, meaning that their geochemical anomalies could be significant	0.68
V9	Zn anomalies in till samples		0.47
V10	Cu anomalies in till samples		0.62
V11	Apparent conductivity	The premise that sulfides are generally conductive warrants the use of this variable	0.78

AUC, Area under the curve; *VHMS*, volcanic-hosted massive sulfide.

most of the known mineral deposits in the area. Given the values of bias and variance, the model is not significantly overfit compared to other models generated for comparative purposes (Fig. 7.5).

7.4 Unsupervised algorithms

Unsupervised learning is commonly used to find structures and trends within a given dataset. These tools are user-friendly and do not require much from the user, making them specifically popular in mineral exploration. Two common forms of unsupervised learning, clustering and dimensionality reduction, are briefly explained herein. Since there are plenty of examples delving into the details of these algorithms, this section merely focuses on their application. Clustering is used for classifying the samples or variables into groups. Indices of

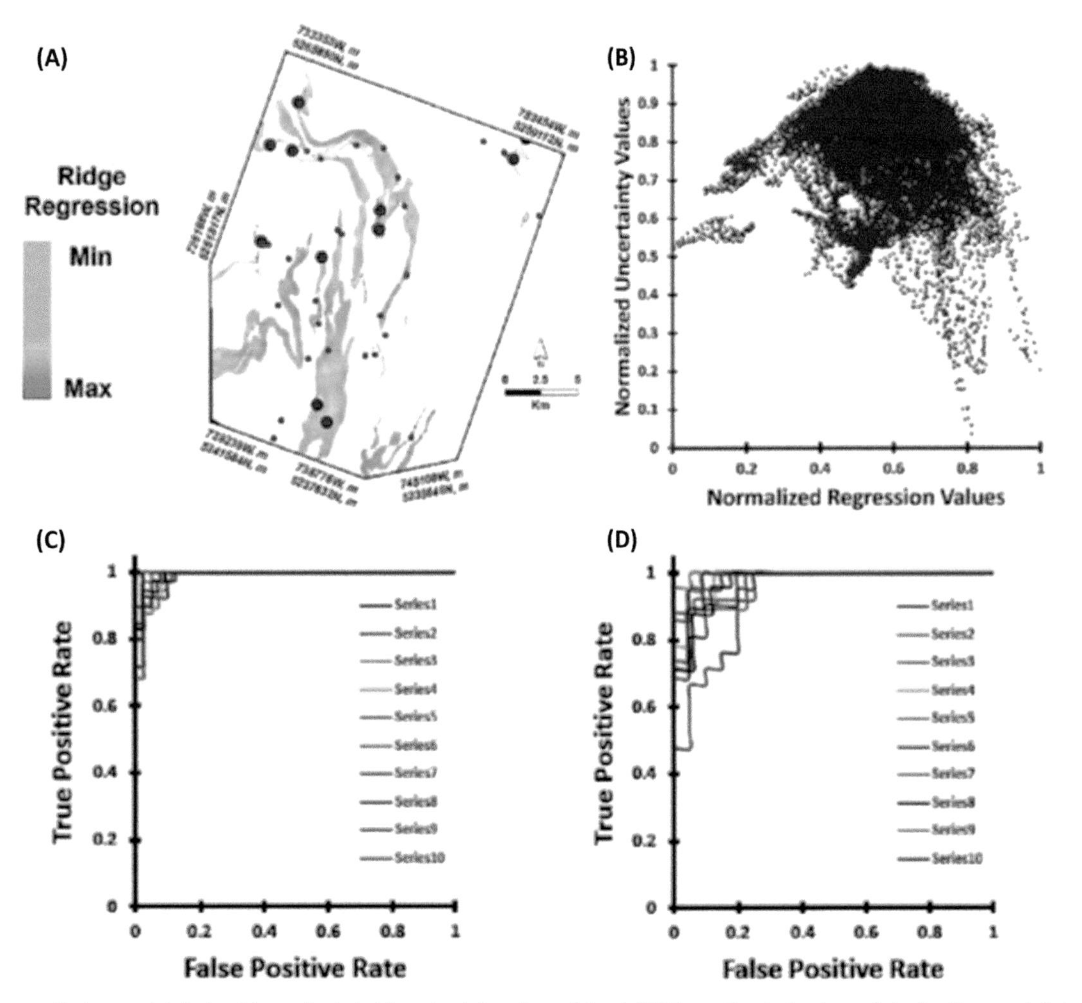

FIGURE 7.4 (A) The predictive model derived from the hybrid methodology is explained. (B) The estimated values of stochastic uncertainty versus the normalized predictive values. ROC curves for individual iteration for calibrating (C) and validating (D) data. *ROC*, Receiver operating characteristic.

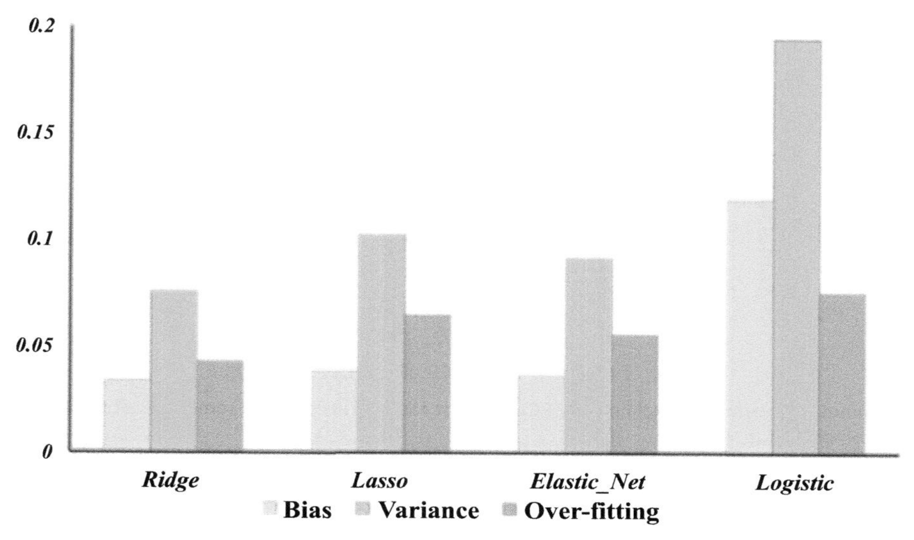

FIGURE 7.5 Plot comparing the performance of the ridge regression model with other models.

similarity, be they Euclidean-based or correlation-based, are employed for recognizing clusters, where samples bearing similarity are grouped into homogenous clusters (Madhulatha, 2012). Some applications of clustering were demonstrated in Chapter 4 for processing and interpreting geochemical data.

Dimensionality reduction is applied to project a large number of variables into a smaller number of variables explaining the variability of datasets. Principal component (PC) analysis (PCA) and factor analysis are two popular methods of dimensionality reduction (Sorzano et al., 2014). PCA summarizes the variables into a number of PCs. The application of this method was also discussed in Chapter 4 with examples of regional-scale geochemical data. PCA can also be applied for preprocessing of datasets before using them in supervised algorithms.

7.4.1 Examples of using unsupervised algorithms in mineral exploration

The following presents an example of using dimensionality reduction techniques in processing geochemical data. In this example, PCA has been applied to the results of chemical analysis for a set of stream sediment geochemical data, including Pb, Zn, Ag, Cr, Ni, Co, V, Sr, Cu, and Ba. The intent of this survey was, in part, to recognize patterns that describe carbonate-hosted Zn–Pb mineral systems in terrain in Western Iran (see Parsa, 2021). Primary commodities linked to these deposits are Pb, Zn, and Cu, appearing in sulfide and nonsulfide minerals, whereas barite is a common gangue mineral observed in these deposits (see Parsa, 2021).

What PCA does is translate the original variables into a secondary set of variables (PCs) that describe the interrelationship between original elements. Usually, the first couple of PCs explains most of the variability inherited in the dataset. In this case, as presented in Fig. 7.6, the first four PCs account for some 80% of data variability, meaning that these PCs can be

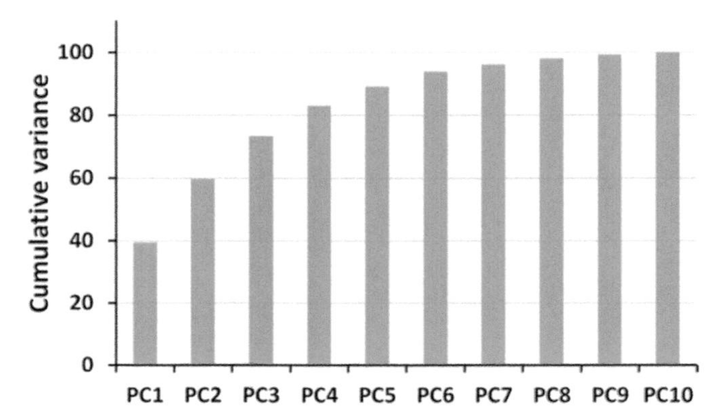

FIGURE 7.6 Cumulative variability explained by PCs. The first four PCs account for some 80% of the total variability. *PC*, Principal component.

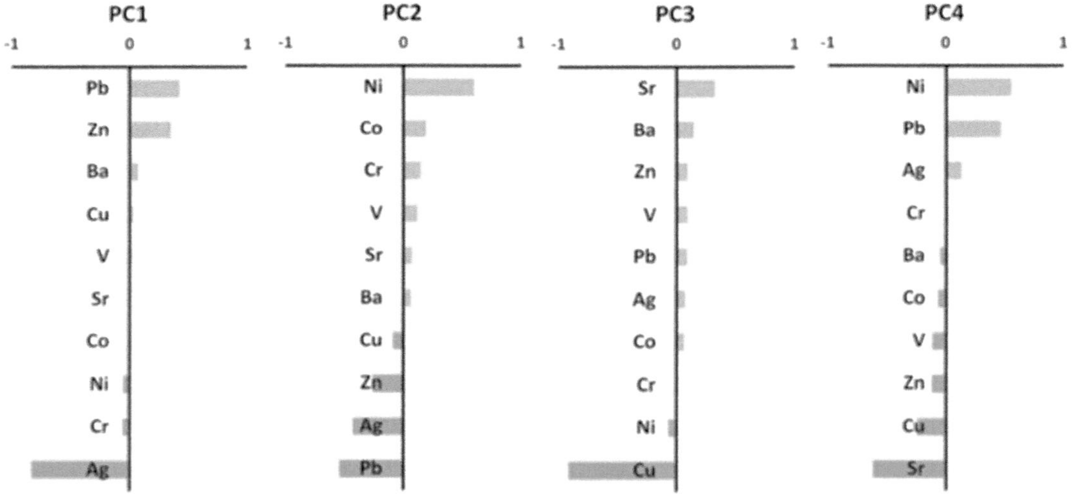

FIGURE 7.7 Loading chart for the first four principal components, indicating positive loadings for ore-related (Zn, Pb, Cu, and Ba) elements in PC1.

employed as a proxy to describe the interrelationship between the original variables (here Pb, Zn, Ag, Cr, Ni, Co, V, Sr, Cu, and Ba).

Using PCA requires extra effort for the interpretation of its results, and this interpretation can be quite subjective. Loadings are employed to understand the relationship between PCs and original variables, whereas scores describe the relationship between samples and PCs. Loadings demonstrate a positive relationship between original variables and PCs. In the case of using geochemical data a positive loading for a given element points to the fact that enriched samples should have higher PC scores and vice versa. Getting back to the earlier example, the first PC, PC1, shows positive loadings for Ba, Cu, Pb, and Zn (Fig. 7.7).

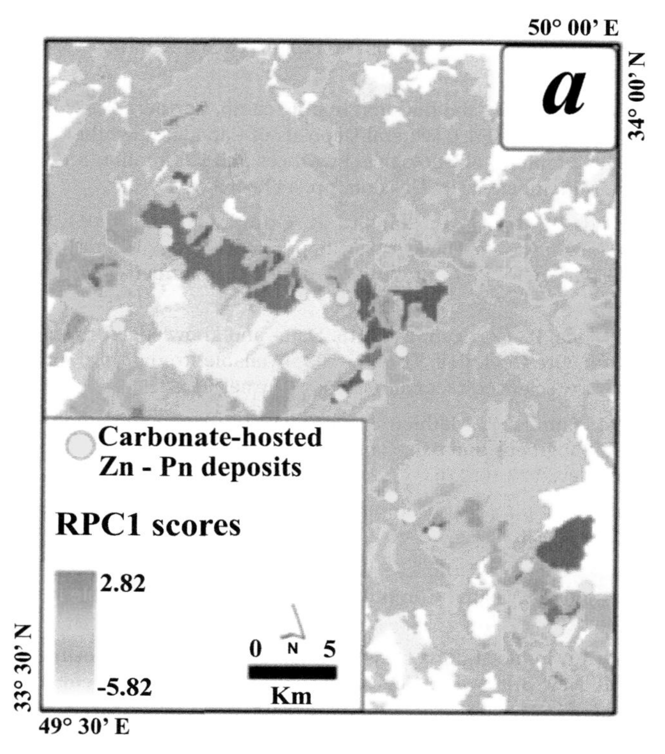

FIGURE 7.8 Scores of PC1. The term RPC refers to the use of a robust version of PCA that resists the effects of outliers. Principal component analysis (PCA). *RPC*, Robust principal component.

These are all ore-related elements, meaning that the first PC can serve as a proxy for tracing carbonate-hosted deposits in this terrain. Therefore the scores of this PC were mapped in this terrain for mineral exploration purposes, as can be seen in Fig. 7.8.

References

Ahrens, A., Hansen, C.B., Schaffer, M.E., 2020. lassopack: model selection and prediction with regularized regression in Stata. Stata J. 20 (1), 176–235. Available from: https://doi.org/10.1177/1536867X20909697; https://journals.sagepub.com/home/stj.

Brandmeier, M., Cabrera Zamora, I.G., Nykänen, V., Middleton, M., 2020. Boosting for mineral prospectivity modeling: a new GIS toolbox. Nat. Resour. Res. 29 (1), 71–88. Available from: https://doi.org/10.1007/s11053-019-09483-8; https://link.springer.com/journal/11053.

Carranza, E.J.M., 2009. Controls on mineral deposit occurrence inferred from analysis of their spatial pattern and spatial association with geological features. Ore Geol. Rev. 35 (3–4), 383–400. Available from: https://doi.org/10.1016/j.oregeorev.2009.01.001.

Carranza, E.J.M., Laborte, A.G., 2015. Data-driven predictive mapping of gold prospectivity, Baguio district, Philippines: application of Random Forests algorithm. Ore Geol. Rev. 71, 777–787. Available from: https://doi.org/10.1016/j.oregeorev.2014.08.010; http://www.sciencedirect.com/science/journal/01691368.

Dong, X., Yu, Z., Cao, W., Shi, Y., Ma, Q., 2020. A survey on ensemble learning. Front. Comput. Sci. 14 (2), 241–258. Available from: https://doi.org/10.1007/s11704-019-8208-z; http://www.springerlink.com/content/2095-2228/0.

Goodfellow, W.D., 2007. Metallogeny of the Bathurst mining camp, northern New BrunswickSpecial Publication In: Goodfellow, W.D. (Ed.), Mineral Deposits of Canada: A Synthesis of Major Deposit-Type, District Metallogeny, the Evolution of Geological Provinces and Exploration Methods, 5. Geological Association of Canada, Mineral Deposits Division, pp. 449–469.

Harris, J.R., Wickert, L., Lynds, T., Behnia, P., Rainbird, R., Grunsky, E., et al., 2011. Remote predictive mapping 3. optical remote sensing – a review for remote predictive geological mapping in northern Canada. Geosci. Can. 38 (2), 49–84. Available from: http://www.gac.ca/membersonly/journals/geoscience/getfile.php?id = 191.Canada.

Harris, J.R., Grunsky, E., Behnia, P., Corrigan, D., 2015. Data- and knowledge-driven mineral prospectivity maps for Canada's North. Ore Geol. Rev. 71, 788–803. Available from: https://doi.org/10.1016/j.oregeorev.2015.01.004; http://www.sciencedirect.com/science/journal/01691368.

Harris, J.R., Naghizadeh, M., Behnia, P., Mathieu, L., 2022. Data-driven gold potential maps for the Chibougamau area, Abitibi greenstone belt, Canada. Ore Geol. Rev. 150, 105176. Available from: https://doi.org/10.1016/j.oregeorev.2022.105176.

Hengl, T., 2006. Finding the right pixel size. Comput. Geosci. 32 (9), 1283–1298. Available from: https://doi.org/10.1016/j.cageo.2005.11.008.

Lawley, C.J.M., Raimondo, S., Chen, T., Brin, L., Zakharov, A., Kur, D., et al., 2022. Geoscience language models and their intrinsic evaluation. Appl. Comput. Geosci. 14, 100084. Available from: https://doi.org/10.1016/j.acags.2022.100084.

Lentz, D.R., 1999. Deformation-induced mass transfer in felsic volcanic rocks hosting the Brunswick No. 6 massive-sulfide deposit, New Brunswick: Geochemical effects and petrogenetic implications. Can. Mineralogist 37 (2), 489–512.

Levinson, A.A., 1974. Introduction to Exploration Geochemistry.

Li, T., Zuo, R., Xiong, Y., Peng, Y., 2021. Random-drop data augmentation of deep convolutional neural network for mineral prospectivity mapping. Nat. Resour. Res. 30 (1), 27–38. Available from: https://doi.org/10.1007/s11053-020-09742-z; https://link.springer.com/journal/11053.

Madhulatha, T.S., 2012. An overview on clustering methods. IOSR J. Eng. 02 (04), 719–725. Available from: https://doi.org/10.9790/3021-0204719725.

McCutcheon, S.R., Luff, W.M., Boyle, R.W., 2003. In: Goodfellow, W.D., McCutcheon, S.R., Peter, J.M. (Eds.), Massive sulfide deposits of the Bathurst mining camp, New Brunswick, and Northern Maine: economic geology monograph, 11. Society of Economic Geologists, pp. 17–35.

Nykänen, V., Lahti, I., Niiranen, T., Korhonen, K., 2015. Receiver operating characteristics (ROC) as validation tool for prospectivity models – a magmatic Ni-Cu case study from the Central Lapland Greenstone Belt, Northern Finland. Ore Geol. Rev. 71, 853–860. Available from: https://doi.org/10.1016/j.oregeorev.2014.09.007; http://www.sciencedirect.com/science/journal/01691368.

Parkhill, M.A., Doiron, A., 2003. Quaternary geology of the Bathurst mining camp and implications for base metal exploration using drift prospecting. Soc. Econ. Geol. Available from: https://doi.org/10.5382/mono.11.28.

Parsa, M., 2021. A data augmentation approach to XGboost-based mineral potential mapping: an example of carbonate-hosted Zn Pb mineral systems of Western Iran. J. Geochem. Explor. 228, 106811. Available from: https://doi.org/10.1016/j.gexplo.2021.106811.

Parsa, M., Carranza, E.J.M., 2021. Modulating the impacts of stochastic uncertainties linked to deposit locations in data-driven predictive mapping of mineral prospectivity. Nat. Resour. Res. 30 (5), 3081–3097. Available from: https://doi.org/10.1007/s11053-021-09891-9; https://link.springer.com/journal/11053.

Parsa, M., Maghsoudi, A., Yousefi, M., 2018a. Spatial analyses of exploration evidence data to model skarn-type copper prospectivity in the Varzaghan district, NW Iran. Ore Geol. Rev. 92, 97–112. Available from: https://doi.org/10.1016/j.oregeorev.2017.11.013; http://www.sciencedirect.com/science/journal/01691368.

Parsa, M., Maghsoudi, A., Yousefi, M., 2018b. A receiver operating characteristics-based geochemical data fusion technique for targeting undiscovered mineral deposits. Nat. Resour. Res. 27 (1), 15–28. Available from: https://doi.org/10.1007/s11053-017-9351-6; http://www.kluweronline.com/issn/1520-7439.

Parsa, M., Carranza, E.J.M., Ahmadi, B., 2022a. Deep GMDH neural networks for predictive mapping of mineral prospectivity in terrains hosting few but large mineral deposits. Nat. Resour. Res. 31 (1), 37–50. Available from: https://doi.org/10.1007/s11053-021-09984-5; https://link.springer.com/journal/11053.

Parsa, M., Lentz, D.R., Walker, J.A., 2022b. Predictive modeling of prospectivity for VHMS mineral deposits, Northeastern Bathurst mining camp, NB, Canada, Using an ensemble regularization technique. Nat. Resour. Res. Available from: https://doi.org/10.1007/s11053-022-10133-9; https://www.springer.com/journal/11053.

Peter, J.M., 2003. Ancient Iron Formations: Their Genesis and Use in the Exploration for Stratiform Base Metal Sulphide Deposits, with Examples from the Bathurst Mining Camp. pp. 145–176.

Porwal, A., Carranza, E.J.M., Hale, M., 2003. Artificial neural networks for mineral-potential mapping: a case study from Aravalli Province, Western India. Nat. Resour. Res. 12, 155–171.

Pour, A.B., Hashim, M., 2012. The application of ASTER remote sensing data to porphyry copper and epithermal gold deposits. Ore Geol. Rev. 44, 1–9. Available from: https://doi.org/10.1016/j.oregeorev.2011.09.009.

Prado, E.M.G., de Souza Filho, C.R., Carranza, E.J.M., Motta, J.G., 2020. Modeling of Cu–Au prospectivity in the Carajás mineral province (Brazil) through machine learning: dealing with imbalanced training data. Ore Geol. Rev. 124. Available from: https://doi.org/10.1016/j.oregeorev.2020.103611; http://www.sciencedirect.com/science/journal/01691368.

Refaeilzadeh, P., Tang, L., Liu, H., 2009. Cross-Validation. Springer Science and Business Media LLC, pp. 532–538. Available from: https://doi.org/10.1007/978-0-387-39940-9_565.

Reimann, C., Filzmoser, P., Garrett, R.G., 2002. Factor analysis applied to regional geochemical data: problems and possibilities. Appl. Geochem. 17 (3), 185–206. Available from: https://doi.org/10.1016/S0883-2927(01)00066-X.

Shives, R.B.K., Ford, K.L., Peter, J.M., 2003. Mapping and exploration applications of gamma ray spectrometry in the Bathurst mining camp, Northeastern New Brunswick. Soc. Econ. Geol. Available from: https://doi.org/10.5382/mono.11.37.

Sinclair, W.D., 2007. Mineral deposits of Canada: a synthesis of major deposit-types, district metallogeny, the evolution of geological provinces, and exploration methods: Geological Association of Canada, Mineral Deposits Division. Spec. Publ. 5, 223–243.

Sorzano, C.O.S., Vargas, J., Montano, A.P., 2014. A Survey of Dimensionality Reduction Techniques.

Swets, J.A., 1988. Measuring the accuracy of diagnostic systems. Science 240 (4857), 1285–1293. Available from: https://doi.org/10.1126/science.3287615.

Templ, M., Filzmoser, P., Reimann, C., 2008. Cluster analysis applied to regional geochemical data: problems and possibilities. Appl. Geochem. 23 (8), 2198–2213. Available from: https://doi.org/10.1016/j.apgeochem.2008.03.004.

van Staal, C.R., Wilson, R.A., Rogers, N., Fyffe, L.R., Langton, J.P., McCutcheon, S.R., et al., 2003. Geology and tectonic history of the Bathurst supergroup, Bathurst mining camp, and its relationships to coeval rocks in Southwestern New Brunswick and Adjacent Maine—a synthesis. Soc. Econ. Geol. Available from: https://doi.org/10.5382/mono.11.03.

Williams, Hd, 1995. Geology of the Appalachian—Caledonian Orogen in Canada and Greenland. Geological Society of America.

Xiong, Y., Zuo, R., 2021. A positive and unlabeled learning algorithm for mineral prospectivity mapping. Comput. Geosci. 147, 104667. Available from: https://doi.org/10.1016/j.cageo.2020.104667.

Zhang, C., Zuo, R., Xiong, Y., 2021. Detection of the multivariate geochemical anomalies associated with mineralization using a deep convolutional neural network and a pixel-pair feature method. Appl. Geochem. 130, 104994. Available from: https://doi.org/10.1016/j.apgeochem.2021.104994.

Zuo, R., 2012. Exploring the effects of cell size in geochemical mapping. J. Geochem. Explor. 112, 357–367. Available from: https://doi.org/10.1016/j.gexplo.2011.11.001.

Zuo, R., 2014. Identification of weak geochemical anomalies using robust neighborhood statistics coupled with GIS in covered areas. J. Geochem. Explor. 136, 93–101. Available from: https://doi.org/10.1016/j.gexplo.2013.10.011.

Zuo, R., 2020. Geodata science-based mineral prospectivity mapping: a review. Nat. Resour. Res. 29 (6), 3415–3424. Available from: https://doi.org/10.1007/s11053-020-09700-9; https://link.springer.com/journal/11053.

Zuo, R., Carranza, E.J.M., 2011. Support vector machine: a tool for mapping mineral prospectivity. Comput. Geosci. 37 (12), 1967–1975. Available from: https://doi.org/10.1016/j.cageo.2010.09.014.

Zuo, R., Wang, Z., 2020. Effects of random negative training samples on mineral prospectivity mapping. Nat. Resour. Res. 29 (6), 3443–3455. Available from: https://doi.org/10.1007/s11053-020-09668-6; https://link.springer.com/journal/11053.

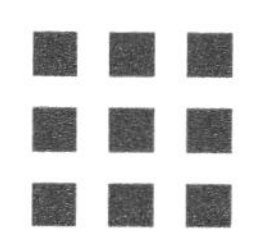

Index

Note: Page numbers followed by "*f*" and "*t*" refer to figures and tables, respectively.

D

E

H

N

9780323956086